STEM Education in the Junior Secondary

Robyn Jorgensen • Kevin Larkin
Editors

STEM Education in the Junior Secondary

The State of Play

Springer

Editors
Robyn Jorgensen
University of Canberra
Bruce, Aust Capital Terr, Australia

Kevin Larkin
School of EPS
Griffith University
Brisbane, Queensland, Australia

ISBN 978-981-10-5447-1 ISBN 978-981-10-5448-8 (eBook)
DOI 10.1007/978-981-10-5448-8

Library of Congress Control Number: 2017948862

Printed on acid-free paper

This Springer imprint is published by Springer Nature
The registered company is Springer Nature Singapore Pte Ltd.
The registered company address is: 152 Beach Road, #21-01/04 Gateway East, Singapore 189721, Singapore

Contents

Chapter 1
Introduction

Kevin Larkin and Robyn Jorgensen

This book has arisen from the current international focus on STEM education. Our specific focus on the Junior Secondary years (or what might be called the middle years) is strategic. This is a little researched area, and yet, we know it is an important grounding area for students and their subsequent choices for the Senior Secondary years of schooling and beyond into tertiary study. Having students engaged, enjoying and becoming confident users and participants in STEM education is critical for sound decision making in later life. Internationally, there is a serious cry for more students to engage with STEM education so that they are able to participate more fully as informed citizens. STEM learning is therefore, in many ways, an equity issue for students as it is highly likely that STEM knowledge and skills will be required by those who wish to be at the cutting edge of future key advances in society, the economy and industry. The Junior Secondary years are foundational in positioning students so that a wide range of STEM study options are available for them in Senior Secondary and beyond.

To create this book, we invited colleagues from the international STEM community who we knew were working in this space. While useful, such a process has its flaws as it is limited by our collective wisdom. As mathematics educators, we were aware of many of the projects being undertaken in our field but wanted to ensure that there was a good representation across the STEM area. Our colleague, Dr. Harry Kanasa who works in science education, was instrumental in helping identify science educators. Larkin's joint interest area is technology, so he was able to identify colleagues in this area as well. Authors were approached and chapters submitted. Chapters were peer reviewed and two were rejected, as they were not suitable for this book. We are very pleased with the chapters that form this collection.

K. Larkin (✉)
School of EPS, Griffith University, Brisbane, Queensland, Australia
e-mail: k.larkin@griffith.edu.au

R. Jorgensen
University of Canberra, Bruce, Aust Capital Terr, Australia

R. Jorgensen, K. Larkin (eds.), *STEM Education in the Junior Secondary*,
DOI 10.1007/978-981-10-5448-8_1

We recognise that there is a wide range of quality work being undertaken in this important area of study that could have been included within the scope of this book. Their non-inclusion is not to be construed as anything other than a very pragmatic decision given publishing constraints. In addition, we have deliberately not set out to construct a book with a specific theoretical, empirical, methodological or reform agenda. STEM education is a contested space; therefore, we wanted to capture some of this complexity with the mixture of chapters within this book.

The collection of chapters that appear in this book are written by authors from across the globe, and therefore these chapters will appeal to an international audience. It was our intent to ensure that we were able to represent, as much as possible, the breadth of work being undertaken in STEM education. Some of the authors have focused on one discipline within STEM – such as mathematics or science or technology – while others have incorporated the breadth of STEM-orientated subjects from either a student or teacher perspective. It is not our intent in this introduction to provide a lengthy overview of the various chapters in the book. Suffice to say here that we have organised the book in three main sections. The three sections of the book are philosophical and theoretical orientations within STEM, STEM in action in Junior Secondary and teacher education contexts, and critiques of STEM education.

There are three chapters in the theoretical section of the book: Roth's chapter focuses on affect in adolescence via a cultural-historical lens; Powell, Alqahtani and Singh outline the importance of a collaborative approach to learning in online environments; and Hubber, Tytler and Chittleborough discuss how "representation construction" can be a vehicle for guided inquiry in science.

The second section, which includes five chapters, focuses on the practical application of STEM in Junior Secondary and adult learning contexts including teacher education and nonschool learning communities. In the first chapter in this section, Rennie, Venville and Wallace explore the importance of motivation and relevance in relation to the creation of STEM curricula. Jorgensen (Zevenbergen) and Alden present an argument for a stronger theory-practice nexus in preservice teacher STEM education, and a multi-faceted approach to STEM teacher education is then outlined by Hobbs, Cripps Clark and Plant. Gates' contribution centres on the importance of visualisation within STEM; and finally, Finger outlines a policy agenda within the technology component of STEM. Collectively, these authors provide valuable insights into the debates, tensions and conflicts when teaching STEM in various educational contexts.

The final section has a focus on critical critiques of STEM education across the globe. Rosa uses an ethnomathematical lens to examine STEM education in the Brazilian context, and Tytler, Symington, Williams and White remind us of the critical importance of School-Community Partnerships in enlivening STEM education. The final chapter of the book by Wolfmeyer, Lupinacci and Chesky takes a socially critical approach to understanding various contemporary viewpoints within the STEM agenda and challenges us to think deeply about the role of STEM in understanding modern society.

Collectively, this selection of chapters provides an overview of the STEM work being undertaken in the middle years of schooling. It was a serious intent of the book to focus on these years of schooling as they are underrepresented in the literature. The chapters provide a sample of the work being undertaken, within different frameworks and with different foci.

Finally, we would like to thank Alanna Grant who has supported us in the management of the chapters, reviews and collation of the various sources of information needed. Alanna has been an enormous help in the finalising of this book, and we duly acknowledge her work and support.

Chapter 2
What Is Unique About Junior STEM?

Robyn Jorgensen and Kevin Larkin

The growth in interest in science, technology, engineering and mathematics (STEM) education has been the catalyst for this book. There has been a worldwide trend to focus on the teaching of STEM across all sectors of schooling, and this is evidenced by the geographical spread of the authors in this book. The initial intent of STEM education was to build strengths in science, technology, engineering and mathematics due to the declining number of students undertaking these courses of study in high school or at university, a perceived decline in the quality of teaching, and an increased recognition that STEM is a key driver in advancing societies (see Han et al. 2015). Most societies have taken the urgency to develop STEM in schools and in the labour market very seriously, with many nations developing productive strategies to boost STEM in schools and in the workplace. While increased emphasis has occurred overall, within the four STEM elements, there is a disparity in the amount of attention each element receives, with science and mathematics remaining the main focus. Daugherty et al. (2014), for example, argue that technology and engineering education continue to struggle to maintain a foothold in secondary education and that, despite some curriculum initiatives in the USA (e.g. Project Lead the Way, Engineering by Design), the "influence of technology and engineering education curriculum at junior high and high schools across America is clearly less than it was just 20 years ago" (p. 45).

Marginson and colleagues (Marginson et al. 2013) have suggested that while many nations have seen an urgency in building STEM, Australia has not been as forthright in its tackling of STEM and that Australia, as a nation, lags behind the USA, East Asia and much of Western Europe in terms of addressing STEM shortfalls. It is recognised that participating in tertiary mathematics and science is

R. Jorgensen (✉)
University of Canberra, Bruce, Aust Capital Terr, Australia
e-mail: robyn.jorgensen@canberra.edu.au

K. Larkin
School of EPS, Griffith University, Brisbane, Queensland, Australia

R. Jorgensen, K. Larkin (eds.), *STEM Education in the Junior Secondary*,
DOI 10.1007/978-981-10-5448-8_2

contingent on achievement in secondary school STEM (Wang, 2013). Students' experiences and successes in school-based STEM influence their decision to enrol in tertiary courses. However, what is less well known is the impact of the years prior to senior study in influencing students to participate in senior secondary and university STEM-related courses. There are a range of ways in which STEM education can be realised – where there may be a focus on one discipline within STEM, such as mathematics or science or technology – and then make connections back to the remaining STEM disciplines, where the focus is on two or more of the STEM disciplines. There is also a focus in the broader literature that the separation into the constitutive subject areas denies the strength of STEM as an integrated and holistic area of study in its own right. The focus on STEM education can be directed at student learning or the professional learning of teachers many of whom may be teaching out of the discipline area, particularly in the junior years of secondary schooling. Finally, there is some critique of STEM education as to whether there is really a crisis in STEM or it is a crisis of convenience to enable (and perhaps perpetuate) the hegemonic power of the STEM discipline(s).

2.1 Junior STEM

So why would we focus on junior STEM? For our purposes, we define junior STEM as the area of secondary schooling that encompasses students aged 12–15 years. In the Australian context, this is referred at as junior secondary as the education systems across Australia are generally divided into two sectors, primary school for students aged 5–12 years and then secondary school for students aged 12–18 years. In other contexts, different signifiers will be used to refer to these years of schooling. The junior secondary years are formative in terms of students' developing a sense of themselves as learners particularly in relation to particular subject areas. This area of schooling is relatively under researched with much of the contemporary STEM research focussing on the bookends of education, i.e. the commencing years of schooling and the final years of schooling. In terms of the early years, research by Piasta, Logan, Pelatti, Capps and Petrill (2015) has found that "children's experiences during early childhood impact their understanding and knowledge of a host of cognitive skills, including math and science concepts" (p. 407). Yet despite the importance of early positive STEM experiences, students are opting out of the STEM areas prior to the final years of schooling. As Larkin and Jorgensen (2016) and Jorgensen and Larkin (2016) have identified in their work on the early and middle years of mathematics education, students in Year 3 are already developing a sense of who they are as learners of mathematics and by Year 6 have a strong sense of whether or not they want to be a part of the mathematics learning community and hence have a trajectory within their curriculum choices to include STEM. In this research on the attitudes towards mathematics of over 275 students, in three different Australian educational jurisdictions, Larkin and Jorgensen (2016) and Jorgensen and Larkin (2016) noted that both psychological and social constructs were instrumental in forming mathematical identities that proved highly resistant to change. This has obvious implications for junior secondary mathematics. There is a strong sense that STEM has a big image problem – it is seen as something for "geniuses, geeks, nerds, and not for girls"(New South Wales Government, 2016, p. 2). This perception of STEM clearly influences the choices students make about participating in, and continuing with, the study of STEM subjects. As we have established in our own work, this sense of mathematical identity is already being formed in the early years of primary school. As such, there is a lot of work to be done, in junior secondary contexts, in terms of shaping positive identities of students who have the confidence and competence to continue in the study of STEM-related subjects.

Likewise, research into elementary school level science learning indicates early formation of science attitudes that prove difficult to change during, and after, adolescence (Kermani & Aldemir, 2015). In addition, research has suggested that integrating science with other disciplines – e.g. technology and mathematics – supports young learners in "the formation of awareness and interest towards science, and eventually affects their overall school performance further down in their education" (p. 1505).

It is therefore apparent in the literature that the practices in schools (elementary and junior secondary) are creating opportunities, often negative, for learners to create a STEM identity which, in turn, will shape their decisions as to whether or not they continue with the study of STEM-related subjects in senior secondary and beyond. Building a strong affiliation with STEM in the junior secondary years is tantamount if students are to continue the study of STEM beyond the compulsory years of schooling.

Widely, and somewhat hysterically, reported in the Australian media at the time of writing this chapter was the demise of Australian students in international testing – TIMMS and PISA– which demonstrated that, comparatively speaking, Australian students are falling behind many other countries. Thomson (2016) reported that Australian students' results are stagnating which has resulted in the international ranking of Australian students dropping due to the successes of students in other countries. Thomson Wernert, O'Grady and Rodrigues (2016), for example, cited that the Year 8 science results are not significantly different from the 1995 TIMMS results suggesting that there has been little to no value adding over the subsequent two decades. Soon after the announcement of the TIMMS and PISA results, the National Assessment Program – Literacy and Numeracy (NAPLAN) results were released, and it was found that, despite considerable investment in education, results across the year levels and areas under discussion here had flatlined (Rice, 2016). Collectively, these results have caused considerable debate and concern concerning the education of our students, including STEM education.

2.2 So Who Participates in STEM and Why?

As a consequence of the data that indicates young people opting out of the enabling sciences, there has been a push to invigorate the study in STEM-related areas as these are the disciplines that enrich and progress societies. This was brought home at the time of writing this introduction when headline news in Australia reported on the brain drain of women undertaking studies in the hard sciences, the high attrition rates of women from the science profession and the implications of these concurrent phenomena (ABC Science, 2016). This becomes a "double-edged" sword: firstly attracting and retaining students in the study of STEM and secondly, their later retention as scientists in the workplace. For example, in terms of engineering, Engineers Australia (2012) reported that only 14% of enrolments in engineering courses were women, and despite efforts to attract women into this profession, this proportion has remained relatively consistent for a decade or more.

There are numerous studies on the patterns of enrolment in the mathematics/science areas in the senior years of schooling that raise concern regarding the declining numbers of students who are taking the "hard sciences". For example, in the Australian context, there has been an increase in the numbers of students remaining to Year 12 Kennedy, Lyons and Quinn (2014) as, in the period from 1992 to 2012, enrolments in Year 12 increased by 16%. This increase did not, however, translate to increased enrolments in the higher levels of sciences and mathematics. Their data indicated that overall enrolments fell in the areas of biology (10%), chemistry (5%), physics (7%), multistrand sciences (5%), intermediate maths (11%) and advanced maths (7%). In contrast, soft enrolments in the "soft sciences" rose – earth sciences (3%) and entry maths (11%). It would appear that this is not the case in other countries; for example, the UK has reported a growth in enrolment in 2010–2015 by 20% (OECD, 2006). However, it was unclear from this data where this growth actually occurred, in the "softer STEM subject options" (i.e. multistrand science or earth sciences) rather than in the "harder STEM options" (i.e. chemistry, physics or advanced mathematics), which have flow on implications into the tertiary education sector. While there are many reasons for this changing profile in subject enrolment in the senior years, there is some concern as to why students are electing to opt out of the more difficult subjects in the STEM areas.

2.3 Teachers of STEM

There is a general recognition that teachers are the key to long-lasting reform and student learning (Hattie, 2003; NSW Government, 2016). There have been ongoing debates about the quality of teachers in terms of entry into the profession, ongoing professional learning and the need to attract more high-performing students into teaching (NSW Government, 2016).

One of the key issues facing the STEM areas in schooling is that there are teachers who are required, due to an unavailability of qualified teachers, to teach outside their discipline area; in many cases these are STEM areas. The Office of the Chief Scientist (2016) reported that 40% of students in Years 7–10 are taught by a teacher who is not trained in mathematics. In addition, the report also indicates that 20% of science teachers are not science trained. While this differs from the ACER report, (Weldon, 2015) where the claim is that only 20% of mathematics teachers were teaching out of their field of expertise, this difference is likely to be a consequence of the way "teaching outside the field" was defined. In the ACER report, the definition was a generous one in which a teacher was deemed to be qualified if he/she had undertaken one semester of study in the second year of a degree in the nominated area of study. This is a very different definition used in most Schools of Education in Australian Universities where it is expected that the teachers would have at least a minor (40 credit points) or a major (60 credit points) in a nominated area. While teacher education is tightly controlled as to what preservice teachers must complete in order to be registered as a teacher, once in the schools, the allocation of teachers to specific subjects and year levels is at the discretion of the principal. As a result, "teaching out of their field" is a commonplace experience particularly in rural and remote secondary schools where it is difficult to attract qualified mathematics and science teachers.

In the junior areas of secondary education, STEM studies are largely compulsory, so students must undertake the study of science, technology and mathematics, and it is only at the end of Year 10 that students can select courses for study in the senior years. For this reason, a collection of work helps us, as a field of education, to better understand the practices in junior STEM that are implicated in the patterns described above and what is being done to enhance enrolments in senior secondary STEM subjects. At the end of 2015, the New South Wales Government hosted a STEM summit at which they discussed the key issues and opportunities for STEM education in Australia (New South Wales Government, 2016). One of the key outcomes was the recognition that there is a clear need for teachers to have STEM content knowledge and confidence in teaching STEM subjects, if they are to be successful teachers in STEM areas. A focus on teacher education is therefore a warranted inclusion in the discussion of STEM education in the junior secondary years of schooling.

2.4 Teaching Out of Field

A serious consideration in all classrooms, but perhaps a more serious issues in junior STEM education, is the capacity of the classroom teacher. As noted earlier, for a range of reasons, many teachers find themselves teaching outside of their area of expertise. This is most common in the area of STEM where specific knowledge and practices are required (e.g. management of science laboratories). Preservice teacher education in STEM requires primary preservice teachers to undertake curriculum study in the STEM area as they will be teaching classes in STEM areas. However, the same requirement is not evident in many secondary education courses although it is likely that they will be required to teach in these areas. This is most notable in the area of mathematics. For example, in the UK, it was reported (Ross, 2015) that there is a decline in the numbers of teachers who are qualified to teach in mathematics from 82.7% in 2013 to 79.8% in 2015. In Eire, a staggering 48% of teachers were teaching mathematics to either lower grades (i.e. junior secondary) or to low-attaining students (Ní Ríordáin & Hannigan, 2011) without being qualified to teach mathematics.

Highly significant in research on teachers working outside their field of expertise is their sense of identity. Grootenboer and Zevenbergen (2008) argued that there is a strong need to consider the mathematical identities of teachers and how this is lived out in their lives as teachers. Hobbs (2013) has argued similarly for science education. In work with German out-of-field mathematics teachers, Bosse (2014) reported that the teachers enjoyed teaching mathematics but framed mathematics as basic or elementary mathematics rather than junior secondary mathematics and that they received little professional development in terms of their mathematics education learning. Similarly, Graven (2004) highlighted the importance of teacher confidence in the teaching of mathematics, again drawing attention to the importance of teachers feeling confident in their curriculum and content knowledge. There is some sense that when teachers have strong mathematical content knowledge, this impacts positively on their identity as a teacher as well as their pedagogical content knowledge (Jorgensen & Lowrie, 2017).

The plethora of findings that has been generated through the research into teachers' pedagogical content knowledge, and their discipline content knowledge, has shown that there are greater gains in students' learning when the teachers have strong content knowledge across the STEM spectrum (e.g. Berry Friedrichsen & Loughran, 2015; Callingham, Carmichael, & Watson, 2016; Ioannou & Angeli, 2015). For example, in mathematics education, studies have consistently shown that there is a very strong relationship between teachers' knowledge of mathematics and student learning (Goos, 2013). Jorgensen and Lowrie (2017) have shown that there is a strong relationship between the content knowledge of teachers and their pedagogical content knowledge. In their study, it was found that engineers who have not undertaken any formal training in education studies, but who possess strong mathematical content knowledge, performed as well as preservice teachers and primary school teachers when their pedagogical content knowledge was assessed. This study highlights the importance of strong mathematical content knowledge. Collectively this corpus of research suggests that there are likely to be issues with teachers teaching out of their field and that professional learning opportunities are needed to support those teachers, particularly around issues of STEM content knowledge. This book, with the focus on teacher professional learning, either at preservice or in-service level, makes a contribution to this knowledge.

2.4.1 STEM Education and Equity

It has been widely recognised that participation and success in the STEM areas has been quite biased, with girls, low-income students, Indigenous students and rural/regional students at greater risk of not participating, or failing, in these areas. There have been ranges of initiatives implemented to redress this.

Some programmes have focussed specifically on the enrolment of Indigenous students in tertiary STEM subjects. The More Aboriginal and Torres Strait Islander Teachers Initiative (MATSITI) has had oversight of the $783,000 project – Excellence and Equity in Maths: Indigenous Student Achievement and Tertiary Aspirations in Mathematics – in which the goal is to support Indigenous students in schools and tertiary settings so as to establish pathways from school to tertiary STEM (MATSITI, 2016). The project has the specific intent of supporting classroom practices in mathematics that assist Indigenous student transition into senior secondary and tertiary mathematics. A more generic project – Teach for Australia – advocates for attracting outstanding professionals and graduates to undertake a fast-tracked teacher education programme in priority areas, including STEM, and to work in educationally disadvantaged contexts (Teach for Australia, 2016). Typically, these schools are unable to attract strong graduates or teachers, so Teach for Australia is seeking to train teachers who are willing to take positions in these schools. Engineers, scientists and mathematicians are amongst the targeted professionals. These people have strong content knowledge and often practical experiences in their professional life, thus creating the potential for quality, grounded teachers to work in educationally disadvantaged contexts and hopefully to open STEM up for students who traditionally have been excluded from participating in this space.

2.5 Jumping on the Bandwagon

The original intent with the STEM initiative was to create a focus on strengthening the "hard sciences" as this was an area of dire need in terms of future employment opportunities. Other disciplines such as the arts and design have been keen to jump on the bandwagon and argue that creative thinking is also a much needed skill and that STEM needs to morph to science, technology, engineering, arts and mathematics (STEAM). Initially championed by the Rhode Island School of Design, it was argued that art and design were needed to drive innovation. Overseas the notion of STEAM rather than STEM is increasingly used in countries such as Korea (see Baek et al., 2011; Kim, Chung, Woo, & Lee, 2012), and in Australia the STEAM agenda is full steam ahead (if you can pardon the engineering pun) in a number of Australian (e.g. Murdoch, Griffith) and US (e.g. San Diego) universities now offering postgraduate degrees in STEAM. In addition, Australian high schools are beginning to market STEAM courses. This is a contentious extension of STEM as it poses two questions: Is it the case that STEM is not able to innovate (or design)? Or it is the case that art and design is seeking to raise its status by badging itself as part of the broader STEM movement? There is some sense that if this is extended, then all curriculum areas could be included and the impetus for STEM has been usurped.

For our purposes, we have opted to remain as STEM as we feel that there is a serious need to re-engage students in the study of science, mathematics, engineering and technology in junior secondary school so that students have a wider range of options in senior secondary and university studies. Furthermore, we argue that the types of skills, dispositions and ways of thinking and working usually associated with the sciences (including mathematics as a science) are quite different from those required in the arts areas. As such, it is our position that the priority for junior secondary education should be building capacity in the STEM area and therefore we have this as a primary focus of this book. The inclusion of other areas of study, such as the arts, has the potential to reduce the primacy of science and mathematics, with the resultant return to the original discourses that have in many ways facilitated the demise of the STEM area. The tension between an integrated curriculum and individual curriculum areas is an "age old" one which can result in more content-specific subjects being diluted. This can result, as has been the case, in subjects like mathematics not being taught well, or not being offered, or only taught in a shallow, tokenistic manner. Over time, this has resulted in many students being denied the rigour and enjoyment that can come from a well-taught science/mathematics/technology classroom.

References

Australian Broadcasting Commission (ABC). (2016). *Hanging up their labcoats: Australia's new brain drain.* Retrieved from http://www.abc.net.au/science/articles/2014/01/16/3926579.htm

Baek, Y. S., Park, H. J., Kim, Y., Noh, S., Park, J. Y., Lee, J., et al. (2011). STEAM education in Korea. *Journal of Learner-Centered Curriculum and Instruction, 11*(4), 149–171.

Berry, A., Friedrichsen, P., & Loughran, J. (Eds.). (2015). *Re-examining pedagogical content knowledge in science education*. New York: Routledge.

Bosse, M. (2014). The practice of out-of-field teaching in mathematics classrooms. In L. Hobbs, G. Törner (Eds.), *Taking an international perspective on out-of-field teaching: Proceedings and Agenda for Research and Action, 1st TAS Collective Symposium*, 30–31 August 2014.

Callingham, R., Carmichael, C., & Watson, J. M. (2016). Explaining student achievement: The influence of teachers' pedagogical content knowledge in statistics. *International Journal of Science and Mathematics Education, 14*(7), 1339–1357. doi:10.1007/s10763-015-9653-2.

Chief Scientist Office. (2016) *Science, technology, engineering and mathematics: Australia's future.* Canberra, Australia: Australian Government. Retrieved from http://www.chiefscientist.gov.au/2016/07/science-and-maths-in-australian-secondary-schools-datasheet/

Daugherty, M., Carter, V., & Swagerty, L. (2014). Elementary STEM education: The future for technology and engineering education? *Journal of STEM Teacher Education, 49*(1), 45–55. Retrieved from http://ir.library.illinoisstate.edu/jste/vol49/iss1/7

Engineers Australia. (2012). *Women in engineering: A statistical.* Barton, Australia: Engineers Australia.

Goos, M. (2013). Knowledge for teaching secondary school mathematics: What counts? *International Journal of Mathematical Education in Science and Technology*, *44*(7), 972–983. Online. doi:org/10.1080/0020739X.2013.826387.

Graven, M. (2004). Investigating mathematics teacher learning within an in-service community of practice: The centrality of confidence. *Educational Studies in Mathematics, 57*(3), 177–211.

Grootenboer, P. J. & Zevenbergen, R. (2008). Identity as a lens to understand learning mathematics: Developing a model. In M. Goos, R. Brown, & K. Makar (Eds.), *Navigating currents and charting directions*. Proceedings of the 31st annual conference of the mathematics education research group of Australasia, Brisbane (Vol. 1, pp. 243–250). Brisbane, Australia: MERGA.

Han, S., Capraro, R., & Capraro, M. (2015). How Science, Technology, Engineering, and Mathematics (Stem) Project-Based Learning (PBL) affects high, middle, and low achievers differently: The impact of student factors on achievement. *International Journal of Science and Mathematics Education, 13*(5), 1089–1113.

Hattie, J. (2003). Teachers make a difference: What is the research evidence? *Interpretation, 36*(2), 27–38.

Hobbs, L. (2013). Teaching 'out-of-field' as a boundary crossing event: Factors shaping teacher identity. *International Journal of Science and Mathematics Education, 11*(2), 271–297.

Ioannou, I., & Angeli, C. (2015). Technological pedagogical content knowledge as a framework for integrating educational technology in the teaching of computer science. In C. Angeli & N. Valanides (Eds.), *Technological pedagogical content knowledge* (pp. 225–237). New York: Springer.

Jorgensen, R., & Larkin, K. (2016). Analysing the relationships between students and mathematics: A tale of two paradigms. *Mathematics Education Research Journal*, doi:10.1007/s13394-016-0183-1.

Jorgensen, R., & Lowrie, T. (2017, April). Pedagogical and mathematical capital: Does teacher education make a difference? In A. Chronaki (Ed.), *Proceedings of the 9th mathematics education and society conference* (pp. 580–591). Volos, Greece.

Kennedy, J., Lyons, T., & Quinn, F. (2014). The continuing decline of science and mathematics enrolments in Australian high schools. *Teaching Science, 60*(4), 34–36.

Kermani, H., & Aldemir, J. (2015). Preparing children for success: Integrating science, math, and technology in early childhood classroom. *Early Child Development and Care, 185*(9), 1504–1527. doi:10.1080/03004430.2015.1007371.

Kim, S. W., Chung, Y. L., Woo, A. J., & Lee, H. J. (2012). Development of a theoretical model for STEAM education. *Journal of the Korean Association for Science Education, 32*(2), 388–401.

Larkin, K., & Jorgensen, R. (2016). 'I hate Maths: Why do we need to do Maths?' using iPad video diaries to investigate attitudes and emotions towards mathematics in year 3 and year 6 students. *International Journal of Science and Mathematics Education, 14*(5), 925–944. doi:10.1007/s10763-015-9621-x.

Marginson, S., Tytler, R., Freeman, B., & Roberts, K. (2013). *STEM: Country comparisons: International comparisons of science, technology, engineering and mathematics (STEM) education. Final report*. Melbourne, Australia: Australian Council of Learned Academies.

MATSITI. (2016). *Excellence and equity in maths for Indigenous students*. Retrieved from http://matsiti.edu.au/amspp/

New South Wales Government. (2016). *STEM education summit*: Sydney, November 15. Retrieved from http://www.educationcouncil.edu.au/site/DefaultSite/filesystem/documents/Summary%20of%20STEM%20Education%20Summit.pdf

Ní Ríordáin, M., & Hannigan, A. (2011). Who teaches mathematics at second level in Ireland? *Irish Educational Studies, 30*(3), 289–304.

Organisation for Economic Cooperation and Development: Global Science Forum. (2006). *Evolution of students' interest in science and technology studies; Policy report*. Retrieved from http://www.oecd.org/sti/sci-tech/36645825.pdf

Piasta, S., Logan, J., Pelatti, C., Capps, J., & Petrill, S. (2015). Professional development for early childhood educators: Efforts to improve math and science learning opportunities in early childhood classrooms. *Journal of Educational Psychology, 107*(2), 407–422.

Rice, S. (2016). NAPLAN results reveal little change in literacy and numeracy performance – here are some of the key takeaway findings. *The Conversation, Dec 13*. Retrieved from https://theconversation.com/naplan-results-reveal-little-change-in-literacy-and-numeracy-performance-here-are-some-key-takeaway-findings-70208

Ross, N. (2015). *School workforce in England: November 2014*. Retrieved on from https://www.gov.uk/government/uploads/system/uploads/attachment_data/file/440577/Text_SFR21-2015.pdf

Teach for Australia. (2016). *In Australia the postcode that you're born in can radically impact on your whole future*. Retrieved from http://www.teachforaustralia.org/

Thomson, S. (2016). Australian schools continue to fall behind other countries in maths and science. *The Conversation, Nov 29*. Retrieved from https://theconversation.com/australian-schools-continue-to-fall-behind-other-countries-in-maths-and-science-69341

Thomson, S., Wernert, N., O'Grady, E., & Rodrigues, S. (2016). *TIMMS 2015: A first look at Australia's results*. Melbourne, Australia: ACER.

Wang, X. (2013). Why students choose STEM majors motivation, high school learning and post secondary context of support. *American Educational Research Journal*, *50*(5), 1081–1121, Online. doi:10.3102/0002831213488622.

Weldon, P. R. (2015, March). The teacher workforce in Australia: Supply, demand and data issues. *Policy Insights, Issue 2.* Melbourne, Australia: ACER.

Chapter 3
STEM and Affect in Adolescence: A Cultural-Historical Approach

Wolff-Michael Roth

Abstract It is well known that by the time students enter adolescence, many of them have lost, or are beginning to lose, interest in STEM subjects. Aversion, anxiety, and other forms of affect toward STEM subjects also turn negative; and what traditional psychology theorizes as motivation also wanes when students pass the crisis—as Vygotsky called it—that takes them into adolescence. Affect tends to be attributed to the individual, but from cultural-historical perspectives, there is much more to it, and there are many social phenomena (e.g., watching a game, being at a party) where the individual is affected by the social effervescence. In this paper, I take a cultural-historical perspective to contribute to building a theory of affect in STEM. I am doing so by drawing on examples from my own teaching experience, where, when students choose the object/motive of activity, motivation also is very high. I discuss two concepts of learning: students engage in *expansive learning* when what they learn increases their agency and control over conditions (e.g., tasks); and they engage in *defensive learning* for the purpose of avoiding negative consequences (e.g., low grades, punishment). One way of approaching affect is through the phenomenon of astonishment. Astonishment is a positive affect, which is not cultured in classrooms; and yet if classrooms offer space for students to submit themselves to astonishment, then there is no need to motivate them by behaviorist means. I conclude by arguing for cultures of affect in STEM classrooms, which are especially important during adolescence, when so many students currently are turned off.

3.1 Introduction

There is a common perception in academic and popular psychology that teaching adolescents is more difficult than teaching younger students (Pickhardt, 2014). Reasons provided gravitate around issues that always involve affective dimensions, such as when young males are said to be more aggressive or when students in early

W.-M. Roth (✉)
University of Victoria, Victoria, BC V8P 5C2, Canada
e-mail: mroth@uvic.ca

R. Jorgensen, K. Larkin (eds.), *STEM Education in the Junior Secondary*,
DOI 10.1007/978-981-10-5448-8_3

adolescence (13–16 years) are said to exhibit active and passive resistance, engage in arguments, and fail to be compliant with the demands of teachers and parents alike. Changes are described from affectionate students who adore their teachers during the elementary school years to the reluctant, procrastinating adolescents who resist not only external demands but also demands that they may have made to themselves. But in the same breath, psychologists might tell stories about very positive student-teacher relations, where the former go considerably beyond what one might expect. This latter form is also the experience I have had in teaching science, mathematics, and computer science. In this chapter, I articulate a cultural-historical approach to affect, which differs in some essential ways from all other approaches, especially in the field of science, technology, engineering, and mathematics (STEM) education (e.g., Roth & Walshaw, 2015). In the following, I begin with several vignettes from my own teaching experience with middle and early high school students, which exemplify some essential invariant features of the phenomenon. The materials from the vignettes are used as resource for concretizing the articulation of a cultural-historical approach that follows. I conclude this chapter by arguing for the culturing of affect in STEM classrooms, both in the sense of growing affective cultures and understanding affect in cultural terms.

3.2 Vignettes of Affect in STEM Teaching

In this section, I present three vignettes from my own teaching. The first vignette derives from the first 2 years of my teaching experience; the second vignette occurred in the school where I completed years 3–5 of my teaching. At the time of the third vignette, I had been teaching for the better part of 11–12 years.

3.2.1 Vignette 1: "I Love Math"

After obtaining a Master's degree in physics, I started teaching without any teacher training in a small, extremely isolated village in Northeastern Canada.[1] Having appreciated independence myself, the students in my mathematics classes got together in groups of approximately equal ability, each of which established a contract with me of how much they would achieve during a 2-week period.[2] One Monday morning, the 13-year-old Earle approached me saying, "Sir, I do not feel like doing math."

[1] There are many opportunities for doing interesting curriculum that comes with village life, which I describe elsewhere (Roth, 2010).

[2] They were assessed according to the progress within their group, which required approximately five different exam forms.

- What do you want to do, I asked.
- Read my novel … from English class.
- Go ahead, I said.

Leaving his group mates who had assembled at their set of desks, he took his chair into a corner of the room and read. The next morning, Earle approached me again.

- Sir, I don't feel like doing math.

We ended up doing the same as on the day before. On Wednesday and Thursday of that week, exactly the same happened, and Earle ended up reading his novel.

On Friday morning, Earle talked to me again. This time, however, it was different.

- Sir, he said, I think I am behind my group mates.
- I think so too.
- I promise you, in 3 weeks I will have caught up.

Two weeks later, Earle not only had caught up with his group mates but also had advanced over them and assisted them when they had difficulties. He ended the year with an A grade, even though he had been a poor mathematics student during the previous year with another teacher.

On the last day of school that year, Earle came to talk to me.

- Sir, he said, I love mathematics. And do you know why?
- Go ahead.
- I love mathematics, he said, because I knew that whenever I did not feel like doing it, I could do something else. I could do mathematics when it felt right.

3.2.2 *Vignette 2: From Dropout to Academic Excellence*

At the time of the events in this second vignette, I was teaching computer science in a rural high school in Newfoundland, with class sizes limited to 12 students because there were only three computers available for teaching the course. One year, Dwayne was in the course. I anticipated that he could mean trouble. He was trouble in all his classes in that school ever since I had started teaching there—often was removed and sometimes suspended from attending school. In computer science, however, things turned out to be different. As in the years before, when my mathematics students worked at their own rate and according to a contract, the computer science students worked according to contracts for 2-week periods at the end of which they would submit the work. Despite the contract arrangements, I felt there was more needed in the case of Dwayne, who often came to school stoned, either having smoked a joint or eaten magic mushrooms picked on the lawn of the vocational school across the street. I entered an arrangement with him that he could do his work anywhere in the school and whenever he pleased. There were two fundamental conditions, though: (a) he would never bother anyone else in class or in school (e.g., when heading for the library), and (b) he must deliver the work to which he had contractually committed.

It turned out that in computer science, Dwayne never was trouble. Only once he did not submit the work by Friday afternoon as per agreement. This led me to call his parents that evening. On the next morning, the work was on the doorsteps of my home. It turns out that Dwayne started coming after school, seeking access to the computers, often staying until 10 pm and even longer (I sometimes drove students home when it had gotten too late).[3] He asked me to help him learn word processing; and he started typing up his assignments for other courses as well, printing them out on the printer we had in the classroom. One night, I had a call from his parents, who inquired about the best model to buy for their son as a gift or his birthday. He was the first student to own a computer.

At the end of the year, and despite some continuing troubles in other courses and teachers, he ended up with the third-highest grade point average of his entire year.

3.2.3 Vignette 3: Different From Everybody Else

Several years after teaching Earle and Dwayne, and following the completion of my doctorate, I was department head of science and physics teacher in a private school. In my physics classes, students spent about 70% of their time doing investigations (experiments) of their own design.[4] During my final year at the school, I shared with students the official curriculum document issued by the Ministry of Education and invited them to design their own curriculum process by means of which they would obtain the stated objectives (i.e., content). For example, in a 6-week unit on electricity, the students spent the first week thinking about, discussing, and designing what they would do and then realize their program over the next four and one half weeks. Presentations and assessment took place in the remainder of the time. We agreed that students would produce presentations or exhibits and that the assessment for the term grade would be shared, 60% coming from the evaluation by their peers in other groups, 5% was coming from self-evaluation, and 35% from the teacher's evaluation.[5]

In one physics class, for example, one group decided to do experiments at room temperature involving regular electrical conductors and semiconductors, and they would complete a series of experiments with superconducting materials, which were their primary interest. I had the superconducting materials, but students

[3] One of the superintendents of schooling started noticing that students came to school in the evening, an astonishing fact given that over 75% of the 18–25-year-olds were unemployed and have little interest or incentive to engage in schooling activity. It was upon his recommendation and with his support that I returned to university to obtain a doctorate.

[4] While there, I conducted a lot of research on student learning (e.g., Roth, 1995).

[5] It turned out that self-evaluation and peer evaluation were more severe than teacher evaluation.

realized that they also needed liquid nitrogen to make the material superconducting. I said that I did not have any but pointed them to the telephone. With some additional pointers, they not only located a company but also the lease for a special container (Dewar flask), transport, and release from all classes on 1 day entirely dedicated to the completion of all experiments planned. Other groups decided to write and design comic strips teaching electricity or write scenarios for a puppet play and to record a performance. One group produced curriculum materials for teaching electricity concepts in a fifth grade (9–10 years) and taught the curriculum.

There was one exception. Richard was opposed to working in groups—a challenge to my philosophy of teaching that was premised on the benefits from working in peers groups. His interests challenged me in another way: he did not want to conduct experiments and investigations, but wanted to do a library research project, which meant he would be outside of my direct supervision. Moreover, he requested access to a university library that was at a distance of 35 kilometers. After a conversation about how I would be affected if he were to "screw up," and a little heavy-heartedly because of my commitment to group work, I let him realize his project.

The entire unit was a success and students in all groups worked hard to realize the unit plans they had designed themselves. More so, the word was spreading in the school so that other students came to see what was happening in the physics course—which always had an open door for visitors and students from different classes working on their project outside of their scheduled times. Other teachers, too, came to see what was happening. Most surprisingly and pleasing at the same time was what had happened to Richard. Being a student who was seen as resistant in the school, always in opposition to what teachers offered, he persisted and worked hard. In his final exhibit to the class, he not only gave a university-level presentation (in terms of content and quality) but also understood the concepts he presented, as I could find out during the question period that followed. After class, Richard told me about how much he had like working in this manner.

3.2.4 Common Aspects Across Vignettes

> Freedom: affect in concept. The grand picture of development of the personality: the way to freedom. Bring Spinozism to life in Marx[ist] psychol[ogy]. The cent[ral] problem of all psychology is freedom. (Vygotsky, in Zavershneva, 2010, p. 66)

There are a number of commonalities across all three vignettes, both in terms of the organization of the classroom experience and the cognitive and affective responses. Although I did not know it at the time, the commonalities relate to the content of the quotation from Vygotsky's personal notebooks written very near the end of his life. These include affect in concept (learning), the development of the whole personality (rather than just abstract mental structures), and freedom. An education that really matters takes into account the way to freedom by affording students to take control of their activities and lives (Roth & Jornet, 2017).

In all cases, there was a good degree of collaboration on the part of students and teacher in the sense that how the curriculum was implemented resulted from agreements that both sides stuck to. Thus, the students framed the rate at which they would progress and what they would accomplish over a certain period of time. This would be the core of the contract, and deviations from it were to be accounted for.

In all three vignettes, the contractual nature gave students the freedom to work at their own rate and time. They were in complete control over what they were doing and when. They could also engage in a discussion with the teacher to renegotiate the group so that they would progress at faster or slower pace to meet their needs. Students such as Dwayne and Richard, who actively resisted in their other classes and sometimes were ejected from classroom life, manifested highly positive affect—e.g., in motivation, willingness to work after hours, and willingness to extend their study beyond requirements.

In all instances, students also were in control over other aspects of classroom life. In contrast to all of their other classes, they did not have to ask for permission to go to the washroom, for, as I argued, as their teacher, I did not have to ask them for permission to do so. It turns out that students were so engaged that they apparently hardly ever felt the need to use attending the washroom as an excuse to escape work. In any event, they did not have to work if they did not feel like. Thus, the classroom doors were always open so that students could come and go as they pleased if they wanted to.

There was considerable freedom in the process of organizing classroom life, especially in the third vignette where students chose very different forms of experiences to learn the mandated set of concepts.

Students tended to be interested to such an extent that many returned to school after dinner. In the context of the second vignette, I often drove students home, which sometimes turned out to be later than 11 p.m.; and student interest was not just internal but remarked by a superintendent. In the school in the third vignette, the school administration had the teacher on duty supervising evening studies close the physics laboratory, "because too many students were spending their time in that facility." But students did so not only to work on the subject areas I was teaching them (computer science, physics) but also to work on other subject areas. There was a culture where learning was associated with positive forms of affect. Here, affect was not hidden somewhere inside but visible to those present. This became clear to me not in the least one evening, when a small group of students came to the door joining my office and the physics laboratory: "Doc, you apparently love to learn." That is, students could see in my behavior a positive affect toward learning, which apparently inspired them to study and learn as well.

In the context of all three vignettes, the measured achievements were high and their success rates in courses after I taught them. For example, students from the school involved in the first vignette had to leave the village to attend the final year of schooling more than 1600 km from home. Over 90% were dropping out within the first year because of mathematics. One formal study in the school of the third vignette, conducted in the similarly taught biology class of a colleague (14–15 years), showed that pairs of these students outperformed preservice science teachers who had already obtained BSc or MSc degrees in terms of their data analysis-related competencies (Roth, McGinn, & Bowen, 1998).

In terms of achievement, it is to be noted that traditionally underperforming students were doing very well. In one study, where my 12-year-old students were learning physics concepts by designing artifacts, the research study revealed that five of the seven students in the first quartile of conceptual understanding were students tagged as "L[earning]D[isabled]" (Roth, McGinn, Woszczyna, & Boutonné, 1999). Some girls were learning to use power tools, others emphasizing their artistic and social knowledgeability.

How might we model and explain what was happening in the situations that I sketch in the vignettes? In the following section, I offer a cultural-historical approach to theorizing affect and intellect that does not reduce one to the other and which allows us to understand the idea I propose below: the culturing of affect in affective cultures.

3.3 Affect in a Cultural-Historical Approach

> Among the most basic defects of traditional approaches to the study of psychology has been the isolation of the intellectual from the volitional and affective aspects of consciousness. (Vygotsky, 1987, p. 50)

In this quotation from the work of Vygotsky, written in the early 1930, the major defect of the discipline of psychology was identified by the way it treated the relationship between affect and intellect. It does not take much work to verify that this separation still exists today, not only in psychology but also in all the disciplines that it informs. That separation manifests itself in different ways, sometimes in models where intellect and affect are external to but affecting (interacting with) each other. This is also the case for constructivist research, both in its Piagetian version—where affect was likened to the gasoline that drives the engine, intellect—and its present day variants—where affect is investigated via the intellect, that is, in terms of what participants say about their feelings, interests, motivations, and so on. But expressions are intellectualizations of what fundamentally is not intellectual, as affects are bodily material and distinct from the affect discourse that is cultural and historical. In Vygotsky's take, affect and intellect are different but the same; they constitute a unity/identity of opposites. They are different because affects are bodily states and sensations, whereas intellect is a cultural-historical phenomenon. They are the same, however, because they are (only) different manifestations of *one substance*, a Spinozist and monist position that Vygotsky (1999) expresses in an incomplete text entitled "The Teaching of Emotions" that was to be the starting point for the development of his own theory. He had started articulating this position in a text on the role of the environment in which he develops the category of *experience* [pereživanie] (Vygotskij, 2001). Pereživanie is the term for a person-acting-in-environment unit, which always has practical, intellectual, and affective shadings even though one or the other shading may be more in the background than the others. That is, from Vygotsky's perspective, every act manifests itself as practical doing, thinking, and feeling. Vygotsky died before he could begin articulating some ideas about the relationship of the three dimensions, but A. N. Leont'ev, a student and collaborator, interested in arriving at the true nature of the human psyche, engaged in this endeavor.

3.3.1 From the Prehistory of the Psyche

Pursuing the same agenda as Vygotsky to establish a truly Marxist psychology, Leont'ev traced the origin of the psyche to the beginning of life when simple organisms were swimming in a life-sustaining fluid medium (Leontyev, 1981). Organisms did not need to move, because they were exposed to it and thus oriented only sporadically. At that time, there were also certain sensitivities, which could have been sensitivities to nutrient gradients or light. As soon as the medium began to differentiate and food no longer was available everywhere but located in space, a new function began to emerge when change in orientation and position came to be associated with access to food. The two possible situations within the organism are "hunger" and "satiation"; the two possible situations in the environment are availability and lack of food. Already in the earliest organisms, mechanisms had to exist to valuate these states positively or negatively, for otherwise the organism population could not have survived. As long as the conditions were such that most of the time food was directly available, moving to get at food was a minor function. It became a major function in the life of the organism when the external contradiction changed such that food was accessible only in some areas but not in others. The "affective" signal went from negative to positive as a consequence of movement (action) that led to the satisfaction of the energetic needs. Here, then the practical aspect of moving to get at food came to be tied to the organism's capabilities to recognize the lack of food and the affective qualities of actually getting to it.

Already at this early stage in evolution, there were two aspects to affect. The first is a valuative measure of the current situation. The second is a sense for the future state of this first dimension that arises from an action. That is, there is an anticipated positive affect that becomes an object/motive for acting in the life of the organism. This sense also works in the reverse, when an action is valuated negatively if the anticipated outcomes lead to a decrease in the quality of life.

3.3.2 Affect and Activity

The basic structure, despite all evolutionary changes in organisms in the course of natural history, never is overturned and lost. More so, it is integral to the way in which society is organized, including its generalized provisions of need (Holzkamp, 1983). Thus, human beings participate in activities even though these do not directly satisfy their basic needs—food and shelter—but rather contribute to the generalized need satisfaction in exchange for the means (i.e., salary, income, profit) that may be used for the purchase of products that meet those needs. This creates a particular tension in the activity of schooling, which claims to educate for the future but which is institutionally organized to produce grades, reports, and certificates that become key, qua symbolic capital, to access other opportunities or to be excluded therefrom (Roth & McGinn, 1998). To articulate the relation between affect and activity requires us first to consider the nature of activity in the context cultural-historical activity theory.

Vygotsky, in this explicitly following K. Marx and F. Engels (1978), recognizes human society as constituting that what sets humans apart from other animals: "the human essence is not an abstraction inherent in the single individual. In its reality, it is the ensemble of societal relations" (p. 6). In other words, the human essence exists in the totality of relations that make society. Thus, in the psychological theory, human reasoning is not created in the head but instead is a particular form of relation among real people. As a consequence, every higher psychological function *was* a social relation with others (e.g., Vygotsky, 1989). Vygotsky does not say that the higher function was *in* the relation, where the individual picks it up to internalize or subjectively construct it. Instead, he writes that the social relation, witnessable by everyone who cares, *is* the higher psychological function. He also writes *psychological* function [*psixologičeskaja funkcija*] rather than "mental" function as this is sometimes claimed. This immediately means that all aspects of the social relation, practical, intellectual, and affective *also* characterize the higher functions. Thus, anything we might eventually ascribe to Earle, Dwayne, or Richard as specific to their personality—whether these are of intellectual, affective, or practical nature—first was a relation with others.

Vygotsky never finished the theory he was on the verge of developing, but one of his students and collaborators (A. N. Leont'ev) would develop a full theory of activity. In this theory, *activity* is a category, a smallest unit that has all the characteristics of human society (Leont'ev, 1978). Examples of such activities are farming (the production of grain), manufacturing tractors (the tools required in farming), or baking (production of food), and, most relevant here, schooling. Each activity is characterized first and foremost by its object/motive, that is, the things that the activity transforms into products that then enter the market place as generalized means for meeting human needs (Roth & Lee, 2007). It is an object/*motive* because the end product already exists in ideal form—i.e., in consciousness—at the beginning of the productive process, e.g., the farmer knows that she will end up with a lot of grain in the elevator. An activity therefore refers to the whole period from the beginning of the productive process where there are only materials to its end when the product is finished. In farming, in the beginning, there is seed grain, and at the end of the season, there is a full harvest of grain in the elevator. Whereas the productions in the three vignettes were still for school purposes, those in the case of the eight-grade student Michelle and her peers that I once taught, who contributed to an open-house event and the education of their municipality generally. Other aspects of an activity are the *means* used in production, such as tractors and irrigation devices in farming.

Rules of engagement, the community of practitioners, and *division of labor* (e.g., between farmer and farm hand or hired harvester operator) are other dimensions.

Once we take the cultural-historical activity theoretic perspective, we may ask this question: “What it is that schooling produces?” It is quite apparent that students, teachers, and principals cooperate—taking their part in the division of labor—in the production of grade reports, which students take away from schooling. Everything else they produce in the course of the year may well end up in the garbage can. Moreover, it does not matter what and how much students learn, as can be seen in a number of facts: (a) students tend to ask, “What d’I get?”; (b) students frequently “cheat” (see below); (c) students can graduate even though they are functionally illiterate; (d) research shows that success at the tertiary level (college, university) and in everyday life tends to be independent of the number of school courses taken and the grades obtained (e.g., Saxe, 1991); and (e) what is learned in schools (e.g., mathematics) often is not useful on the job, and job competence is irrelevant in school (e.g., Roth, 2014). On the other hand, when Michelle and her peers produce an exhibit for an environmentalist open house, they contribute something to the common good, including the data they had collected that entered a database to which environmentalists, university students, and scientists also contributed; and it contributed to the common good in educating visitors of the open-house event.

The activity theoretic perspective makes apparent the relationship between the motive of activity and one dimension of affect: The distance between the activity in its current state and the motive, which is an image of the final product. Another dimension is the “payoff” that the product brings with itself. Most importantly, the cultural-historical approach thereby makes unnecessary the concept of motivation from traditional psychology. As the vignettes show, Earle, Dwayne, Richard, Michelle, and their peers did not need to be motivated externally. This situation arises from the fact that motivation, the pursuit of the object/motive, cannot be segregated from its content (Holzkamp, 2013). They did not need to be motivated because being motivated requires agents to be in a position to anticipate that their control over the conditions, life possibilities, and quality of life is enhanced. Such an enhancement does not depend on the individual or collective agent but is an objective feature of the motive of activity. In contrast, motivation theorized in traditional psychology is subject to strong criticism from cultural-historical perspectives because the concept is used to theorize how to make others (students, laborers) do what they *do not inherently want to do* (Holzkamp-Osterkamp, 1975, 1976). Traditional motivation psychology thereby occludes the object/motive of activity, which may actually not be in the interest of the person. As a result, motivation theories are tools for those intending to control (generally capital and middle/upper class) and, thereby, subject others (generally working and lower class). When the social mechanisms have been individualized, then traditional motivation psychology speaks of self-motivation, which in fact is individualized coercion.[6] But such

[6] In Vygotsky’s theory, where any higher psychological function *was* a social relation first, self-motivation would have its genetic origin in social relations where coercion is external; and when the relation is individualized, then this form of coercion becomes self-motivation.

external control—and its individualized version—is unnecessary when the agents (students, but also teachers) themselves not only have control but also increase control over their conditions, especially through efforts that increase their agency, a situation that is articulated in the following subsection as *expansive learning*. Cultural-historical activity theory makes the whole concept of motivation superfluous. Perhaps unsurprisingly, the control of affect, through motivational strategies, historically had its origin in industrial psychology, where it was intended to get workers to be more productive; and motivational psychology has a tremendous field of application in schools, where the conditions are not unlike those in the factories where many students would be working (e.g., Roth & McGinn, 1998). It is well known that adolescents in particular attempt to escape external control, a part of their effort to become themselves, independent persons. This is why it is so important to involve adolescents specifically in gaining control over the object/motive of their activity. This was the case in all of the vignettes described above.

3.3.3 Expansive and Defensive Learning

In the course of production, we may realize that there are things we can do that will expand our agency. Thus, it was apparent that Dwayne treated knowing word processing as an expansion of his power to act and control over his conditions; he would be able to do what he could not do before, including easy corrections and revisions and still deliver a perfectly clean product without having to rewrite an entire essay. In the course of completing the work he had contracted, he identified in word processing an expansion of his control over the activity. It was in his interest to learn word processing, because it expanded what he could do and therefore the control over his condition. This is an instance of *expansive learning* (Holzkamp, 1993). The concept inherently includes affect, for the expansion of agency and control over condition is associated with positive valuation and affect. The students who had decided to do experiments with superconductors expanded their power to act by finding suppliers of liquid nitrogen and Dewar flasks and then organized when to do the experiments, delivery, release from other courses, and laboratory space, as well as the experiments themselves.

In all vignettes, we observe examples of control over conditions, which would explain why the environment was associated with positive affective value. In framing the contents of the contracts, the students controlled what they would produce over a certain period of time—which varied from context to context between 1 and 2 weeks. Students also controlled how they would work together; and especially salient of current trends in the industry, they controlled when they would do the work outlined in their contracts. Such control over the condition of production, in contrast to lack of control and the experience of "being pushed around," tends to have positive value. There were a few exceptions, most of which were tied to students from Asia. These students, pursuing the object/motive of a school leaving certificate with a grade point average sufficiently high to make it into a school of their choice, expressed the wish to have their tasks specified such that their work habits that led to (positively experienced) success in the past also would work in the classes I was teaching.

The expansive nature of learning also may explain why students returned to school in the evening. Engagement at that time led to an expansion of agency in a context where others also worked toward expansion of their personal agency and, therefore, where there was a culture with positive affective values: students learned because the associated expansion of agency was experienced in a positive manner. They also returned to school in the evenings because the successful production itself was imbued with positive affect; and the collective effervescence during those evening hours was experienced positively. The school administrators associated with the third vignette perceived the situation as one in which "students do too much physics"; they failed to ask themselves why that might have been so and what kind of arrangements would make it such that the students would have similar forms of affective experiences in all of their other work.

Both during the day—in the classroom with its open doors so that other students, teachers, and visitors of the school could enter to observe, participate, or do their own work—and in the evenings, the positive affective climate was palpable and, as such, objectively existing. All students were working on their contracts, or whatever else they had to do at the instant, and nobody was waiting for the teacher. Students were working even when, for one or another reason, I was late for a scheduled class. That is, we had cultured affect by creating a culture of positive emotions. As cultural sociologists show and theorize, such affect may infect others (Collins, 2004). A good example are games and parties: Even if we experience an emotional low, going to watch a game or going to a party may lead to an emotional high when we get swept up by the collective effervescence.

In expansive learning, there is a promise of a "payoff"; and it is toward this payoff that labor is directed. This is so even if there is work to be completed and even if this work is hard. Thus, even if the process toward the payoff itself is tinged negatively in affective terms, we engage in the work because of the affective gains that we experience in the end. Of course, if the work itself comes with affectively positive qualities even though it is "hard," e.g., for athletes, even training hard comes with positive affect once they enter "the zone."

When the current affective state has negative value, giving up actually does not change. Instead, any hope of overcoming the negative affective state is to actually do the work. In one study we showed how a student (Mario) in mathematics class, even though he was very frustrated—as apparent from his physical demeanor—and verbally expressed failing to understand was continuing (Roth & Radford, 2011). Even the teacher became frustrated when she tried to assist Mario to comprehend the task and the steps he had to undertake to complete it. But eventually there was a breakthrough, and in the end, Mario said with apparent satisfaction in his voice, "Me, I now understand." There is a contradiction in the sense that to get out of the frustration, the student had to continue, even though each act may have (initially) felt hard. Actually, initially the continuation only aggravated the situation, as Mario manifested increasing frustration. Only subsequently—and despite the negative affective qualities associated with acting when one does not know and understand—did the affective quality of the work change. A turnaround was observable when an action apparently led to a result that the teacher marked as correct, and then, rapidly, after another correct action, the student engaged no longer expressing frustration until he clearly manifested satisfaction.

Defensive learning occurs when students develop competencies that allow them to complete some task without necessarily—and frequently despite—learning what the task is to teach. In such instances, students do and learn whatever is required to get the job done to avoid negative repercussions—which may come in the form of criticism, a low grade, or punishment. It is well known that many students learn to "cheat," that is, find ways in which they complete assignments to get a good or passing grade without actually learning. Thus, for example, another student in Mario's group was equally frustrated. She pounded the table repeatedly, threw herself against the back of the seat, and began to disengage by resting her head on her arms on the desk. The reverse occurred from what we perceive in Mario's case. She did not engage, and her situation did not change (Roth & Walshaw, 2015). In the end, she copied what others had written on their worksheets. It appeared as if she had done the task, when in fact the video shows that she had not. None of the adolescents in the vignettes above had to feign learning because they were in control over when, where, how, and on what to work or whether to work at all. In fact, even those students learn who get better at cheating; it is just that they get better at something that in the context of schooling is not appreciated and is sanctioned—though in the work world, drawing on all the resources possible to get a job done would be the norm, including talking to others or getting help.

Defensive learning tends to occur when the students do not or cannot see why they are doing (have to do) what they are asked to do in STEM classes. Teachers and school counselors might tell them that they need to know science or mathematics to make a good living. Students, however, inherently know this to be a lie, for they can see many people, even those living on their own street, leading good lives without knowing any or much science and mathematics. Thus, there is no real incentive telling adolescents that they need science or mathematics. Tricking students into completing tasks by "motivating" them is but another strategy that does not have to lead to learning because the motive of the task now is obtaining gratification (e.g., a star, candy, or whatever else teachers promise) rather than learning what is to be learned. In contrast, as shown in the vignettes, when students are and experience themselves in control over conditions, gaining greater control is inherently rewarding in affective terms.

3.3.4 Affect and the Whole Person

> The inevitable consequence of the isolation of these functions [affect, intellect] has been the transforming of thinking into an autonomous stream. Thinking itself became the thinker of thought. Thinking was divorced from the full vitality of life, from the motives, interests, and

> inclinations of the thinking individual. Thinking was transformed either into a useless epiphenomenon, a process that can change nothing in the individual's life and behavior, or into an independent and autonomous primeval force that influences the life of consciousness and the life of the personality through its intervention. (Vygotsky, 1987, p. 50)

Entering adolescence—Vygotsky calls it a critical phase in the transformation of the person that occurs around the age of 12 with the onset of the biological transformation into puberty—means radical transformation in the consciousness of the person. That is, there is a decisive change in the way we experience ourselves as persons, and a tremendous shift in personality. Readers will be able to affirm that most STEM research does not even consider affect, being concerned with what and how students construct knowledge. Affect, if treated at all, is conceptualized as another factor in learning (constructing), that is, inherently as something outside of and somehow mediating intellect. Vygotsky, as the quotation shows, takes a different approach. Not only does he ask us to consider intellect and affect as two manifestations of one and the same, but also he encourages us to consider the two in the context of the "full vitality of life" of an individual. Only by considering the two in this context will we avoid making thinking an epiphenomenon. To understand thinking, it has to be theorized in the context of other manifestations of life activity, that is, the motives, interests, and inclinations of the thinker. Focusing on mathematics anxiety, or science interest, or preferences for computers will not do the trick. How are we then to think about affect through a lens that considers the *whole person*, not just the thinker in a science, technology, engineering, or mathematics class? For a response, Vygotsky already provides a lead in the quotation: We need to take a look at intellect and affect through the lens of *personality* and the *whole* life, which always is societal life. Again, it was Vygotsky's student Leont'ev (1978) who made personality a category, tying it to that of activity. That is, personality is the smallest unit that contains all features of society as a whole and is pertinent to describing the individual person.

Every day, we participate in many different societal activities, each time being part of the division of labor characteristic of that activity. For example, I make a living as a professor, engaging both in the production of knowledge and in the reproduction of society through the training of graduate students. I am also a husband, a hobby gardener year-round producing all vegetables that my family eats, an amateur athlete, a hobby craftsperson renovating the house, a shopper for groceries, a consumer of television news, and so on. All of these *societal* activities are characterized by object/motives, to the production and transformation of which I contribute in and through my participation. Who I am is defined by these forms of participation and in the way that they are connected for me. This is so because "[h]ow individuals express their lives, so they are. What they are coincides with their production, as much as with *what* they produce as with *how* they produce. What individuals are depends on the material conditions of their productions" (Marx & Engels, 1978, p. 21).

All of the adolescents in my vignettes—Earle, Dwayne, Richard, and Michelle—are real persons, in flesh and blood, who do many things besides going to school. Unless we take into account all of these other things, we do not understand what they are doing in school and how they are doing it. The proposal therefore is to theorize personality by means of all the object/motives that we contribute to realizing (Leont'ev, 1978). In this way, our personality reflects a part of society, which is a network of activities. That is, personality is a concrete realization of part of the network. In this, personality is societal through and through, as each activity is societal in character, reflecting all characteristics that society has a whole. But there is a highly individualistic aspect, which makes it that each person differs from everyone else: within the person, there is a highly personal hierarchy of the activities so that even if two individuals were engaging in precisely the same activities, they would not be the same because of the different organization of the network of object/motives. Thus, for someone, the job may be the most important aspect of daily life and shopping or engaging in sports may be at the bottom, whereas for another person the relationships and intensities thereof between different object/motives will be different.

Adolescence is characterized by increasing participation in an increasing number of activities and a coincident sense of independence and control over conditions. Thus, considering adolescents and STEM, we must orient not simply to doing this or that in science or mathematics class to "motivate" students, for we would not understand how schooling generally and science or mathematics particularly relate to everything else that the student does. Teachers inherently cannot know *all* the activities in which each student engages. Thus, I now realize that already in my early teaching, having students participate in the decision-making processes allowed them to take control in ways that are appropriate to their personality. Michelle cared little about the science experiments her male peers did, for example, floating oranges in the creek to measure speed, which they correlated with aspects of the cross section. Instead, she was concerned with the social aspects of life and with contributing to the education of others. Giving students the opportunity to define the object/motive takes into account their interests, motivations, and inclinations, allowing them to determine what to do and how to do it. It takes into account where in their development they are at the moment so that they can expand their agency and control based on who they are at the time. It is out of such engagement with things and phenomena already familiar that qualitative changes in consciousness arise (Vygotsky, 1997). That is, development (i.e., jumps to qualitatively different forms of being and consciousness), as distinct from incrementally increasing competencies (i.e., learning), nevertheless arises in learning where the revolutionary changes in personality and psychological functions are provoked.

Taking a whole-life perspective on adolescents in STEM also means transcending the perspective of students as they are now, today, in our class. We need to think beyond the tasks of today's STEM class. Instead, a whole-life perspective considers the future and the past; we need to consider STEM learning with respect to the whole life span (Roth & van Eijck, 2010): where will Earle, Dwayne, Richard, and Michelle be in 5 or 10 years down the road, and who will they be?[7] I have frequently

---

[7] Only 2 years before the writing of this chapter, 32 years after teaching Earle, I was contacted by one of his former classmates, who had become manager of operations in an oil mining camp. He

heard teacher colleagues comment that the "hormones are taking over" or that a student "is dreaming away," or make comments that go in the same direction. In school staffrooms, we can hear many comments about loss of interest and lack of motivation on the part of adolescents with respect to STEM courses. But all we have to do is go outside the school, and we may see the same students engage for hours in some pursuit, be it in some sport or other form of activity. That is, those adolescents are not *inherently* unmotivated and disinterested. The root cause of any such assessment is school itself. Schooling is the problem, and, therefore, STEM education may have to be rethought in terms of deinstitutionalization (Roth, 2015). We experienced an interesting case in what we thought to be a very open science curriculum for adolescents—the one in which Michelle also participated. There was one (indigenous) student to whom participation in environmentalism did not appeal at all, even under the conditions where he was in the position of deciding what he wanted to do and how. It turned out that he was interested in filming generally and in filming (as we were doing) what his peers were doing particularly. We then framed his activity as one in which he would produce a documentary of what his peers were doing, including following them around in the watershed and interviewing them.

In the physics courses of the third vignette, as the experimental curriculum focusing on environmentalism, students chose what to do and how to do it in terms of larger motives, in the context of their lives. Thus, there were students with interests in theater and envisioning themselves working toward such careers; they realized the electricity curriculum by writing plays and by acting them using puppets. Richard was interested in becoming a medical doctor, which allows us to see why doing research on the functioning of electricity in the brain might be of interest. The 13-year-old Michelle in the environmentalism curriculum envisioned herself as journalist, so she included in her project photo series featuring the poor environmental health of a local creek, recorded reports live from different spots along the creek, and interviewed First Nations elders and community politicians about their understandings of and actions toward improvement of water quality in the creek. Interviewed 1 year later by an independent researcher, she would say, "It was fun … I think that every student should do it … all students [in the school] … so more people would get educated about it." What these adolescents were doing as part of schooling already was aligned with possible futures and careers. Their work in science and computer science classes already was aligned with other activities of their life high on their personality constituting hierarchy.

---

thanked me for the affective support, which allowed him and his family to make the decision to continue high school far from home despite his learning disability, which had been made visible in all his other courses.

3.3.5 Affect and Affect Discourse

Without doubt, affect is an important feature of adolescence, especially because the radical nature of the revolutionary transformations occurs in individuals entering adolescence, transformations to qualitatively new forms of biological, cognitive, pragmatic, and affective life. Whereas much research does not even consider affect, those studies that do tend to fail distinguishing the biological and cultural dimensions. Thus, even though the biological and the cultural-historical dimensions are manifestations of one and the same, and even though these dimensions intermingle and cannot be taken apart, they can be distinguished (Vygotsky, 1997).

I therefore maintain that affect (in activity) and talking affect are very different phenomena in the same way as thinking (in activity) is different from thinking (talking about) thinking. In fact, this position is a consequence of Vygotsky's Spinozist orientation according to which one manifestation of life—here, language and consciousness—cannot express the truth of another manifestation—here affect. Moreover, from the cultural-historical activity theoretic perspective, this is so because the two belong to different forms of activity and, therefore, come with different forms of consciousness. Thinking and affect in the course of activity—e.g., producing a report on electricity in the brain or producing an exhibit for an environmentalist open house—have in view the specific outcomes (object/motives) related to *that* activity. Talking affect, for example, in the context of an interview with a researcher has the production of interview text focusing on affect as its motive. This production and participation therein is characterized by its own object/motive and, thereby, by its corresponding versions of intellect and affect. Responding to researchers interested in affect, adolescents will draw on whatever affect discourse is available for the purpose of helping the researcher. It is well known in practice theory that (a) there may be no relation between what practitioners say they do (and feel) and what they do and feel in the activity and (b) practitioners are no better getting at the heart of a praxis-related phenomenon than researchers (Bourdieu 1980). If we are interested in affect among adolescents *while* participating in STEM tasks and activities, then we are better off to make observations directly, especially in situations where affect manifests it such that everyone present notices it, including the vicariously observing researchers. Thus, in our research in workplaces and STEM classes alike (e.g., Roth & Tobin, 2010), analytic techniques were used to show affect at work without having to ask participants—e.g., by using voice analysis software over periods of time that involves different manifestations of affect. Analysis of movements among other participants—e.g., rhythmic behavior—shows when there is solidarity with respect to the situation and, therefore, the degree to which others empathize with the person expressing strong emotions.

A second dimension of importance is the fact that affect discourse is shaped by the available language. Thus, what adolescents (or any other type of person for that matter) say about the affective dimensions of their lives is a function of this language not a function of the individual. Therefore, it is not the individual student that will elucidate the nature of affect in STEM classes, but instead it will be the language itself. The unit of analysis of affect discourse, therefore, is not the individual student but language itself. Different students merely realize this or that aspect of the available discourse. Just as the adolescent discourse about technology and the environment reflects the discourse available in society (Zeyer & Roth, 2009), we will find that the discourse about the STEM-related interests, needs, inclinations, or motivations merely reflects the discourse available in the society as a whole (Hsu & Roth, 2012). This does not mean, however, that there is no relation at all between affect talk and affect in activity specifically and affect and intellect generally.

In the cultural-historical approach, intellect and affect, together with practice, are taken to be manifestations of an overarching whole. Manifestations inherently are one-sided. They therefore cannot be used to constitute the whole by such means as interactions or mediators. The relationship exists at the overarching level, and this is where mutual determinations occur. Thus, affect and intellect change together, for example, in the course of accomplishing some task. In the description and analysis of Mario's work in the mathematics class, we observe how practical and intellectual engagement in the task are associated with changing affect; and, of course, changing affect may be associated with changing levels of intellectual and practical engagement, as seen in the case of Mario's peer, who completely disengaged (Roth & Walshaw, 2015). But we also know that simply saying, "I am not afraid" when I am afraid or stating, "I am not nervous" prior to a mathematics exam when I am nervous, does not change our state. Talk does not change our affective state—engagement in activity does, as we know when we forget about being nervous or anxious as soon as we are fully engaged in the taking of the mathematics exam. Thus, rather than trying to tell students to talk themselves into changing affect, it may be more productive to offer up opportunities for participating in events in the course of which affect changes—which may include providing intellectual, practice, and affective support when the going is tough. But, as shown throughout all the vignettes I provide here from my own teaching, when students are in control over a task, and thereby feel in control over it, they also are in a position to make the kind of changes required to manage their current affective states—e.g., by reading a novel, as Earle did when he did not feel like doing mathematics—and then return to the task when they are ready, affectively, intellectually, and practically speaking.

3.4 Culturing Affect, Affective Cultures

In the preceding sections, I present vignettes from my own teaching and then use these to exemplify a cultural-historical approach to affect in adolescence. Invariant across all the vignettes is the existence of a culture of learning premised by a lot of student control over the conditions, including determining the object/motive and the means of attaining them. In the process, students engaged in expansive learning activities, which are characterized by the increasing room to maneuver and power to act. Increasing power to act, agency, or room to maneuver is reflected by positive affective valence in the agents, who also have greater likelihood to reach their goals and successfully complete what they set out to produce. Affect and the affective-motivational aspects of activity are especially important in adolescence, which is a period in which personality begins to take shape while the individual moves from childhood to adulthood. In the vignettes, adolescents are supported by contexts that culture affect and, in so doing, create affective cultures. How, especially pragmatically oriented educators focusing on applicability of ideas to the classroom might ask, may we go about culturing affect? Although there cannot ever be generic answers to such questions, because the solutions depend on all participants including the students, I offer in the following one avenue. This avenue is organized around *astonishment*, a form of affect that already was of interest to the philosopher of reason I. Kant, who used examples from mathematics to illustrate the concept.

The verb to "*astonish* can imply a dazing or silencing or it may mean to surprise so greatly as to seem incredible … or sometimes merely unusual" (Merriam-Webster, 1984, p. 804). In contrast to surprise, which may be anywhere along the continuum from positive or negative, astonishment is shaded positively similarly to what we experience in wonderment. Astonishment, therefore, comes with both intellectual and affective characteristics. Associated with affect and being-affected by the unexpected, astonishment requires us to think differently about the subject—who neither has some stable identity nor is exclusively agential. Using the geometrical object of a circle, Kant (1957) suggests that it is a source of continued wonderment. For example, he proposes a theorem that can be shown to be valid *in every case considered:* any two lines going through a point located within a circle and intercepting the circumference *intersect each other in proportion* (Fig. 3.1a). As readers can see from the materials provided in my proof account (Fig. 3.1b), the theorem is valid *for every case*, which is a source of astonishment and wonderment for those inclined. Kant states several other theorems (without proof), all involving the circle. He suggests that it is really astonishing how a simple geometrical figure such as the circle can harbor so many amazing truths: "indeed, one is surprised and justifiably put into admiration by such a curious combination of the manifold arising from such fruitful rules and such a poor and simple-minded thing, as is the circle" (Kant, 1960, p. 656). Although I personally was as astonished as Kant was—as I found out doing all the proofs for the theorems Kant states because he does not provide them—we cannot expect that every adolescent will be interested in pursuing the phenomenon of astonishment here. I merely make the point that whenever the object is STEM, classrooms that foster astonishment inherently culture affect and, thereby, create affective cultures. In other words, teachers might orient their organizational actions with the vision to get their adolescents to exclaim, "This is amazing! I would not have thought that this would happen (is the case, I could do that)." Such an orientation will have to begin with the beginning of schooling so that by the time students enter adolescence, expansive learning, control over condition, and creating opportunities to experience astonishment are features of their classroom culture as well as features of who they are and become.

Throughout my teaching and research career, I have found that the most difficult part of the enterprise is that of getting such a culture created and going. But I also experienced that once established, the culture sustains itself. When students get to

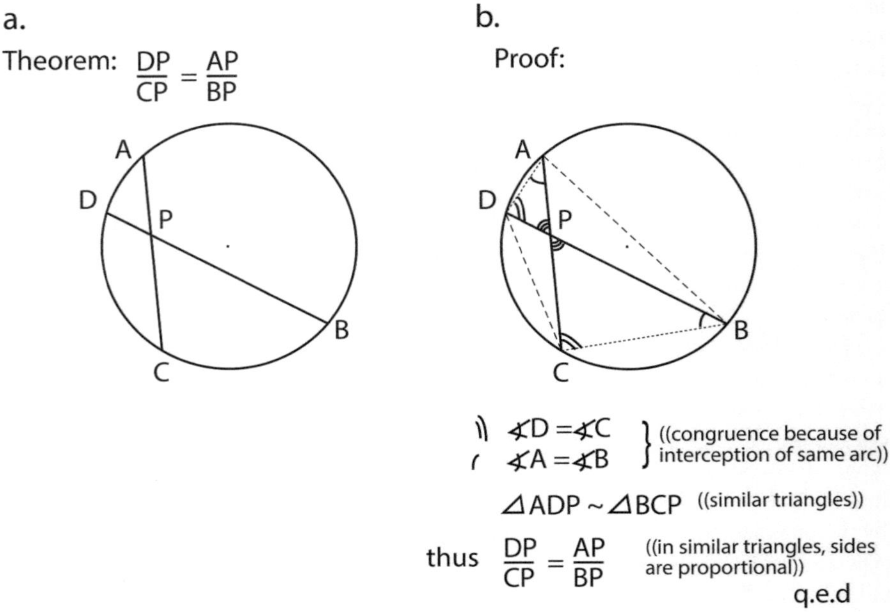

Fig. 3.1 (**a**) Kant's theorem (**b**) proof account for the theorem

share what they have done and produced, others often become interested themselves and take up object/motives or the means of production that they have come to know about. It is for this reason that I personally prefer multiage or multilevel classrooms, where only some students leave each year replaced by a similar number of new students. The culture is sustained because one half or two thirds of the students remain, doing what they have done before. I know that means other things than just the intellectual aspects of culture. For example, I know that it was important for my homeroom students when I spent time with them during their lunch period. During that time, they talked to me a lot about their real concerns. For example, they told me that it is better getting stoned by smoking a joint rather than getting drunk consuming alcohol, because they get more easily caught with a case of beer. Just as the psychologist suggests to whom I refer in the beginning of this chapter, students affectively bond when they "have the attentive middle school science teacher with whom some of her students choose to eat lunch a couple times week. 'We just hang out, but it's no big deal'"(Pickhardt, 2014, para. 8). It is not big deal, perhaps, but, as Pickhardt realizes, it is a big deal "for young people hungry for meaningful communication with an adult." The affective cultures I advocate here on cultural-historical grounds have a lot to offer to turn STEM classrooms into environments where it is safe and feels good to grow into adulthood. That is, we do not have to create afterschool science clubs, such as those created in Montreal for largely immigrant students (Rahm, 2016), for every student to become part of an affective culture. Instead, STEM classrooms can offer such safe and feel-good environments.

This is especially true for those students, who, as I have, come from lower-class backgrounds and poverty—which is one of the reasons why their well-being has always been at the core of my educational endeavors.

References

Bourdieu, P. (1980). *Le sens pratique* [The logic of practice]. Paris: Les Éditions de Minuit.

Collins, R. (2004). *Interaction ritual chains*. Princeton, NJ: Princeton University Press.

Holzkamp, K. (1983). *Grundlegung der Psychologie* [Foundations/Founding psychology]. Frankfurt/M, Germany: Campus.

Holzkamp, K. (1993): *Lernen: Subjektwissenschaftliche Grundlegung* [Learning: Foundation in a science of the subject]. Frankfurt/M, Germany: Campus.

Holzkamp, K. (2013). In E. Schraube & U. Osterkamp (Eds.), *Psychology from the standpoint of the subject: Selected writings*. Basingstoke, UK: Palgrave Macmillan.

Holzkamp-Osterkamp, U. (1975). *Grundlagen der psychologischen Motivationsforschung 1*. Frankfurt/M, Germany: Campus.

Holzkamp-Osterkamp, U. (1976). *Grundlagen der psychologischen Motivationsforschung 2*. Frankfurt/M, Germany: Campus.

Hsu, P.-L., & Roth, W.-M. (2012). Understanding beliefs, identity, conceptions, and motivations from a discursive psychology perspective. In B. J. Fraser, K. Tobin, & C. McRobbie (Eds.), *Second international handbook of science education* (pp. 1435–1447). Dordrecht, The Netherlands: Springer.

Kant, I. (1957). *Werke Band V* (Works Vol. 5). Wiesbaden, Germany: Insel.

Kant, I. (1960). *Werke Band I* (Works Vol. 1). Wiesbaden, Germany: Insel.

Leont'ev, A. N. (1978). *Activity, consciousness and personality*. Englewood Cliffs, NJ: Prentice Hall.

Leontyev, A. N. (1981). *Problems of the development of mind*. Moscow: Progress.

Marx, K., & Engels, F. (1978). *Werke Band 3* [Works Vol. 3]. Berlin, Germany: Dietz.

Merriam-Webster. (1984). *Dictionary of synonyms*. Springfield, MA: Author.

Pickhardt, C. E. (2014, October 27). Teaching adolescents. *Psychology Today*. https://www.psychologytoday.com/blog/surviving-your-childs-adolescence/201410/teaching-adolescents. Accessed 7 Jan 2016.

Rahm, J. (2016). Videomaking projects in STEM after-school clubs: A space-time analysis of one production and its implicit dialogue among identities, ways of knowing and doing in and with science. In L. Avraamidou & W.-M. Roth (Eds.), *Intersections of formal and informal science* (pp. 69–81). New York: Routledge.

Roth, W.-M. (1995). *Authentic school science: Knowing and learning in open-inquiry science laboratories*. Dordrecht, The Netherlands: Kluwer Academic Publishers.

Roth, W.-M. (2010). Local matters, ecojustice, and community: Opportunities of village life for teaching science. In M. Müller & D. Tippins (Eds.), *Cultural studies and environmentalism: The confluence of eco-justice, place-based (science) education, and indigenous knowledge* (pp. 51–82). Dordrecht, The Netherlands: Springer.

Roth, W.-M. (2014). Rules of bending, bending the rules: The geometry of conduit bending in college and workplace. *Educational Studies in Mathematics, 86*, 177–192.

Roth, W.-M. (2015). Schooling is the problem: A plaidoyer for its deinstitutionalization. *Canadian Journal for Science, Mathematics, and Technology Education, 15*, 315–331.

Roth, W.-M., & Jornet, A. (2017). *Understanding educational psychology: A late Vygotskian Spinozist approach*. Dordrecht, The Netherlands: Springer.

Roth, W.-M., & Lee, Y. J. (2007). "Vygotsky's neglected legacy": Cultural-historical activity theory. *Review of Educational Research, 77*, 186–232.

Roth, W.-M., & McGinn, M. K. (1998). Science education: /lives/work/voices. *Journal of Research in Science Teaching, 35*, 399–421.
Roth, W.-M., & Radford, L. (2011). *A cultural-historical perspective on mathematics teaching and learning*. Rotterdam, The Netherlands: Sense Publishers.
Roth, W.-M., & Tobin, K. (2010). Solidarity and conflict: Aligned and misaligned prosody as a transactional resource in intra- and intercultural communication involving power differences. *Cultural Studies of Science Education, 5*, 805–847.
Roth, W.-M., & van Eijck, M. (2010). Fullness of life as minimal unit: STEM learning across the life span. *Science Education, 94*, 1027–1048.
Roth, W.-M., & Walshaw, M. A. (2015). Rethinking affect in education from a societal-historical perspective: The case of mathematics anxiety. *Mind, Culture, and Activity, 22*, 217–232.
Roth, W.-M., McGinn, M. K., & Bowen, G. M. (1998). How prepared are preservice teachers to teach scientific inquiry? Levels of performance in scientific representation practices. *Journal of Science Teacher Education, 9*, 25–48.
Roth, W.-M., McGinn, M. K., Woszczyna, C., & Boutonné, S. (1999). Differential participation during science conversations: The interaction of focal artifacts, social configuration, and physical arrangements. *The Journal of the Learning Sciences, 8*, 293–347.
Saxe, G. B. (1991). *Culture and cognitive development: Studies in mathematical understanding*. Hillsdale, NJ: Lawrence Erlbaum Associates.
Vygotskij, L. S. (2001). *Lekcii po pedologii* [Lectures on pedology]. Izhevsk: Udmurdskij University.
Vygotsky, L. S. (1987). *The collected works of L. S. Vygotsky, Vol. 1: Problems of general psychology*. New York: Springer.
Vygotsky, L. S. (1989). Concrete human psychology. *Soviet Psychology, 27*(2), 53–77.
Vygotsky, L. S. (1997). *The collected works of L. S. Vygotsky, Vol. 4: The history of the development of higher mental functions*. New York: Springer.
Vygotsky, L. S. (1999). *The collected works of L. S. Vygotsky, Vol. 6: Scientific legacy*. New York: Springer.
Zavershneva, E. I. (2010). The way to freedom. *Journal of Russian and East European Psychology, 48*(1), 61–90.
Zeyer, A., & Roth, W.-M. (2009). A mirror of society: A discourse analytic study of 14-15-year-old Swiss students' talk about environment and environmental protection. *Cultural Studies of Science Education, 4*, 961–998.

Chapter 4
Supporting Students' Productive Collaboration and Mathematics Learning in Online Environments

Arthur B. Powell, Muteb M. Alqahtani, and Balvir Singh

Abstract Digital technologies provide a wide range of tools and functions that can support students' learning of mathematics as well as the development of their mathematical and collaborative practices. Bringing such technologies to mathematics classrooms often do not parallel students' previous classroom experiences, especially when collaborative practices are emphasized. When facilitating mathematics learning, discrepancies between students' previous classroom experiences and their expected engagement with new collaborative technologies result in challenges to which teachers need to attend. In this chapter, we describe how a high school mathematics teacher engaged his students in an online collaborative environment, Virtual Math Team with GeoGebra (VMTwG), and how he addressed students' technological and collaborative challenges to support growth in their geometrical understanding. From a cultural historical perspective, we present a model of how teachers can support students' instrumentation of collaborative environments and mathematical understanding. In our model, during a mathematical activity, teachers progressively decentralize their role and, simultaneously, support students' development and performance of collaborative practices. This model informs the theory of instrumental orchestration (Trouche L, Interact Comput 15(6):783–800, 2003; Trouche L, Int J Comput Math Learn 9(3):281–307, 2004; Trouche L, Instrumental genesis, individual and social aspects. The didactical challenge of symbolic calculators. Springer,

This chapter is based upon work supported by the National Science Foundation (NSF), DRK-12 program, under award DRL-1118888. The findings and opinions reported are those of the authors and do not necessarily reflect the views of the funding agency.

A.B. Powell (✉)
Rutgers University-Newark, Newark, NJ, USA
e-mail: powellab@newark.rutgers.edu

M.M. Alqahtani
State University of New York, Cortland, NY, USA
e-mail: muteb.alqahtani@cortland.edu

B. Singh
Burlington City High School, Burlington, NJ, USA
e-mail: balvir.singh@gse.rutgers.edu

R. Jorgensen, K. Larkin (eds.), *STEM Education in the Junior Secondary*,
DOI 10.1007/978-981-10-5448-8_4

New York, pp 197–230, 2005) by providing a pedagogical intervention trajectory that supports students' instrumental genesis (Rabardel P, Beguin P, Theor Issues in Ergon Sci 6(5): 29–461, 2005) of collaborative mathematical environments and shifts students' focus from their teacher to their peer collaborators.

4.1 Introduction

The functionalities and tools of Web 2.0 applications offer potential support for mathematics learning by providing virtual spaces for individuals to perform collaborative and mathematical practices. Mathematical practices (Common Core State Standards Initiative, 2010) can be performed and made visible with dynamic mathematics software such as dynamic geometry environments (DGEs). These environments afford learners' abilities to construct, visualize, and manipulate geometric objects and relations and dependencies. These affordances support empirical explorations and theoretical justifications or proofs (Christou, Mousoulides, & Pittalis, 2004). In DEGs, empirical explorations are experienced immediately, while the need to formulate proofs is latent and to be realized requires either learners' disposition toward justification or pedagogical intervention. Pedagogically motivated transitions from empirical explorations to theoretical justifications depend on carefully designed tasks, teacher guidance, and classroom climates that support conjecturing and deductive justifications (Öner, 2008).

Conjecturing and deductive reasoning or formal proofs have been regarded as the pinnacle of geometry education (Wu, 1996). Students taking a formal geometry course at the high school level are expected to construct (in Euclidean sense) geometric objects and use the relations among objects (or parts of objects) to prove why certain properties or relations are true (Common Core State Standards Initiative, 2010). In contrast, at the middle school level, students are primarily expected to solve basic geometric problems (numerical and algebraic) using given formulas, and at best, they may be expected to describe or verify properties through experiments (Common Core State Standards Initiative, 2010). Noticeably, students are not expected to provide arguments for the properties and relations; however, at the high school level, this expectation changes dramatically. Without prior experiences justifying mathematical statements, this dramatic change causes difficulties for students to understand basic tenets of mathematical proofs (Miyazaki, Fujita, & Jones, 2016). One objective of STEM education concerns helping students develop meaningful use of tools to investigate phenomena and construct viable arguments. To address this objective in mathematics education, teachers need to support students' explorations and thinking about mathematical objects and relations among them. DGEs, uniquely designed to promote explorations, can be used to transition middle school students from the current geometric-properties focused learning to relational reasoning focused learning (Jones, 2000). This relational understanding is what enables students to move away from empirical explanations toward deductive arguments. In addition to geometric constructions, DGEs provide seamless access to both graphical and algebraic representations as well as present immediate, visual feedback. Appropriate and strategic use of DGEs as vehicles for representing mathematical situations supports STEM education in mathematics classrooms.

Learning environments that support conjecturing and deductive reasoning can be virtual as well as presential, focused on the individual or collaborative groups. Support for social conjecturing and justification can occur in computer-supported collaborative learning (CSCL) environments (Öner, 2008; Silverman, 2011). Longitudinal investigations suggest that learners' dispositions toward conjecturing and deductive reasoning can emerge from collaborative interactions among learners in online environments (Alqahtani, 2016; Alqahtani & Powell, 2016, 2017; Stahl, 2015). However, in such CSCL settings, mathematics education researchers and mathematics teachers remain unsure of how to orchestrate students' instrumentation of collaborative environments so as to support students' mathematical practices and movement between exploration and deductive justifications. Knowing how to orchestrate and promote this movement will enable mathematics education researchers and mathematics teachers to realize the potential of DGEs to improve geometry learning and of CSCL environments to engage learners in developing mathematical ideas through online collaboration that parallel the real-world online, collaborative work of mathematicians, including Fields Medal recipients (Alagic & Alagic, 2013).

In this chapter, reporting from a larger iterative project,[1] informed by design-based research (Cobb, Confrey, diSessa, Lehrer, & Schauble, 2003), we address practical and theoretical challenges concerning the orchestration of students' collaborative mathematical interactions in an online environment. After positioning our work in the literature and presenting our conceptual framework, we describe an online environment for collaborative learning, called Virtual Math Team with GeoGebra (VMTwG). Following these, we illustrate the case of a teacher, working with early high school students (15-year-olds), whose pedagogical orchestrations shape students' movement between exploration and deductive justification by focusing on students' collaborative practices. We understand pedagogical orchestrations to be instructional actions initiated by teachers that precede, invite, sustain, monitor, or reflect on students' activity. By movement between exploration and deductive justifications, we mean discursive, recursive trajectories in which students are motivated by mathematical relations that they notice while manipulating mathematical objects to develop and communicate convincing arguments about the relations that satisfy their peers. Finally, we propose a model of how teachers can support students' instrumentation of collaborative environments and mathematical understanding. In our model, during a mathematical activity, teachers progressively decentralize their role and, simultaneously, support students' development and performance of collaborative practices. This model informs the theory of instrumental

[1] The project—Computer-Supported Math Discourse among Teachers and Students—is an NSF-funded collaboration among researchers affiliated with The Math Forum at the National Council of Teachers of Mathematics (NCTM) and Rutgers University-Newark.

orchestration (Trouche, 2003, 2004, 2005) by providing a pedagogical intervention trajectory that supports students' instrumental genesis (Lonchamp, 2012; Rabardel & Beguin, 2005) of collaborative mathematical environments and shifts students' focus from their teacher to their peer collaborators.

4.2 Positioning Within the Literature

Our online environment, VMTwG,[2] is an interactional, synchronous space, containing support for chat rooms with collaborative tools for mathematical explorations, including a multiuser, dynamic version of GeoGebra. This dynamic geometry environment within VMTwG provides affordances typical and beyond most DGEs. From different perspectives and foci, how DGEs influence learning has been the object of research. Some studies focused on affordances of DGEs and how learners use them, while others discussed how DGEs mediate mathematical activity and shape mathematical understanding. Early research noticed differences between pencil-and-paper geometric constructions and dynamic geometry constructions. Laborde (1993) distinguished between *drawing* and *figure* in DGEs to emphasize these differences. A drawing refers to the perceptual image as drawn on paper, while a figure is the theoretical object, constructed in DGE, and whose defining properties remains invariant under the drag test. Focusing on the dragging affordance of DGEs, researchers investigated how learners understand and use dragging and identified different dragging modalities that shape learners' interactions with the environment and their mathematical understanding (Alqahtani & Powell, 2016, 2017; Arzarello, Bairral, & Danè, 2014; Arzarello, Olivero, Paola, & Robutti, 2002; Baccaglini-Frank & Mariotti, 2010; Hollebrands, 2007; Hölzl, 1996; Lopez-Real & Leung, 2006). Measurement affordance of DGEs was also investigated to understand how it influences learners' mathematical understanding (González & Herbst, 2009; Hollebrands, 2007; Olivero & Robutti, 2007; Sinclair, 2004). In DGEs that provide multiple representations of objects such as GeoGebra, Alqahtani and Powell (2017) found that the analytical information offered in Algebra view provided additional support for learners' discussion of properties and relations of geometric figures. These studies of affordances of DGEs show that learners' cognitive processes relate to how learners use these affordances.

Other researchers studied how DGEs mediate learners' activities to justify and prove mathematical propositions. With DGEs, learners justify and prove relations

[2]The environment, Virtual Math Teams (VMT), has been the focus of years of development by a team led by Gerry Stahl, Drexel University, and Stephen Weimar, The Math Forum at the National Council of Teachers of Mathematics (NCTM) (formerly, The Math Forum @ Drexel University), and the target of considerable research (see, e.g., Powell & Lai, 2009; Stahl, 2008; Stahl, 2009b). This chapter is part of a recent body of investigations centered on an updated VMT with a multiuser version of GeoGebra (see, for instance, Alqahtani & Powell, 2016, 2017; Grisi-Dicker, Powell, Silverman, & Fetter, 2012; Powell, 2014; Powell, Grisi-Dicker, & Alqahtani, 2013; Stahl, 2013, 2015).

using empirical and deductive reasoning (Hadas, Hershkowitz, & Schwarz, 2000; Jones, 2000; Lachmy & Koichu, 2014; Leung & Lopez-Real, 2002; Mariotti, 2000, 2006, 2012; Marrades & Gutiérrez, 2000; Powell & Pazuch, 2016). DGEs allow learners to identify properties of mathematical objects, notice relations and dependencies among them, make and test conjectures, and develop proofs. In addition, DGEs provide systems of tools, such as dragging and trace, and signs associated with these tools that learners can internalize and use to build mathematical meaning (Falcade, Laborde, & Mariotti, 2007; Mariotti, 2000). This internalization influences teachers' and students' mathematical discourse and activity (Alqahtani & Powell, 2015a; Powell & Alqahtani, 2015; Sinclair & Yurita, 2008; Stahl, 2015).

Some studies that investigated how DGEs support learning of mathematics used collaborative settings in which learners share and discuss their ideas as they manipulate and construct objects (Alqahtani & Powell, 2016, 2017; Arzarello et al., 2014; Baccaglini-Frank & Mariotti, 2010; Jones, 2000; Lachmy & Koichu, 2014; Leung & Lopez-Real, 2002; Mariotti, 2000, 2012; Marrades & Gutiérrez, 2000; Stahl, 2015). However, few studies attended to collaborative practices that learners develop while working in small groups with DGEs (Alqahtani & Powell, 2016, 2017; Stahl, 2015). Affordances of DGEs enrich learners' mathematical discourse when learners are working collaboratively (Oner, 2008, 2013; Wei & Ismail, 2010).

In the literature, some studies attend to how teachers organize instruction to support students' learning with digital technologies. Using technology in their classroom, teachers often have to manage new pedagogical situations and develop "a new repertory of appropriate teaching practices for these technology-rich settings" (Drijvers et al., 2014, p. 190). To understand these situations and practices for a given instructional setting, Trouche (2004) introduced the construct of instrumental orchestration that explains how teachers organize available artifacts and engage students with them. Later, Drijvers, Doorman, Boon, Reed, and Gravemeijer (2010) and Drijvers (2012) highlighted the complexity of teaching processes and further developed Trouche's construct. Adding to his two components of instrumental orchestration, didactic configuration and exploitation mode, they distinguish a third component, didactical performance. The didactical configuration concerns how teachers arrange learning environment such as tools, materials, and seating. In the exploitation mode, they plan how to engage students with tasks and tools and in discussions. The third component of instrumental orchestration captures how teachers make instructional decisions in real time under changing circumstances. Together, these three components focus on the design, the didactical context, and the use of the technological tools in a classroom. Drijvers et al. (2010) identified various instrumental orchestrations for whole-class teaching and orchestrations for settings in which students work individually or in pairs with technology, distinguishing between teacher-centered and student-centered orchestrations.

Several studies examined how teachers use technological tools in their classrooms. Using instrumental orchestration, researchers investigated how teachers support students' instrumentation of technological tools in classroom (Alqahtani & Powell, 2015b; Drijvers et al., 2010; Erfjord, 2011). Others analyzed teachers' pedagogical interventions in classrooms to support students' mathematical understanding while working with digital technologies (Biza, 2011; Dove & Hollenbrands, 2014; Laborde, 2007; Sutherland, Olivero, & Weeden, 2004). Teachers support students' learning by making available appropriate tools and materials and by engaging students with tasks that enhance their mathematical understanding.

In our review of the literature, we found that studies investigated how DGEs shape mathematical understanding and how teachers organize learning environments to support students' use of digital technologies. Learners' interactions with DGEs influenced their mathematical understanding. Teachers' different instructional configurations supported students' learning with digital technologies. Among these studies, we found only one study that investigated how teachers support students' learning with collaborative DGE (Alqahtani & Powell, 2015b). This suggests a need to further understand how teachers' orchestration of mathematics classrooms that use synchronous, collaborative digital technologies.

4.3 Theoretical Framework

To understand how teachers use collaborative technologies to support students' mathematics learning with these technologies, we draw on several theoretical foundations for our design and analysis. We employ a cultural historical perspective that encourages learners to collaborate with each other and communicate their ideas. Using Vygotsky's ideas about tool-mediated activity and the role of signs and tools in human development, we view learners' interactions in VMTwG as mediated activity through which students develop their understanding of mathematics and the VMTwG environment. We explain how learners develop their understanding of VMTwG and its different functions using Rabardel and Beguin's (2005) notion of instrumental genesis. It allows us to describe how users appropriate tools and use them as instruments to solve mathematical problems. Finally, we use the construct of instrumental orchestration (Trouche, 2004) to describe how teachers organize and support students' learning of mathematics while using technological tools.

To understand how learners use technological tools to collaborate in solving mathematics problems, we draw on Vygotsky's perspective on the role in human development of cultural signs and tools. He believed that material tools, which are developed historically in cultures, influence human's cognitive behavior and development. In addition to tools, he included signs (e.g., written and spoken language, number systems) in human activity. The "alloy of speech and action has a very specific function in the history of the child's development" (Vygotsky, 1978, p. 30). This perspective on the role of signs and tools informs our conceptual view of how learners use technological tools (online collaborative environments and dynamic mathematics software) and cultural signs (natural language and symbols) to construct together geometric figures and solve jointly geometrical problems. While performing mathematical activities, learners interact with each other to work on shared tasks using available environmental tools and signs. The mediational role of the tools and signs supports learners' cognitive development through a process of internalization. In it, learners transform external activities that are linked to tools into internal activities that are linked to signs (Mariotti, 2000; Vygotsky, 1978). The external actions learners perform with technological tools transform into signs that learners use to think about and communicate mathematical ideas.

The link between external actions and signs in the internalization process indicates the significance of human interactions. A major implication of Vygotsky's theory is the importance of social interactions in learning and human development. During an activity that is directed toward an object, learners employ "tools, speech directed toward the person conducting the experiment or speech that simply accompanies the action" to achieve their goal (Vygotsky, 1978, p. 30). Learners' engagement with these actions supports the internalization process that transforms social phenomena into psychological phenomena (mental functions) (Wertsch, 1985). This perspective emphasizes the importance of collaboration among students during mathematical activities. Collaboration with others gives learners opportunities to reflect on their own thinking and on thinking of others (Daniels, 2001) as well as improves mathematics achievement (Springer, Stanne, & Donovan, 1999).

Building on Vygotsky's work, researchers have developed other constructs to explain how learners build knowledge while interacting with others through technological tools. Rabardel and Beguin introduced the notion of instrumental genesis (Lonchamp, 2012; Rabardel & Beguin, 2005), which theorizes how learners interact with tools that mediate their activity. Learners transform tools into instrument by developing their own knowledge of how to use them. The instrument then mediates activities between learners and a task. In the activities, learners perform actions upon an object (matter, reality, object of work, etc.) in order to achieve a goal using a tool (technical or material component). Rabardel and Beguin (2005) emphasize that the instrument is not just the tool or the artifact, the material device or semiotic construct, it "is a composite entity made up of an artifact component and a scheme component." (p. 442). To transform the tool into an instrument (appropriation), learners develop their own utilization schemes through two important dialectical processes that account for potential changes in the instrument and in learners, called instrumentalization and instrumentation. Instrumentalization is "the process in which the learner enriches the artifact properties" (Rabardel & Beguin, 2005, p. 444). Instrumentation is about the development of the learner side of the instrument; the learner assimilates an artifact to a scheme or adapts utilization schemes. When engaging students with different technological tools in mathematics classrooms, instrumentation plays a significant role in how students build their knowledge about using the tools and how these tools support and shape students' mathematical knowledge.

In mathematics classroom settings, how teachers support learners' instrumental genesis is multifaceted. To understand how to support students' instrumentation of technological tools, Trouche (2004, 2005) introduces "instrumental orchestration" to describe how teachers plan and implement mathematics lessons that integrate technological tools. The instrumentation process is an important multidimensional process, including individual as well as social dimensions (Trouche, 2005). Since instrumental genesis mainly accounts for the individual dimension, instrumental orchestration accounts for the social dimension of the instrumentation process. It describes the arrangements of artifacts in the environment, *didactical configurations*, and teacher and student move within these configurations, *exploitation modes*

(Trouche, 2004, 2005). A third component was added by Drijvers et al. (2010), *didactical performance*, to describe teachers' instructional decisions responding to circumstances during mathematical lesson. There are different combinations of didactical configurations, exploitation modes, and didactical performance that teachers use to support their students' instrumentation. The didactical configurations concern the "layout of the artifacts available in the environment" (Trouche, 2004, p. 296) for the students to interact with during the mathematical lesson. The exploitation modes represent actions that teachers choose for students to perform based on their lesson's objects. The didactical configurations and the exploitation modes work together to support students' achievement of lesson's objects. Combinations of didactical configurations and the exploitation modes act on three levels: artifacts, instruments, and students' relationship with the instruments (Trouche, 2005). Within these levels, teachers attend to the tools as artifacts (before instrumentation) and after students appropriate the tools. After students appropriate the tools, teachers' decisions involve how to guide students' interactions with the instrument and support their mathematical understanding.

4.4 Online Environment for Collaborative Learning

The online environment, VMTwG, is an interactional, synchronous space. It contains support for chat rooms with collaborative tools for mathematical explorations, including a multiuser version of GeoGebra, where team members can construct dynamic objects and drag elements to visually explore relationships (see Fig. 4.1). VMTwG records users' chat postings and GeoGebra actions, which teachers can review and even replay at various speeds. The research team designed dynamic geometry tasks to encourage participants to discuss and collaboratively manipulate and construct dynamic geometry objects, notice relations and dependencies among the objects, make conjectures, and build justifications.

The data for this study come from the second course and concern the work of a high school mathematics teacher, Mr. S. He engaged his class in VMTwG in small teams of three to four students each. The class worked in a computer lab, and Mr. S. encouraged students to communicate only through VMTwG. To understand his instrumental orchestration, we analyze qualitatively four sources of data: (1) the tasks he used with his students, (2) the modifications he made to the tasks after reviewing teams' work, (3) the logged VMTwG interactions of two teams of his students on the tasks, and (4) his reflections on their work, which he wrote after each class session. We chose to analyze two teams, team 1 and team 6, since Mr. S. considered those teams to be most collaborative.

On each of the four data sources, we performed conventional and directed content analysis (Hsieh & Shannon, 2005). We were particularly interested in coding and categorizing both Mr. S.'s pedagogical interventions and the deductive justifications of two teams of his students. The data drives our analysis, and we interpret them using the theories of instrumental genesis and orchestration whenever there are links between the data and the theories.

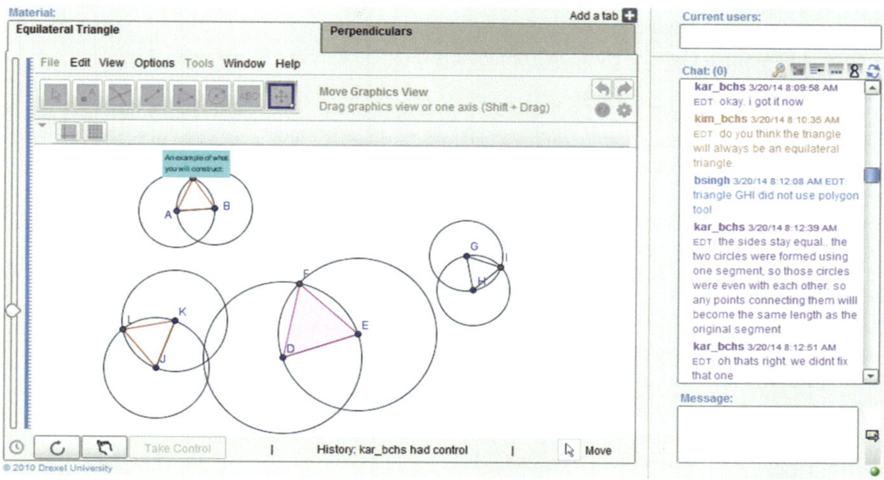

Fig. 4.1 Screenshot of VMTwG environment with the work of Mr. S.'s students

4.5 Pedagogical Setting: Teachers Learning

The work of the high school mathematics teacher who engaged his students in an online collaborative environment, Virtual Math Team with GeoGebra (VMTwG), to extend their geometric knowledge was informed by a particular pedagogical setting. The setting is a professional development project, "Computer-Supported Math Discourse among Teachers and Students," that involves middle and high school teachers in two 15-weeklong, technology-focused online courses. The first course engages teachers, working synchronously in small groups, in interactive, discursive learning of dynamic geometry through collaborating in VMTwG to solve 55 tasks that involve constructing geometric figures and solving open-ended geometric problems. In addition, the teachers realize and, in writing, reflect on their mathematical and collaborative practices; read and discuss synchronously and asynchronously articles about technology and pedagogy (Battista, 2002; Mishra & Koehler, 2006; Stahl, 2009a), lesson types with technology (McGraw & Grant, 2005), collaboration and discourse (Mercer & Sams, 2006; Michaels, O'Connor, & Resnick, 2007; Resnick, Michaels, & O'Connor, 2010), and mathematical practices (Common Core State Standards Initiative, 2010, pp. 6–8); and collaboratively plan the content and means to implement what they learn in the course in lessons with their students.

The Implementation Plan is a major component of the 15-weeklong professional development course. Completed over 10 of the 15-weeklong professional development course, it is divided into the following five phases:

Phases of the Implementation Plan
Phase 1: Develop a plan for garnering school and district support and for technology availability.
Phase 2: Select a focus for VMTwG lessons that is of interest to you individually as a teacher. Discuss your focus during your team's second synchronous session, and post your focus to the Blackboard Discussion forum. Do other teachers within your team or in other teams share the same focus?
Phase 3: Define your focus statement including a clear set of pedagogical goals. Collaborate within your VMT Team, and also exchange ideas with teachers from other teams through Blackboard Discussion forums. It is acceptable if several teachers share a focus statement, but it is not required.
Phase 4: Develop an activity list of collaborative dynamic mathematics activities to foster a developmental trajectory aligned with your focus statement. It is acceptable if several teachers share activities or all or part of a developmental trajectory, but it is not required.
Phase 5: Develop a coherent set of scripted VMTwG sessions, and decide how you might implement this curriculum next term. It is acceptable if several teachers share some or all scripted VMTwG sessions, but it is not required.

Each phase is further explained in an implementation document that the teachers receive. Over the course of the 10 weeks, each teacher uploads to the discussion board space of an online course management system (Blackboard™) their response to each phase of the plan and receives feedback from other teachers in the course as well as the course facilitators. This collaborative development of the teachers' plan is a substantial way in which the course supports each teacher's implementation efforts with their students.

A second major support for teachers' classroom implementation is the second course, which is a reflective practicum. In the course, teachers post their planned lessons to Blackboard, receive constructive feedback and exchange ideas, post a reflection about each lesson that includes information about their students' learning and about challenges and triumphs and to these receive feedback. In the teachers' reflection on each of their lessons they write about the goals of the lesson, how whether the students achieved the goals, what worked and did not work, and what support the teacher provided suggest changes for further revisions to improvement; from the discursive and inscriptive, data highlight the collaborative and mathematical practices; comment on each teachers' reflection.

4.6 Pedagogical Setting: Teachers Supporting Student Learning

The teachers engage their students in at least 10 h of class sessions to learn dynamic geometry through the use of VMTwG to work on construction and problem-solving tasks. This study examines a teacher's and his students' initial engagement with the VMTwG program in an urban high school in southern New Jersey. The high school has a diverse student body, where 52% of the students identify as Black, 35% as White, 10% as Hispanic, and 3% as Asian. Twenty-one percent of the students have a classified disability, while 57% live in economically disadvantaged households.

For the 2012–2013 school year, the high school's suspension rate was 48%, and its 2014 graduation rate was 82%.

This mathematics teacher, Mr. S., has taught at this high school since the beginning of his teaching career and, at the time of this study, had taught there for 6 years. During regular class time, from two different 10th and 11th grade classes, he engaged a total of 31 students in tasks in VMTwG. Nineteen of his students were females and 12 were males. Academically, they had performed below or at the average on statewide standardized assessments. Specifically, statewide assessment data for 25 of the 31 students were available, and of them, 19 passed the 7th grade assessment and still fewer, 17, passed it in the 8th grade. At the time of this study, eight of the 31 students were concurrently enrolled in a separate mathematics remediation course. All 31 students neither had prior experience with dynamic geometry nor previously worked in a computer-supported collaborative learning environment.

For working in VMTwG, the students worked in a computer lab and divided themselves into teams. Teams were formed based on students' already established social groupings since students chose their teammates according to with whom they normally socialized during class. They formed eight teams, seven teams with four students in each and one team of three. The teacher did not have students in regularly assigned seats. In the computer lab, the computers were arranged on pentagonal-shaped tables. There was a teacher station that was connected ceiling projector and two large wall-mounted whiteboards. The teacher provided his students with log-ins and assigned each student team to a VMTwG chat room. Students were not able to enter other teams' chat rooms.

4.6.1 Teacher's Instrumental Orchestration

Based on our analyses of Mr. S.'s implementation of the project design, his instrumental orchestration was directed at supporting three categories of students' actions: collaborative practices, mathematical reasoning, and the use of technology. In addition, the analysis reveals that Mr. S. followed a trajectory of pedagogical interventions focused on his students' discursive interactions and their emerging knowledge of dynamic geometry. In his reflections on his students' work, Mr. S. expresses an overall goal that, within their teams, students manipulate and construct dynamic geometric objects and notice and discuss relations among them, particularly relations of dependency. To achieve this goal, Mr. S.'s didactical configurations had students work in small groups in a computer lab and communicate online through VMTwG. His pedagogical interventions focused on how the teams of students collaborate. Having given his students a task designed to promote collaboration, Mr. S. expressed concern in his weekly reflection that the teams did not collaborate successfully. He reported that to ensure successful collaborative sessions, he subsequently discussed with his class features of successful collaborations and presented examples of what he considered good collaborative moves. To underscore his advice, he distributed a list of behaviors that he judged could help to ensure successful collaboration and called it "The Pledge." It contained statements of behaviors such as "Include everyone's ideas" and "Ask what my team members think and what their reasons are."

These pedagogical interventions and ones that we present below focused on collaboration. They reveal that Mr. S. choose exploitation modes (instructional decisions) that encourage students to be reflective of their work within their teams. His pedagogical interventions are mostly focused on the second and third level of his instrumental orchestration. Those levels are concerned, respectively, with the instrument and the students' relations with the instrument. Mr. S. used collaboration as a vehicle to orchestrate his students' appropriation of VMTwG artifacts and movement toward deductive justifications. In his weekly reflections, he assessed his students' reasoning by tracking their collaborative practices and their use of mathematical language.

Closely following Mr. S.'s interventions concerning his students' collaborative practices, he then focused on aspects of their use of the technological environment. This focus is at the first level—artifact level—of his instrumental orchestration. In his weekly reflections, he reported that during his students' engagement in VMTwG, he "monitored progress and resolved some tech issues." He helped students gain insights into the use of particular GeoGebra commands by modifying tasks and directing his students to view specific YouTube GeoGebra clips.

As Mr. S.'s teams of students increased their effective collaborative interactions, he shifted his pedagogical interventions more explicitly toward supporting their mathematical reasoning. He discussed with his class the concept of dependency in dynamic geometry to contrast it with dependency in other mathematical domains and modified the tasks to explicate particular mathematical ideas. He posed detailed questions to foreground mathematical discourse. For example, he found that the tasks' original questions were not specific enough to elicit mathematical reasoning, so he included the following questions in one of the tasks, "constructing an equilateral triangle":

1. What kinds of triangles can you find here?
2. Drag the points. Do any of the triangles change kind? Discuss this in the chat.
3. Are there some kinds [of triangles] you are not sure about?
4. Why are you sure about some relationships?
5. Does everyone in the team agree?

These questions prompted his students to attend to particular objects and relations in the construction and to discuss the behavior of these objects and relations.

4.6.2 Students' Work in VMTwG

Mr. S.'s instrumental orchestration and his other pedagogical interventions contributed to his teams of students' instrumentation and movement toward greater collaboration and deductive justifications. For example, according to Mr. S. and our analyses, a team of three students (team 6) improved their collaboration, explorations, and mathematical reasoning. In their third session, the task asked them to construct an equilateral triangle, find the relationships among objects in their construction, and justify their claims. The students first dragged a preconstructed figure of an equilateral triangle (see triangle ABC in Fig. 4.1 above) to explore elements of the construction and their behavior. Afterward, they each constructed a similar figure (see Fig. 4.1) and dragged their construction vigorously to validate and justify their construction. Below, an excerpt[3] of their discussion shows how a team of students articulated a valid justification of why their constructions were of equilateral triangles.

18 kar_bchs: looks like we both got it [both successfully construct and drag the figures vigorously]
19 kim_bchs: yay, it seems like for a second one of the circles appeared much larger. but that was my imagination.
20 kar_bchs: oh. lol. why is the third point dependent on the distance between the first two points? (number 7)
21 kar_bchs: it just connects the points and the circles. making them all one piece
22 kim_bchs: as the segments change sides so does the radius of the circle. However, the triangle remains an equilateral triangle
23 bsingh: [the teacher] be sure to read directions, ALL, and make the pledge
24 kim_bchs: triangle
25 kar_bchs: yea. even though the sizes of the sides change, the fact that it is an equilateral triangle doesn't
26 kar_bchs: each side has the same distance in between it. even when you move the points
27 kim_bchs: i notice that point d and e are on the circumference of one circle. while point f is an intersetion of both circle. making it dependent on both points.
28 kar_bchs: if you try and move the intersected point (F and I), it won't move. but yea you're right, the intersecting point depends on the segment that was made
29 kim_bchs: *point f is an intersect of both circles
30 bsingh: [the teacher] there is something missing, are you reading the directions
31 bsingh: [the teacher] we are only doing tab 1 today
32 kar_bchs: i didnt use the polygon tool.. that's missing in mine
33 kim_bchs: i just notice that.
34 kar_bchs: can i try?
35 kar_bchs: okay. i got it now
36 kim_bchs: do you think the triangle will always be an equilateral triangle.
37 kar_bchs: the sides stay equal.. the two circles were formed using one segment, so those circles were even with each other. so any points connecting them will become the same length as the original segment
. . .
. . .
. . .

[3]This and the next excerpt in this chapter are from students engaged in chat communication in VMTwG. In this setting, at the same time that they write informally, without focusing on conventions of academic writing, students direct their attention to communicating quickly their mathematical ideas to themselves and their teammates. For this reason, we have chosen not to correct their orthography or any other aspect of their writing. We feel that it is important to honor and understand their authentic expressions.

50 kim_bchs: the radius of a circle is the same distance. segment AB is Sure. the radii of both circles and Segment AC and BC are also radii of both circles. hence, the triangle should be equilateral.
51 kar_bchs: the circles are equal. making the circumference of each equal to one another

This team of students noticed that the equilateral triangle depended on the relationship between the two circles that they created. They discussed their constructions and the relationships they noticed (lines: 18–29). Both students noticed that the construction maintains the triangle equilateral as vertices are dragged (lines 22 and 25). They tried to explain how the intersection points of the circles are dependent on the centers of the circles (lines 27–29). In line 36, kim_bchs asks whether the triangle is always equilateral. In response, kar_bchs states that the sides of the triangle are equal and mentions that the two circles are "even" or congruent. In line 50, it seems that kim_bchs builds on kar_bchs's observation and notes that the radii of both circles are equal and that imply that the triangle is equilateral and, in line 51, that the circumferences of the two circles are equal. The students successfully build on each other's ideas and justify why their constructions yield equilateral triangles and justify other equivalences that they notice. They also note that the congruence of their circles depends on the segment that they share (line 37: "the two circles were formed using one segment, so those circles were even with each other") and that two sides of the given triangle are dependent on segment AB (line 50: "the radius of a circle is the same distance. segment AB is Sure. the radii of both circles and Segment AC and BC are also radii of both circles. hence, the triangle should be equilateral."). This provides further evidence that these students are justifying mathematical relations, moving themselves toward deductive justification. This also indicates that this student team transformed artifacts of the technological environment such as chat, dragging, and tools involved in constructing equilateral triangles into instruments.

4.7 Discussion

Mr. S's students' actions led to their transformation of technological artifacts of VMTwG into instruments of their knowledge building. In this process, they accomplished movement between visual and dragging explorations and discursive deductive justifications. Their movement toward deductive justifications was evidenced in their discursive, interaction motivated by their perception of mathematical properties and relations that they notice while manipulating mathematical objects to develop and communicate convincing arguments about the mathematical relations that satisfy their team members.

These student knowledge-building actions were supported by Mr. S.'s pedagogical orchestrations. As we defined it earlier, such orchestrations are instructional actions initiated by teachers that precede, invite, sustain, monitor, or reflect on students' activity. Initial actions in the trajectory of Mr. S.'s pedagogical orchestrations began with a focus on supporting teams of his students to have effective collaborative interactions. The extension of their collaborative practices evidence collaborative learning, as Jeong and Hmelo-Silver (2016) suggest: "a group of people engage[s] in activities toward a shared goal. They may divide the tasks in the process of working together, but the ultimate goal is to produce an outcome that collectively advances the knowledge of individuals as well as the collectives" (p. 248). Once Mr. S. was satisfied that, within teams, students were listening to each other and building on each other's ideas, he shifted to focus his instructional interventions around ideas of mathematical reasoning and justifications. Our analysis of his weekly reflections, his later analysis of his students' work, and our analysis of his students' work indicate that, in parallel with his trajectory, his students progressed toward more pointed justifications of geometric relations that they noticed, including, for dynamic geometry, mathematically significant relations of dependencies (Stahl, 2013; Talmon & Yerushalmy, 2004).

Mr. S.'s pedagogical orchestrations not only shaped his students' transformation of technological artifacts of VMTwG into instruments for knowledge building but also inform the theory of instrumental orchestration (Trouche, 2003, 2004, 2005). His instructional actions undergird a model of pedagogical orchestration, the purpose of which is to support students' instrumental genesis (Rabardel & Beguin, 2005) of collaborative mathematical environments such as VMTwG. The didactical configuration involves a technological environment specifically designed to support collaborative knowledge building among small teams of interlocutors, interacting in coordinated discursive (chat) and inscriptive (GeoGebra) spaces. Another aspect of the didactical configuration is the open-ended, collaborative, and discourse-provoking nature of the mathematical tasks that the teacher chose and modified. The choices of technological environment and tasks are instructional moves that shift students' focus in the classroom from their teacher to their peer collaborators.

In the exploitation mode, teacher and student moves vary significantly from Mr. S.'s students' previous mathematical experiences in school. His students' school experiences in mathematical classrooms neither include little to any work in collaborative teams nor with technological or dynamic geometry. They had no experience working on open-ended tasks in which they were expected to build their own geometrical knowledge. This expectation that students were to build their own geometrical knowledge as they collaborative resolved open-ended tasks and constructed geometric figures rather than being told what they were to learn served to decentralize further the teacher's role. As a consequence, the teacher's pedagogical orchestration meant that there were minimal opportunities for Mr. S. to intervene contemporaneously while he was interacting in VMTwG. After the students' sessions, he reviewed their interactional work and made decisions in the realm of didactical performance to respond to circumstances that had occurred in his students' online mathematical work. His whole-class discussion of collaborative moves, including "The Pledge," is how he addressed the challenge of supporting his students' development of productive collaborative practices.

The example of Mr. S. provides evidence of pedagogical orchestration that supports students' instrumentation. He models how teachers can support students' instrumentation of collaborative environments and the extension of their mathematical understanding. In this model, during students' mathematical activity, teachers progressively decentralize their role and, simultaneously, support students' development and performance of collaborative practices. This model augments the theory of instrumental orchestration (Trouche, 2003, 2004, 2005) by providing a pedagogical intervention trajectory that supports students' instrumental genesis of collaborative mathematical environments and shifts students' focus from their teacher to their peer collaborators. In general, this model contributes to an understudied area of DGEs, teacher practice (Sinclair et al., 2016).

References

Alagic, G., & Alagic, M. (2013). Collaborative mathematics learning in online environments. In D. Martinovic, V. Freiman, & Z. Karadag (Eds.), *Vis ual Mathematics and Cyberlearning* (Vol. 1, pp. 23–48). New York: Springer.

Alqahtani, M. M. (2016). *Interactions between teachers' use of collaborative, dynamic geometry environment and their geometrical knowledge.* Unpublished doctoral dissertation Ph.D., Rutgers University, New Brunswick, NJ.

Alqahtani, M. M., & Powell, A. B. (2015a). Co-action and dynamic geometry knowledge. In K. Beswick, T. Muir, & J. Wells (Eds.), *Proceedings of 39th Psychology of Mathematics Education Conference* (Vol. 2, pp. 17–24). Hobart, Australia: PME.

Alqahtani, M. M., & Powell, A. B. (2015b). Teachers' support of students' instrumentation in a collaborative, dynamic geometry environment. In N. Amado & S. Carreira (Eds.), *Proceedings of the 12th international conference on technology in mathematics teaching* (pp. 268–276). Faro, Portugal: University of Algarve.

Alqahtani, M. M., & Powell, A. B. (2016). Instrumental appropriation of a collaborative, dynamic-geometry environment and geometrical understanding. *International Journal of Education in Mathematics, Science and Technology, 4*(2), 72–83.

Alqahtani, M. M., & Powell, A. B. (2017). Teachers' instrumental genesis of a dynamic geometry environment and their geometrical understanding. *Digital Experiences in Mathematics Education, 3(1)*, 9–38.

Arzarello, F., Bairral, M. A., & Danè, C. (2014). Moving from dragging to touchscreen: geometrical learning with geometric dynamic software. *Teaching Mathematics and its Applications, 33*(1), 39–51. doi:10.1093/teamat/hru002.

Arzarello, F., Olivero, F., Paola, D., & Robutti, O. (2002). A cognitive analysis of dragging practises in Cabri environments. *International Reviews on Mathematical Education (ZDM), 34*(3), 66–72.

Baccaglini-Frank, A., & Mariotti, M. A. (2010). Generating conjectures in dynamic geometry: The maintaining dragging model. *International Journal for Computers in Mathematical Learning, 15*, 225–253.

Battista, M. T. (2002). Learning geometry in a dynamic computer environment. *Teaching Children Mathematics, 8*(6), 333–339.

Biza, I. (2011). Students' evolving meaning about tangent line with the mediation of a dynamic geometry environment and an instructional example space. *Technology, Knowledge and Learning, 16*(2), 125–151.

Christou, C., Mousoulides, N., & Pittalis, M. (2004). Proofs through exploration in dynamic geometry environment. *International Journal of Science and Mathematics Education, 2*(3), 339–352.

Cobb, P., Confrey, J., diSessa, A., Lehrer, R., & Schauble, L. (2003). Design experiments in educational research. *Educational Researcher, 32*(1), 9–13.

Common Core State Standards Initiative. (2010). *Common core state standards for mathematics*. National Governors Association Center for Best Practices, Council of Chief State School Officers. Retrieved from http://www.corestandards.org/assets/CCSSI_MathStandards.pdf

Daniels, H. (2001). *Vygotsky and pedagogy*. London: Routledge.

Dove, A., & Hollenbrands, K. (2014). Teachers' scaffolding of students' learning of geometry while using a dynamic geometry program. *International Journal of Mathematical Education in Science and Technology, 45*(5), 668–681.

Drijvers, P. (2012). Teachers transforming resources into orchestrations. In G. Gueudet, B. Pepin, & L. Trouche (Eds.), *From text to 'lived' resources: Mathematics curriculum materials and teacher development* (pp. 265–281). Berlin, Germany: Springer.

Drijvers, P., Doorman, M., Boon, P., Reed, H., & Gravemeijer, K. (2010). The teacher and the tool: instrumental orchestrations in the technology-rich mathematics classroom. *Educational Studies in Mathematics, 75*(2), 213–234.

Drijvers, P., Tacoma, S., Besamusca, A., van den Heuvel, C., Doorman, M., & Boon, P. (2014). Digital technology and mid-adopting teachers' professional development: A case study. In A. Clark-Wilson, O. Robutti, & N. Sinclair (Eds.), *The mathematics teacher in the digital era* (pp. 189–212). New York: Springer.

Erfjord, I. (2011). Teachers' initial orchestration of students' dynamic geometry software use: Consequences for students' opportunities to learn mathematics. *International Journal of Computers for Mathematical Learning, 16*(1), 35–54.

Falcade, R., Laborde, C., & Mariotti, M. A. (2007). Approaching functions: Cabri tools as instruments of semiotic mediation. *Educational Studies in Mathematics, 66*(3), 317–333.

González, G., & Herbst, P. G. (2009). Students' conceptions of congruency through the use of dynamic geometry software. *International Journal of Computers for Mathematical Learning, 14*(2), 153–182.

Grisi-Dicker, L., Powell, A. B., Silverman, J., & Fetter, A. (2012). Addressing transitional challenges to teaching with dynamic geometry in a collaborative online environment. In L. R. Van Zoest, J.-J. Lo, & J. L. Kratky (Eds.), *Proceedings of the 34th annual meeting of the North American Chapter of the International Group for the Psychology of Mathematics Education* (pp. 1024–1027). Kalamazoo, MI: Western Michigan University.

Hadas, N., Hershkowitz, R., & Schwarz, B. B. (2000). The role of contradiction and uncertainty in promoting the need to prove in dynamic geometry environments. *Educational Studies in Mathematics, 44*(1–2), 127–150.

Hollebrands, K. F. (2007). The role of a dynamic software program for geometry in the strategies high school mathematics students employ. *Journal for Research in Mathematics Education, 38*(2), 164–192.

Hölzl, R. (1996). How does 'dragging' affect the learning of geometry. *International Journal of Computers for Mathematical Learning, 1*(2), 167–187.

Hsieh, H.-F., & Shannon, S. F. (2005). Three approaches to qualitative content analysis. *Qualitative Health Research, 15*(9), 1277–1288.

Jeong, H., & Hmelo-Silver, C. (2016). Seven affordances of computer-supported collaborative learning: How to support collaborative learning? How can technologies help? *Educational Psychologist, 51*(2), 247–265. doi:10.1080/00461520.2016.1158654.

Jones, K. (2000). Providing a foundation for deductive reasoning: students' interpretations when using dynamic geometry software and their evolving mathematical explanations. *Educational Studies in Mathematics, 44*(1–3), 55–85.

Laborde, C. (1993). The computer as part of the learning environment: The case of geometry. In C. Keitel & K. Ruthven (Eds.), *Learning from computers: Mathematics education and technology, NATO ASI Series* (Vol. 121, pp. 48–67). Berlin/Heidelberg, Germany: Springer.

Laborde, C. (2007). The Role and Uses of Technologies in Mathematics Classrooms: Between Challenge and Modus Vivendi. *Canadian Journal of Science, Mathematics, and Technology Education, 7*(1), 68–92.

Lachmy, R., & Koichu, B. (2014). The interplay of empirical and deductive reasoning in proving "if" and "only if" statements in a dynamic geometry environment. *The Journal of Mathematical Behavior, 36*, 150–165.

Leung, A., & Lopez-Real, F. (2002). Theorem justification and acquisition in dynamic geometry: A case of proof by contradiction. *International Journal of Computers for Mathematical Learning, 7*(2), 145–165.

Lonchamp, J. (2012). An instrumental perspective on CSCL systems. *International Journal of Computer-Supported Collaborative Learning, 7*(2), 211–237.

Lopez-Real, F., & Leung, A. (2006). Dragging as a conceptual tool in dynamic geometry environments. *International Journal of Mathematical Education in Science and Technology, 37*(6), 665–679. doi:10.1080/00207390600712539.

Mariotti, M. A. (2000). Introduction to proof: The mediation of a dynamic software environment. *Educational Studies in Mathematics, 44*(1), 25–53.

Mariotti, M. A. (2006). Proof and proving in mathematics education. In*Handbook of research on the psychology of mathematics education: Past, present and future* (pp. 173–204). Rotterdam, The Netherlands: Sense.

Mariotti, M. A. (2012). Proof and proving in the classroom: Dynamic Geometry Systems as tools of semiotic mediation. *Research in Mathematics Education, 14*(2), 163–185.

Marrades, R., & Gutiérrez, Á. (2000). Proofs produced by secondary school students learning geometry in a dynamic computer environment. *Educational Studies in Mathematics, 44*(1–3), 87–125.

McGraw, R., & Grant, M. (2005). Investigating mathematics with technology: Lesson structures that encourage a range of methods and solutions. In W. J. Masalski & P. C. Elliott (Eds.), *Technology-supported mathematics learning environments, Vol. Sixty-Seventh Yearbook* (pp. 303–317). Reston, VA: National Council of Teachers of Mathematics..

Mercer, N., & Sams, C. (2006). Teaching children how to use language to solve maths problems. *Language and Education, 20*(6), 507–528.

Michaels, S., O'Connor, C., & Resnick, L. (2007). Deliberative discourse idealized and realized: Accountable talk in the classroom and in civic life. *Studies in Philosophy and Education, 27*(4), 283–297.

Mishra, P., & Koehler, M. J. (2006). Technological pedagogical content knowledge: A new framework for teacher knowledge. *Teachers College Record, 108*(6), 1017–1054.

Miyazaki, M., Fujita, T., & Jones, K. (2016). Students' understanding of the structure of deductive proof. *Educational Studies in Mathematics*, 1–17. doi:10.1007/s10649-016-9720-9.

Olivero, F., & Robutti, O. (2007). Measuring in dynamic geometry environments as a tool for conjecturing and proving. *International Journal of Computers for Mathematical Learning, 12*(2), 135–156.

Öner, D. (2008). Supporting students' participation in authentic proof activities in computer supported collaborative learning (CSCL) environments. *Computer-Supported Collaborative Learning, 3*, 343–359.

Oner, D. (2013). Analyzing group coordination when solving geometry problems with dynamic geometry software. *International Journal of Computer-Supported Collaborative Learning, 8*(1), 13–39.

Powell, A. B. (2014). Construção colaborativa do conhecimento tecnológico, pedagógico e do conteúdo de professores de matemática [Mathematics teachers' collaborative construction of technological pedagogical content knowledge]. *Boletim GEPEM, 64*, 88–103.

Powell, A. B., & Alqahtani, M. M. (2015). Tasks promoting productive mathematical discourse in collaborative digital environments. In N. Amado & S. Carreira (Eds.), *Proceedings of the 12th international conference on technology in mathematics teaching* (pp. 68–76). Faro, Portugal: University of Algarve.

Powell, A. B., Grisi-Dicker, L., & Alqahtani, M. (2013). Letramento matemático: Desenvolvendo as práticas colaborativas, matemáticas, e discursivas com tecnologia [Mathematical literacy: Development of collaborative, mathematical and discusive practices with technology] *XI Encontro Nacional de Educação Matemática, Educação Matemática: Retrospectivas e Perspectivas [XI National Conference of Mathematics Education*, Mathematics Education*: Retrospectives and Perspectives*. Curitiba, Paraná.

Powell, A. B., & Lai, F. F. (2009). Inscription, mathematical ideas, and reasoning in VMT. In G. Stahl (Ed.), *Studying virtual math teams* (pp. 237–259). New York: Springer.

Powell, A. B., & Pazuch, V. (2016). Tarefas e justificativas de professores em ambientes virtuais colaborativos de geometria dinâmica. *Zetetiké, 24*(2), 191.

Rabardel, P., & Beguin, P. (2005). Instrument mediated activity: From subject development to anthropocentric design. *Theoretical Issues in Ergonomics Science, 6*(5), 429–461.

Resnick, L. B., Michaels, S., & O'Connor, C. (2010). How (well-structured) talk builds the mind. In*Innovations in educational psychology: Perspectives on learning, teaching and human development* (pp. 163–194). New York: Springer.

Silverman, J. (2011). Supporting the development of mathematical knowledge for teaching through online asynchronous collaboration. *The Journal of Computers in Mathematics and Science Teaching, 30*, 61–78.

Sinclair, M. (2004). Working with accurate representations: The case of preconstructed dynamic geometry sketches. *The. Journal of Computers in Mathematics and Science Teaching, 23*(2), 191–208.

Sinclair, N., Bartolini Bussi, M. G., de Villers, M., Jones, K., Kortenkamp, U., Leung, A., & Owens, K. (2016). Recent research on geometry education: An ICME-13 survey team report. *ZDM Mathematics Education, 48*(5), 691–719. doi:10.1007/s11858-016-0796-6.

Sinclair, N., & Yurita, V. (2008). To be or to become: How dynamic geometry changes discourse. *Research in Mathematics Education, 10*(2), 135–150.

Springer, L., Stanne, M. E., & Donovan, S. S. (1999). Effects of small-group learning on undergraduates in science, mathematics, engineering, and technology: A meta-analysis. *Review of Educational Research, 69*(1), 21–51.

Stahl, G. (2008). Social practices of group cognition in virtual math teams. In S. Ludvigsen, A. Lund, & R. Säljö (Eds.), *Learning in social practices: ICT and new artifacts—transformation of social and cultural practices*. Pergamon, Turkey.

Stahl, G. (2009a). The VMT vision. In G. Stahl (Ed.), *Studying virtual math teams* (pp. 17–29). New York: Springer.

Stahl, G. (Ed.). (2009b). *Studying virtual math teams*. New York: Springer.

Stahl, G. (2013). *Translating Euclid: Designing a human-centered mathematics*. San Rafael, CA: Morgan & Claypool.

Stahl, G. (2015). *Constructing dynamic triangles together: The development of mathematical group cognition*. Cambridge, UK: Cambridge University.

Sutherland, R., Olivero, F., & Weeden, M. (2004). Orchestrating mathematical proof through the use of digital tools. In*Proceedings of the 28th Conference of the International Group for the Psychology of Mathematics Education* (Vol. 4, pp. 265–272). Bergen, Norway: Bergen University College.

Talmon, V., & Yerushalmy, M. (2004). Understanding dynamic behavior: Parent-child relations in dynamic geometry environments. *Educational Studies in Mathematics, 57*(1), 91–119.

Trouche, L. (2003). From artifact to instrument: mathematics teaching mediated by symbolic calculators. *Interacting with Computers, 15*(6), 783–800.

Trouche, L. (2004). Managing the complexity of human/machine interactions in computerized learning environments: Guiding students' command process through instrumental orchestrations. *International Journal of Computers for Mathematical Learning, 9*(3), 281–307.
Trouche, L. (2005). Instrumental genesis, individual and social aspects. In*The didactical challenge of symbolic calculators* (pp. 197–230). New York: Springer.
Vygotsky, L. S. (1978). *Mind in society: The development of higher psychological processes* (M. Cole, V. John-Steiner, S. Scribner, & E. Souberman, Trans.). Cambridge, MA: Harvard.
Wei, C. S., & Ismail, Z. (2010). Peer interactions in computer-supported collaborative learning using dynamic mathematics software. *Procedia – Social and Behavioral Sciences, 8*, 600–608.
Wertsch, J. V. (1985). *Vygotsky and the social formation of mind*. Cambridge, MA: Harvard.
Wu, H. H. (1996). The role of Euclidean geometry in high school. *The Journal of Mathematical Behavior, 15*(3), 221–237. doi:10.1016/S0732-3123(96)90002-4.

Chapter 5
Representation Construction: A Guided Inquiry Approach for Science Education

Peter Hubber, Russell Tytler, and Gail Chittleborough

Abstract This chapter outlines a guided inquiry approach, called representation construction, which was successfully developed within an Australian Research Council (ARC) project that links student learning and engagement with the knowledge production practices of science. This approach involves challenging students to generate and negotiate the representations (text, graphs, models, diagrams) that constitute the discursive practices of science, rather than focusing on the text-based, definitional versions of concepts. The representation construction approach is based on sequences of representational challenges which involve students constructing representations to actively explore and make claims about phenomena. It thus represents a more active view of knowledge than traditional structural approaches and encourages visual as well as the traditional text-based literacies. The approach has been successful in demonstrating enhanced outcomes for students, in terms of sustained engagement with ideas, and quality learning, and for teachers enhanced pedagogical knowledge and understanding of how knowledge in science is developed and communicated. This chapter draws on specific examples of how the approach was implemented in a variety of topics, such as energy, forces, astronomy and ideas about matter within junior secondary science classrooms. It will also draw on the issues associated with the adoption of the approach in laptop/tablet classrooms where part of the curriculum is delivered in the cloud.

5.1 Introduction

This chapter describes an approach to inquiry teaching and learning in science that has been developed and trialled over a 10-year programme of research, which is based on students actively constructing representations in response to structured challenges (Tytler, Prain, Hubber, & Waldrip, 2013). The approach has its basis in a number of practical concerns and theoretical insights. There is mounting concern that traditional teacher-centred approaches to science are failing to engage students

P. Hubber (✉) • R. Tytler • G. Chittleborough
Deakin University, Burwood, VIC, Australia
e-mail: phubber@deakin.edu.au; russell.tytler@deakin.edu.au; gail.chittleborough@deakin.edu.au

R. Jorgensen, K. Larkin (eds.), *STEM Education in the Junior Secondary*,
DOI 10.1007/978-981-10-5448-8_5

and, in particular, are not developing the inquiry and problem-solving skills, and creativity, needed by citizens engaged in the twenty-first-century workforce (Chubb, 2014). Despite a history of curriculum advocacy, inquiry approaches have failed to take hold:

> *Four decades after* Schwab's (1962) *argument that science should be taught as an 'enquiry into enquiry', and almost a century since John Dewey (1916) advocated that classroom learning be a student-centred process of enquiry, we still find ourselves struggling to achieve such practices in the science classroom.* (Osborne 2006, p. 2)

A decade after Osborne made this point, the situation in Australia has not changed much (Goodrum, Druhan & Abbs, 2012), despite growing evidence that inquiry and open problem-solving approaches lead to more robust learning in science (Chi, 2009; Furtak, Seidel, Iverson, & Briggs, 2012), as they also do in mathematics (Kapur, 2008).

Allied to the call for inquiry, there is increasing advocacy that school science should better represent the epistemic practices by which knowledge is built in science (Prain & Tytler, 2012). Recent research in science studies has yielded fresh insights into the way that representational work is central to discovery processes and the increasingly pervasive role of the diverse representational work in generating and communicating knowledge (Elkins, 2011). Increasingly we come to understand that scientific knowledge is built by more complex processes than rational, logical reasoning applied to hypothesis generation and testing for evidence. Scientific discovery involves imaginative and often communal processes of creation of models and representations as new tools that mediate our understandings of the world. Inscriptions such as graphs and diagrams, 3D models and, increasingly, digital images and simulations created by sophisticated software and hardware that itself mediates and transforms data, provide new conceptual tools for interpreting the world. Latour (1999) was an early commentator on laboratory work and the complex processes by which teams of scientists generated representations to guide and make sense of data generation. Studying the process by which two scientists studied the encroachment of agricultural land into the Amazon forest, Latour (1999) charted the process of representational redescription, through 'circulating representations', from ordered and labelled soil container arrays, to measurements of soil characteristics, to tables and finally graphs that were transported to Paris then transformed into the abstracted text that was the final published paper. Gooding (2004, 2006) analysed Michael Faraday's notebooks to show the key role played by visual representations in Faraday's developing thinking on the relationships between magnetism and electric current. Gooding identified a fundamental pattern of dimensional transformation, from 2D to 3D to 4D and back to 2D as processes were abstracted and communicated. He argued that complex informal, visual reasoning through a mix of inscriptions and artefacts was a fundamental but unacknowledged characteristic of scientific discovery processes.

These ideas are central to new understandings of how students learn in science classrooms. Lemke (1990, 2004) identified the key role of representational work in learning science, as students are introduced to the complex multimodal representations through which scientific explanatory work is pursued, often is quick succession from text to diagram to symbol to graph and often is without acknowledgement of the representational conventions and complexities of coordination that underpin deeper understanding. Often, in fact, knowledge is thought of in terms of appreciation of the abstracted textual forms in which curricula and textbook conclusions are framed, without due acknowledgement of the representational practices that underpin this knowledge and its use (Tytler, Haslam, Prain, & Hubber, 2009). With growing realization of the importance of representational work, there has been a strong strand of research in cognitive science focused on the role of different representational modes and how these might best be coordinated to support learning (Ainsworth, 2006, 2008). Sociocultural theorists (Lemke, 2004; Moje, 2007) have characterized representational work as central to the development of scientific disciplinary literacy through which students come to know and achieve competence in the discursive practices that characterize science.

Our work sits within an inquiry tradition of research with students actively generating representations rather than being taught to interpret teacher-generated representation. Researchers in this tradition argue that students benefit from opportunities to explore, elaborate, redescribe representations and coordinate them across multiple modes and to negotiate their meaning with support from teachers (Cox, 1999; Greeno & Hall, 1997; Hubber, Tytler & Haslam, 2010; Lehrer & Schauble, 2006a, 2006b; Tytler, Peterson, & Prain, 2006; Waldrip, Prain & Carolan, 2010). Different forms of representation support different insights, and students need to explore the advantages and limitations of particular representational forms and modes for reasoning about phenomena (Greeno & Hall, 1997; Cox, 1999). In theorizing the power of representation construction, we have developed a model (Prain & Tytler, 2012) through which we link classroom inquiry practices with those of science. We argue that each representation has a partial and approximate relation to scientific phenomena, with understanding involving accessing and coordinating multiple, multimodal representations. We argue that the value of each representation can be understood in terms of its affordances (Gibson, 1977) understood as productive constraints on thinking. Thus, a drawing achieves its affordance through its visual and spatial specificity, such as in speculative drawings of particle representations of macro phenomena or in selection and abstraction processes involved in representing complex ideas such as animal diversity or movement (Tytler et al., 2009).

From our perspective, representations actively mediate and shape reasoning such that the targets of classroom activities are on the representational resources needed to support scientific problem-solving and explanatory practices, rather than the establishment of abstracted concepts or mental models. In traditional accounts, representations are often cast as efficient and effective ways to introduce and illustrate abstracted concepts such as waves, chemical bonds or ecological interactions that are considered distinct from the representations through which they are generated and communicated. From our perspective, however, representations are the reasoning tools *through which* we imagine and visualize these concepts and model phenomena. This view is fundamentally Vygotskian, characterizing representations as the disciplinary language tools that mediate or frame our thinking and knowing (Moje, 2007).

5.2 The Development of the Representation Construction Approach to Teaching and Learning Science

This section outlines a guided inquiry approach, called representation construction, which was successfully developed and implemented within three Australian Research Council (ARC) projects that link student learning and engagement with the knowledge production practices of science.

Within the first of the projects, 'The Role of Representations in Learning Science (RiLS; 2007–2010)', the researchers collaborated with Middle Years teachers in several schools in exploring the role of representation in teaching whole topics of science. This exploratory work on representations led to the development of a set of pedagogical principles (detailed below) based on representations and came to be known as representation construction. Within the second project, 'The Role of Representations in Learning Science (RiLS; 2007–2010)', the principles were refined and trialled in several more Middle Years classrooms. The research was extended to involve the delivery of the representation construction approach in blended learning classroom environments in the third project, 'Developing digital pedagogies in inquiry science through a cloud-based teaching and learning environment' (iSTELR; 2014–2016).

This approach involves challenging students to generate and negotiate the representations (text, graphs, models, diagrams) that constitute the discursive practices of science, rather than focusing on the text-based, definitional versions of concepts. The representation construction approach is based on sequences of representational challenges which involve students constructing representations to actively explore and make claims about phenomena. It thus represents a more active view of knowledge than traditional structural approaches and encourages visual as well as the traditional text-based literacies.

Central to the representation construction approach is the view that understanding and practising science involve coordinating and reasoning with multimodal representations. These include verbal and written language (including topic- and process-specific vocabulary), drawing, three-dimensional modelling, mathematical (graphs, tables, equations) and gestural language. In learning these particular literacies of science, students are learning how to invest these representations with appropriate meaning as part of learning how to reason and communicate in this subject. The teacher's task in scaffolding conceptual understanding thus becomes, importantly, about representational processes and products. Whilst students have to learn how to interpret and critique authorized scientific representations, a focus on teacher-guided student construction and justification of their own representations can (a) develop conceptual understanding and reasoning capacities in this subject and (b) enable students to participate in knowledge production methods aligned with scientific practice. Given the teacher's role is to lead students to develop an understanding of the authorized scientific representations, the representation construction approach is considered a guided inquiry pedagogy.

The set of principles that underpin the representation construction approach (Tytler et al., 2013, p. 34) are described as:

1. *Sequencing of representational challenges involving students generating representations to actively explore and make claims about phenomena:*
   (a) *Clarifying the representational resources underpinning key concepts*: Teachers need to clearly identify big ideas, key concepts and their representations, at the planning stage of a topic in order to guide refinement of representational work.
   (b) *Establishing a representational need:* The sequence needs to involve explorations in which students identify the problematic nature of phenomena and the need for explanatory representation, before the introduction of the scientifically accepted forms.
   (c) *Coordinating/aligning student-generated and canonical representations:* There needs to be interplay between teacher-introduced and student-constructed representations where students are challenged and supported to refine and extend and coordinate their understandings.
2. *Explicitly discussing representations*: The teacher plays multiple roles, scaffolding the discussion to aim at student self-assessment as a shared classroom process:
   (a) *The selective purpose of any representation:* Students need to understand that a number of representations are needed for working with multiple aspects of a concept.
   (b) *Group agreement on generative representations*: There needs to be a guided process whereby students critique representations to aim at a resolution.
   (c) *Form and function:* There needs to be an explicit focus on representational function and form, with timely clarification of parts and their purposes.
   (d) *The adequacy of representations:* There needs to be ongoing assessment (by teachers and students) of student representations.
3. *Meaningful learning*: Providing strong perceptual/experiential contexts and attending to student engagement and interests through choice of task and encouraging student agency;
   (a) *Perceptual context:* Activity sequences need to have a strong perceptual context (i.e. hands on, experiential) and allow constant two-way mapping between objects and representations.
   (b) *Engagement/agency:* Activity sequences need to focus on engaging students in learning that is personally meaningful and challenging, through affording agency and attending to students' interests, values and aesthetic preferences and personal histories.
4. *Assessment through representations*: Formative and summative assessment needs to allow opportunities for students to generate and interpret representations. Students need to be supported to extend and demonstrate learning through developing explanations that involve coordinating and re-representing multiple modes.

The following sections provide illustrations of practice taken from case studies from the ARC projects that adopted the representation construction approach. Examples are also provided from a successful statewide professional learning programme, Switched on Secondary Science Professional Learning (SOSSPL 2010–12), funded by the Victorian Department of Education that introduced representation construction to over 300 teachers across the state which they then trialled in their schools (Hubber, Tytler, Chittleborough, Campbell, & Jobling, 2012).

5.3 Introducing Ideas About the Representation Construction Approach

In enacting a representation construction approach, importance needs to be given at the planning stage of the key concepts that underpin the topic to be taught (refer to ***Principle 1a*** above). These concepts need to be expressed as statements of understanding that are couched at a level of language that is readily understood by the students. For example, 'Object like the Earth and Moon spin, or rotate, on an axis, and revolve, or obit, other object' [Year 7 Astronomy] or, 'The temperature of an object is related to the average kinetic, or motion, energy of the particles that make up the object' [Year 9 Energy].

Teachers who are initially introduced to the representation construction approach readily understand that whilst a planning document for a topic might include a series of concepts as statements of understanding, the regurgitation of such statements, say in a topic test, does not necessarily imply understanding. Understanding a concept implies an ability to make links between multiple modes of representations. Figure 5.1 gives examples of a representational challenge given to groups of 3–4 secondary science teachers in which the group was given the challenge to represent their understanding of the concept of 'temperature' on a mini whiteboard.

The group-generated representations of Fig. 5.1 illustrate multiple modes of representation from those associated with everyday experiences to the more formalized canonical forms of representation that might be found in a science textbook. Discussions with teachers who have undertaken this task, from a pedagogical perspective, usually generate a view that their role is one of guiding students in linking the everyday representations they have about a concept, and which they bring to the classroom, with the canonical representations. This representational challenge has also been seen by the teachers as a useful activity they might employ in the classroom, particularly as a formative assessment task designed to elicit students' understanding of a concept at the beginning of a topic (***Principle 4***). The use of a mini whiteboard was seen as beneficial to use instead of butcher's paper as it allows for editing by those who generate the presentation.

Given the definition of a representation as something that explains some aspect of nature and the means by which we understand and communicate our science understandings, teachers readily provide graphic and physical representations such

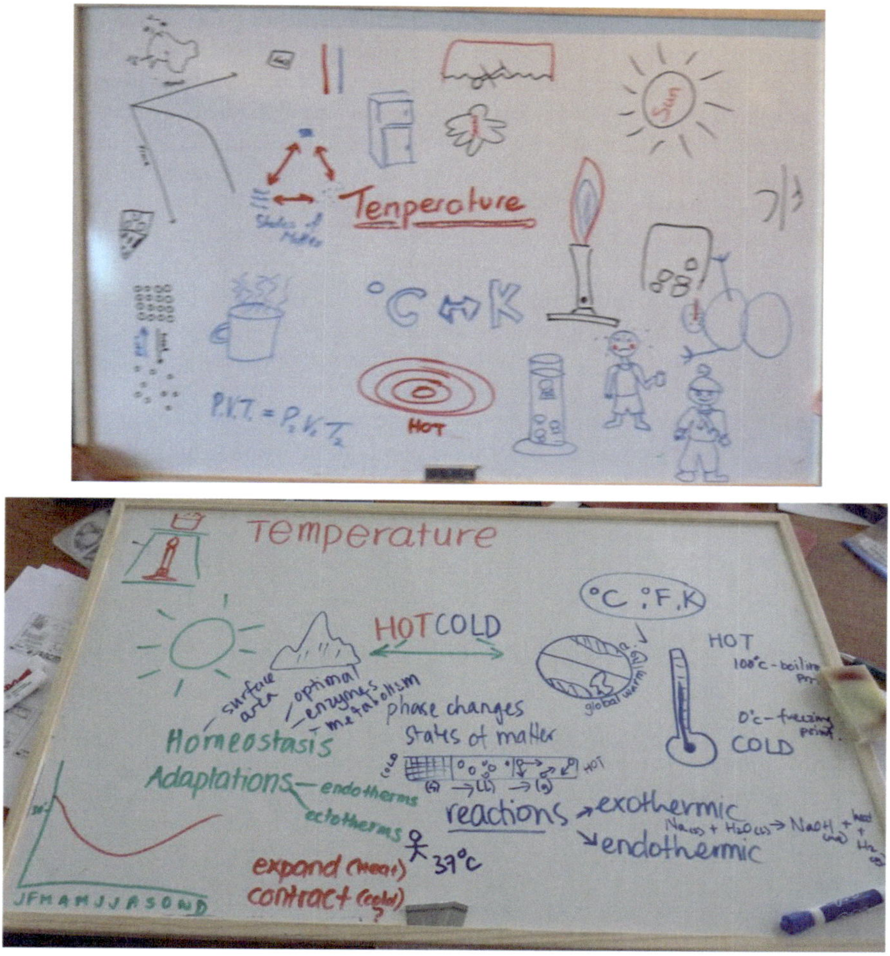

Fig. 5.1 Group challenge to represent the concept of temperature

as diagrams and 3D models. It takes some teasing out for them to recognize that language both in verbal and written forms is a key representational mode they use in the classroom. Apart from thinking about the affordances of individual representational forms, there needs to be some thought to the ways in which different representations can be linked. For example, embodied representations in the form of gesture can provide a link between representational forms. In introducing an idea to a class represented in a diagram, a teacher might provide a verbal explanation whilst at the same time pointing to various parts of a diagram or gesturing to represent movement or spatial relations. Individuals given the representational challenge of explaining how a snake moves through the grass find the task quite difficult as they are not allowed to use gesture to accompany their verbal explanation.

A key element of the representation construction pedagogy for teachers to know and for students to learn is that any one representation is only partial in its explanatory power of the target phenomena, idea or process (***Principle 2***). To illustrate this point, consider that the target concept is *the human heart*. Figure 5.2 provides three representations of the heart. It is important that when presenting a representation to students that discussions with them not only involve those features of the target concept that are shown by the representation but there are also discussions about what features of the target concept are not shown by the representation. For example, Table 5.1 lists some features of the human heart that are shown in the Fig. 5.2 representations and some features that are not shown. Table 5.1 illustrates that collectively the three representations provide more insights into providing an understanding of the target concept than any single representation can possibly provide. In addition, representations, such as those shown in Fig. 5.2, are not things that are readily understood by all those who view them, for example, the significance of the coloured arrows in Fig. 5.2b or mechanical pump as a metaphor for the function of the heart. Such representations need to be interpreted with accompanying text that might be given in a textbook or verbal explanations given by the teacher in the classroom.

5.4 Enacting Representation Construction in the Classroom

The following classroom examples relate to the topic of energy taught at Year 9 in a blended learning environment and a Year 8 class. Most of the examples are digital in nature and reflect both the teachers' use of representations in teaching ideas about energy and students' use of representations in learning ideas about energy.

5.4.1 Representational Challenges with Student-Generated Word Clouds and Mindmaps

The following two examples relate to our iSTELR project where Year 9 students were given cloud-based challenges (***Principle 1***) related to the topic of energy given as initial tasks in the topic of energy. The first challenge asked students to create a word cloud (Fig. 5.3) that represented their understanding of energy and then upload their representation to their cloud-based learning platform (STILE https://www.stileeducation.com/). The function and form of word clouds were explicitly discussed with the students prior to the task (***Principle 2c***). The word clouds in Fig. 5.3 not only show the words the students associate with energy, but they also show which words the students considered as more important (greater font size). The second challenge for the students was to construct a mind map (Fig. 5.4) to represent how different forms of energy connect with their daily lives (***Principle 3b***).

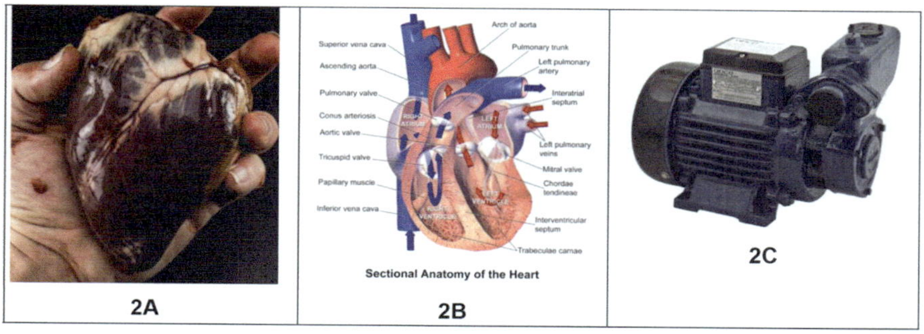

Fig. 5.2 Three representations of the heart

Table 5.1 Some features of the human heart that are shown and not shown in the Fig. 5.2 representations

Representation A		Representation B		Representation C	
What it shows	What it does not show	What it shows	What it does not show	What it shows	What it does not show
Shape	Placement in the body	Internal structure with names of parts	Placement in the body		Shape
Surface feature on one side	Internal structure	Direction of blood flow	Colour	Key function of the heart as a pump	Internal structure
Colour					

It is important to note that a key element of all representational challenges is to have some evaluation of the student-generated representations. Challenges in the classroom usually lead to evaluative discussions amongst the students or in class discussions by the teachers. In the cases illustrated in Figs. 5.3 and 5.4, students shared their word clouds/mind maps in small groups. The teacher had access to all students' representations through the STILE platform which she used to inform her subsequent teaching and to initiate class discussion by projecting selected students' word clouds or mindmaps for the whole class to view.

The technical features word clouds and mindmaps are different and afford and constrain the representations that are constructed. Knowledge of these features by the students allows them to make certain decisions in what they wish to express in their representation. It is the role of the teacher to ensure that students gain such knowledge not only for the specific task as described above but to add to the students' kitbag of representational forms they might draw on in the future (***Principle 2***). In this way teachers have a role to play in developing students' meta-representational competence (diSessa, 2004). According to diSessa (2004) meta-representational competence includes:

- The ability to invent novel representations
- The ability to critique existing representations
- Knowledge of the functions that representations perform
- Knowledge that facilitates the rapid learning of new representation

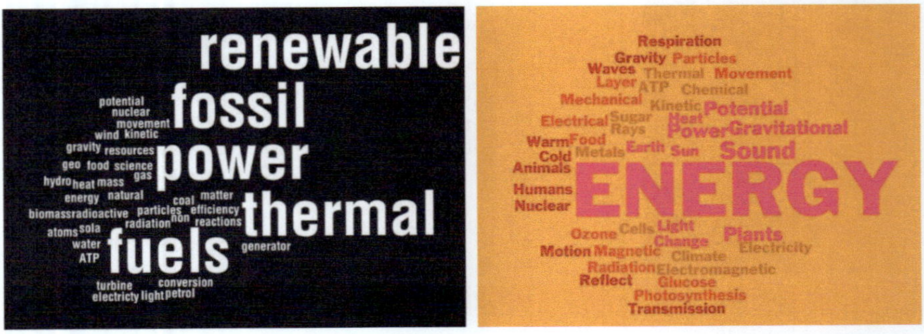

Fig. 5.3 Year 9 students' word clouds of the concept of energy

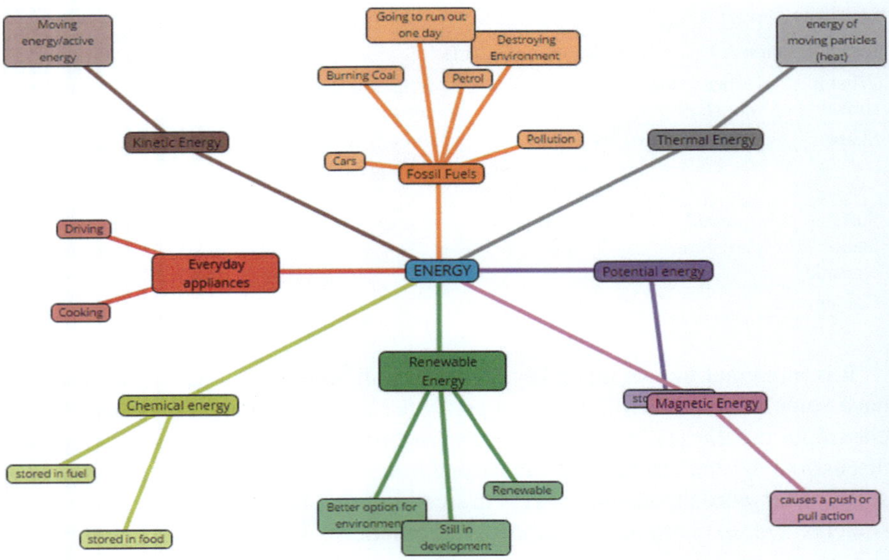

Fig. 5.4 Year 9 student's mind map connecting energy forms to their daily lives

5.4.2 *Opinion Polls and Surveys*

The use of a cloud-based platform accessible to all students in the class provides the teacher with ways to quickly gain formative data that can be put into various representational forms that can be fed back to the class to initiate discussion and inform the direction of further teaching (***Principle 4***). The following two examples relate to the use of opinion poll and survey strategies within the STILE platform.

The purpose for the opinion poll was to elicit Year 9 students' attitudes to climate change (***Principle 3b***) with questions that were part of an annual survey of Australian attitudes to climate change conducted by the Australian Government Commonwealth

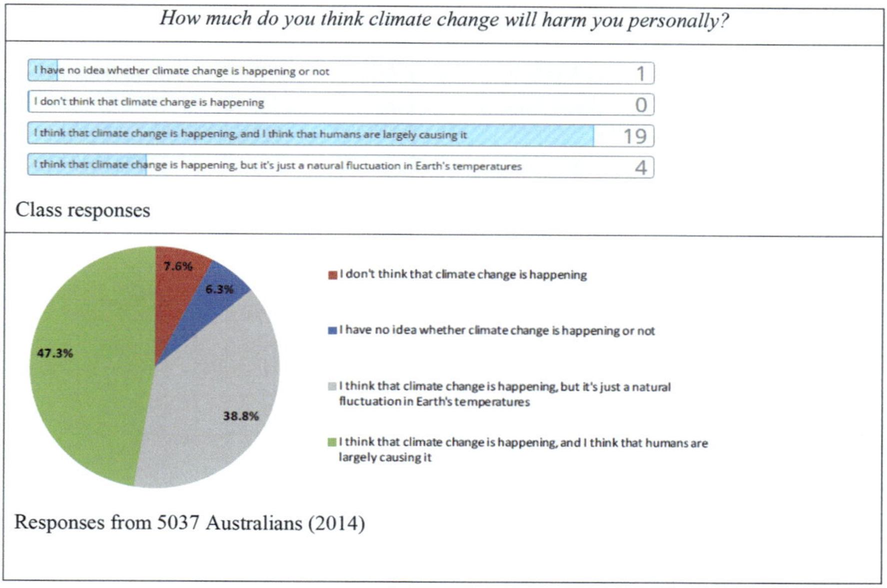

Fig. 5.5 Year 9 students' and Australian responses to an opinion poll/survey of attitudes to climate change (Leviston et al. 2014)

Scientific and Industrial Research Organisation (CSIRO) (Leviston, Price, Malkin, & McCrea, 2014). Questions included:

- Is climate change happening?
- What best describes your thoughts about climate change?
- How worried are you about climate change?
- How much do you think climate change will harm you personally?

After the class responded to a question, a graphical representation of the class results, automatically generated by the STILE platform, was shown to the students alongside a graphical representation of individual responses from the CSIRO survey. Figure 5.5 shows an example of the class and national results to one of the questions. The teachers reported that the class critique and analysis of graphical data were highly valuable in generating discussion about climate change from a personal and national perspective.

An online survey was used to determine the Year 9 students' prior knowledge of the particle model to inform the teacher in planning to teach energy transfer processes such as conduction and convection. Figure 5.6 shows a list of statements for which the student was to respond as either true or false. As with the opinion poll, the STILE platform immediately generated a graphical representation of the class results. A colour code is used in the representation with red indicating a non-scientific response and green indicating a scientifically correct response. The key affordance of this representational form was that the teacher gained instant feedback

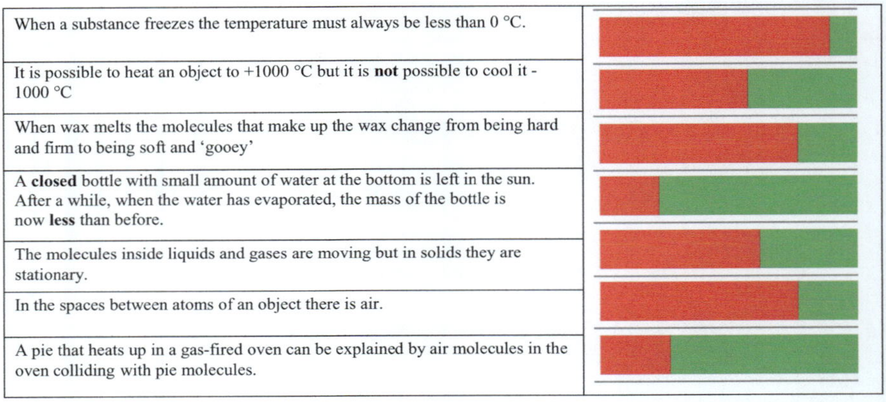

Fig. 5.6 Year 9 students' responses to a survey to elicit views about particle model

on students' thinking across several areas of the particle model at the same time. The teacher also used the representational form as stimulus to generate classroom activities and discussion about a topic the students had been taught in previous years.

5.4.3 Representational Challenges Employing Particle Ideas About Matter

Following the results from the survey (Fig. 5.6), the teacher engaged a review of the particle model and its role in science as a representation to explain properties of matter. Students were given the representational challenge to use particle ideas to represent the following properties of matter:

1. A lump of plasticine holds its shape.
2. A lump of plasticine can be changed into a different shape.
3. A piece of chalk can't change shape; it breaks easily (brittle).
4. A rubber band can stretch and return to its original shape.
5. Red cordial and water mix easily.
6. An iron cube is much heavier than an aluminium cube of the same size.

Students had the option of either creating a digital representation using a drawing tool embedded in the STILE platform, or they could draw on paper and take an image to then upload to the STILE platform. Figure 5.6 provides some examples of student-generated representations in response to the challenge. The evaluation of the student-generated representations was initially undertaken at the class discussion level with the teacher making references to specific representations.

The decision as to whether a representation is suitable depends on the nature of the property of the matter that is to be explained. For example, one might consider

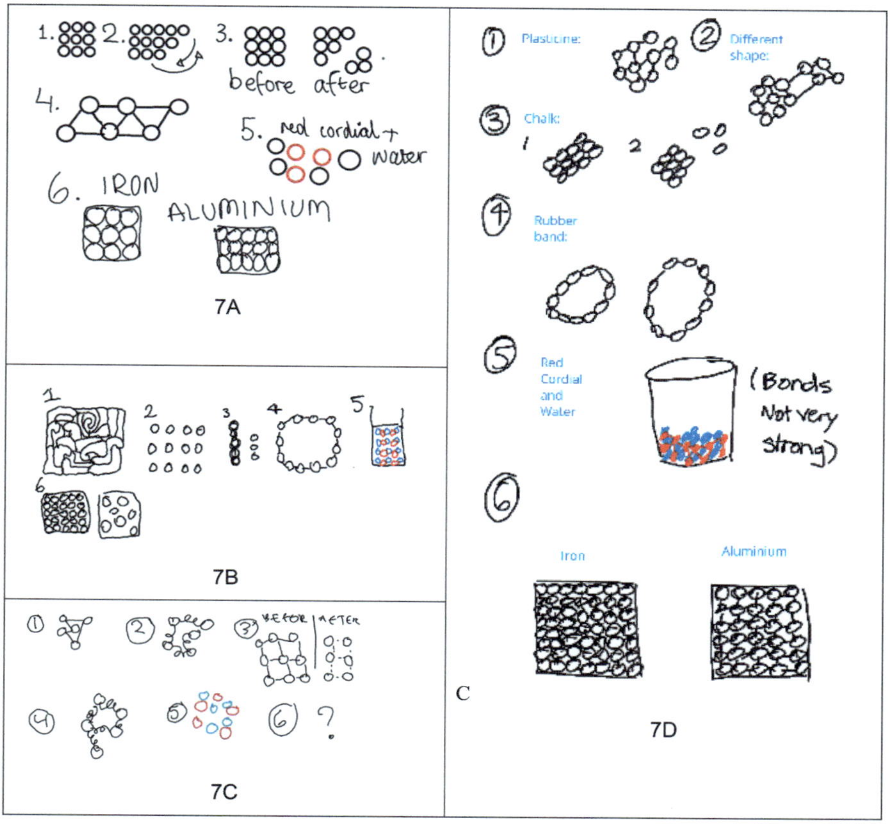

Fig. 5.7 Year 9 students' explanations of properties of matter using particle ideas

that a suitable representation for Challenge 1 listed above requires that the representation needs to show particles that are connected in some way to explain stability of shape of a lump of plasticine. Figure 5.7b–d successfully show this, whereas Fig. 5.7a does not show connectedness of particles. For Challenge 4 the representation needs to show particles, and connections between them can be maintained even though the particles move apart as occurs when the rubber band stretches. Figure 5.7d shows this with a before and after image, whereas Fig. 5.7c shows this through representing the connections as springlike.

These representation challenges described students engaging in the modelling process. Students are expected to take elements of the particle model of matter to generate in a drawing a model that explains the given property of matter. This replicates the work of scientists in creating scientific models to explain the observations they make of the world.

Having considered using the particle model to explain some physical properties of solids, the teacher introduced temperature and the idea that the temperature of an

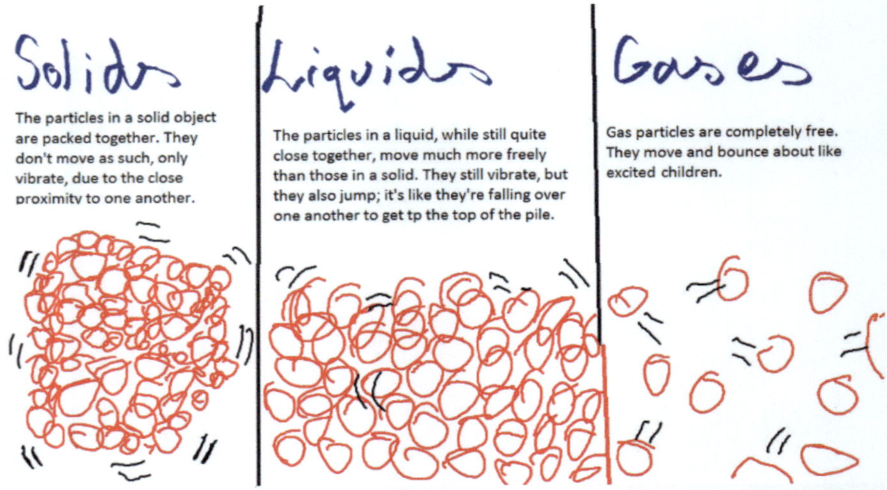

8A

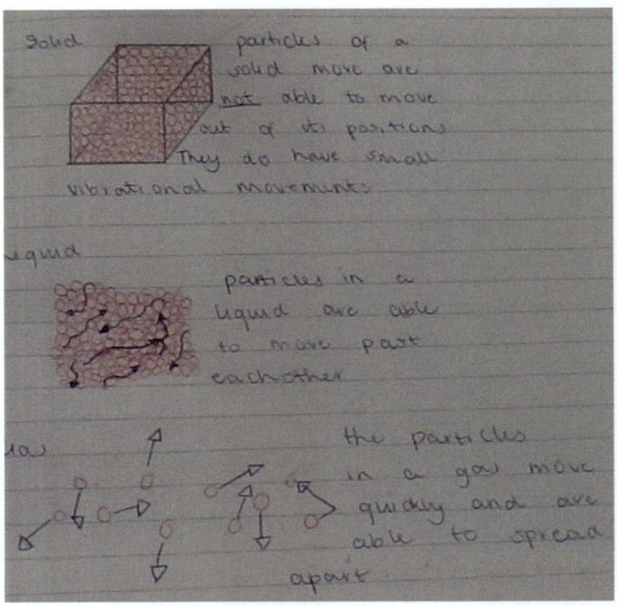

8B

Fig. 5.8 Year 9 students' re-representations of particle movement

object relates to the average kinetic, or motion, energy of the particles that make of the object. This led to the presentation of animations of particles in each state. The representational challenge was for the students to describe the motion of the particles in each of the states. Refer to Fig. 5.8 for student-generated representations to this task. Figure 5.8a describes very well the movement of particles in each of the states, whereas Fig. 5.8b makes effective use of arrows in representing movement of particles represented in a liquid state (curved arrows) and particles represented in a solid state (straight arrows).

For one of the classes that undertook this challenge, discussions of the student-generated representations led to a class convention as how particles might be represented diagrammatically. The class agreed that when representing matter in a solid state, particles would have bracket-type symbols to represent the movement (vibrations); when representing matter in a liquid state, particles would have curved arrow symbols to represent movement (moving around each other); and when representing matter in a gas state, particles would have straight arrows to represent movement (moving in straight lines). Coming to a class consensus view replicates the manner in which scientists undertake their work through generating and validating standard representational forms as a way to explain and communicate evidenced-based findings.

5.4.4 Interactive Simulations and Animations to Represent Dynamic Processes

One of the constraints associated with representations based on drawing is that it is sometimes difficult to show dynamic processes such as the movement of particles of matter in a particular state. In addition many of the dynamic processes that drive phenomena occur at the submicroscopic domain, and so animations and, in particular, simulations can assist the learner in developing an understanding of the phenomena. Animations and simulations can provide more insights into a dynamic process than drawings can, and so it can be beneficial for student learning if teachers use animations and develop students' skills in creating them (***Principle 2***). The following two examples relate to the teaching and learning of energy as part of the iSTELR project.

The first example relates to the use of an interactive simulation by the teacher. Figure 5.9 shows a snap shot image from an interactive simulation representing an energy system involving the sun, solar panel, heater and tank of water. The simulation is part of a high-quality set of freeware digital resources developed by the University of Colorado, Boulder (USA), and is known as Physics Education Technology (PhET) interactive simulations (https://phet.colorado.edu/).

The PhET simulation was used by the teacher as part of a class discussion to explore key ideas associated with energy. As mentioned early in these chapter discussions about what the simulation shows about energy transfer in a system were had alongside discussions about what the simulation does not show. A key feature of the simulation is the way in which energy is visually represented which, from a

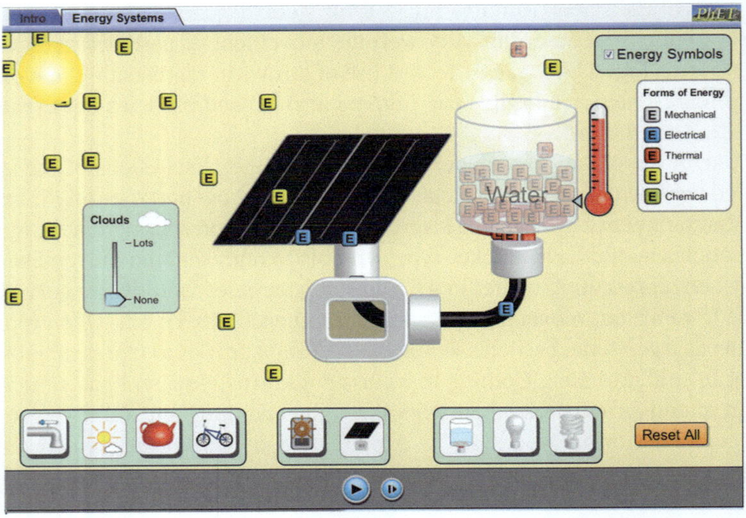

Fig. 5.9 PhET animation of an energy system (*Source*: PhET Interactive Simulations University of Colorado http://phet.colorado.edu)

scientific perspective, is a very abstract concept. Through the energy symbols interacting with the parts of the system, the student gets meaning to such ideas as:

- Energy is manifest in different forms – represented by different coloured symbols.
- Energy transformation – symbols change colour as they move through the various parts of the system.
- Conservation of energy – symbols might change colour but do not disappear as the simulation is run.
- Dissipation of energy – the energy symbols eventually spread out into the surrounding environment.
- Heating arises through the absorption of thermal energy – the simulation links the macroscopic observations of heating the water which are represented as rising scale of the thermometer and steam rising off the water surface with thermal energy symbols filling the water container.

Some aspects of energy transfer through the system that are not shown in the simulation include:

- The mechanism by which energy is transformed from one form to another such as light energy transforms into electrical energy at the solar cell.
- The heater attached to the solar cell is not explicitly represented.
- The simulation only shows electrical energy transformation to thermal energy at the water container.

The PhET simulation was also embedded in the STILE platform and therefore accessible to the students. The interactive nature of the simulation meant that students could explore energy transfer in other systems with components shown at the bottom of the simulation page (Fig. 5.9). In interrogating other systems, students were given a task to create energy flow diagrams.

The second example relates to a representational challenge given to the students that involved them creating an animation using PowerPoint that uses a particle representation to represent how solar radiation from the sun can interact with Earth in different ways. They were to choose one of several scenarios to represent as a PowerPoint animation. Two such scenarios include:

- Solar radiation enters Earth's atmosphere, reflects off Earth's surface and returns to space.
- Solar radiation enters Earth's atmosphere and gets absorbed at the surface. Thermal radiation is emitted that either:
  - Enters atmosphere passes through into space
  - Enters atmosphere, gets absorbed by carbon dioxide and radiates back to Earth to then be reabsorbed by Earth and then reradiated to the atmosphere then into space

Students were given a template, shown in Fig. 5.10a–c, to construct their PowerPoint animation. Through the STILE platform, they were provided with a short video screen cast that outlines the procedures to construct an animation.

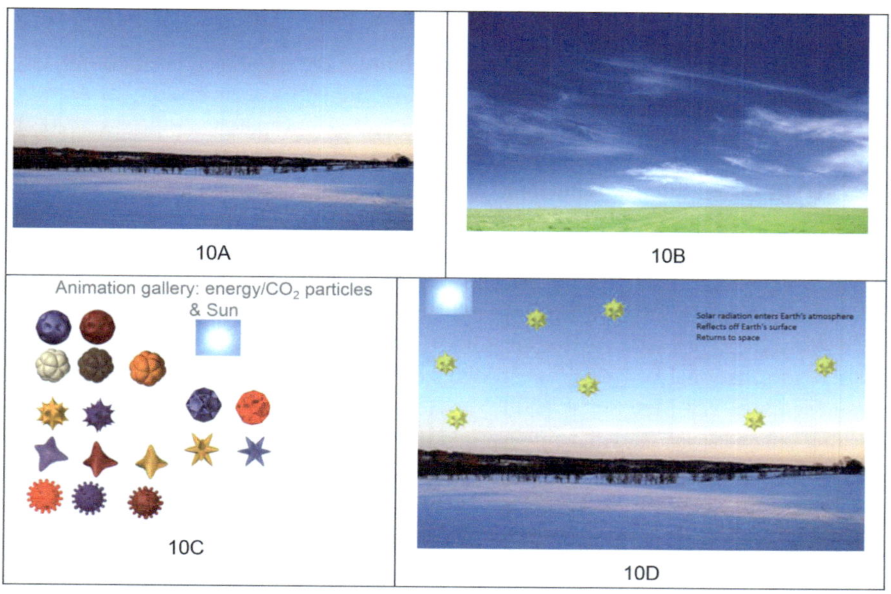

Fig. 5.10 Year 9 challenge to create an animation representing the Sun's radiant energy interactions with the Earth

Figure 5.10d is a snap shot of a student-generated animation representing one of the first scenarios listed above. The animation challenge explored key ideas that formed the basis of understanding the warming of Earth through the greenhouse effect.

5.5 Drawing to Learn in Science

As will be apparent from the illustrations of student work in the previous sections, much of the representation construction involves student drawings of a variety of types, either iconic or symbolic. We have found that drawing has much to offer in terms of student engagement, focus on learning and collaborative construction of explanation. Much of the research on the impact of drawing on learning has come from the cognitive science literature and characterizes drawing in terms of its support of cognitive processes. There is, however, growing interest from a sociocultural perspective in drawing as an important discursive practice mediating learning (Vygotsky, 1981). A recent review of the literature that combines both traditions (Ainsworth, Prain & Tytler, 2011) identified five reasons for a renewed focus on drawing as a classroom activity:

1. Drawing to enhance engagement
2. Drawing to learn to represent in science
3. Drawing to reason in science
4. Drawing as a learning strategy
5. Drawing to communicate

Researchers argue that students, when generating their own representations such as line graphs, can come to a deeper understanding of the conventions of specific representations, as a form of meta-representational competence (diSessa, 2004; Gilbert, 2005). Other researchers have focused on the benefits of collaborative drawing, in which the explicitness of drawings provides opportunities for students to exchange and clarify ideas (Schwartz, 1995). The public sharing of representations, a key component of our own approach, has been argued to allow students to productively critique the clarity, coherence and content of drawings (Linn, Lewis, Tsuchida & Songer, 2000). Hackling and Prain (2005) found, in an evaluation of a large-scale literacy-based Australian science programme, that teachers perceived a positive motivational benefit when students drew to explore and justify understandings in science. Van Meter et al. (2006) argue that the strength of drawing to support learning occurs when students are required to translate between modes, thus enriching understanding.

Other studies (Stieff, 2011; Zhang & Linn, 2008) have shown marginal effects of coupling drawing with dynamic visualizations of complex molecular structures in high school chemistry. Similarly Stieff and DeSutter (2016) found only marginal learning gains when students made observational, then reflective sketches of a dynamic simulation of molecular behaviour. Van Meter and Garner (2005), in a review of the literature, argue that interest in research on drawing has fallen off because of a history of inconsistent results, which relate to variation in the way drawing is conceived of and used in different studies.

In our own research, we position student drawing as an important aspect of students generating representations in response to challenges. For us, the key question is how the act of constructing and collaboratively negotiating ideas through drawing can support enriched reasoning and learning as the task is engaged with. In order to further explore how students collaboratively reason through the construction and coordination of multimodal representations, we conducted research in a specially constructed classroom with multiple cameras and radio microphones, in order to capture a comprehensive record of student activity, including talk, gesture, experimental exploration, drawing and modelling and embodied interactions, in response to representational challenges. Within the analysis we were able to focus particularly on students' drawing activity, and in this chapter we provide illustrations of some of the conditions, and learning affordances, associated with learning through drawing.

The research involved the planning, execution and analysis of six single science lessons on the topics of levers, toys and energy, plant reproduction and astronomy, conducted in the Science of Learning Research classroom at the University of Melbourne. The Year 7 classes (students age 12) were taught by their own science teachers, with the activities developed jointly by the research team and the teacher based on the representation construction inquiry approach. The broad lesson outline involved an introduction to the topic, a preliminary challenge engaged with by pairs of students, reporting back and then a more advanced challenge tackled first by pairs of students and then shared within groups of four students to negotiate resolution. Most activities involved drawing, either using pen and paper or using markers on a portable whiteboard, or both in sequence. The analysis of the video record was undertaken first by selection of groups to illustrate a variety of levels of engagement and production and then transcription of the audio and video record to identify sequences that provided insight into processes of collaborative reasoning through exploration and representation construction. From these analyses a number of principles were constructed concerning the roles of drawing in supporting reasoning and learning (Tytler, Ferguson, Aranda, Gorur, & Prain, 2016). In this chapter we describe a number of these insights into the way drawing supports reasoning and learning in science.

1. *Drawing can play an active role in framing student exploration and reasoning*

The lever task required students to explore, using a set of small weights and a see-saw constructed from a ruler with an attached fulcrum in the centre, what combinations of weights would create a balance. They were to draw a representation of their findings.

Students used the drawings sometimes as predictive and then tested their hypothesis using the see-saw and in other cases first tested using the see-saw and then noted through drawing (Fig. 5.11). The drawings thus ranged from generative of ideas to consolidating, but the distinction is not clear-cut. The drawings in either

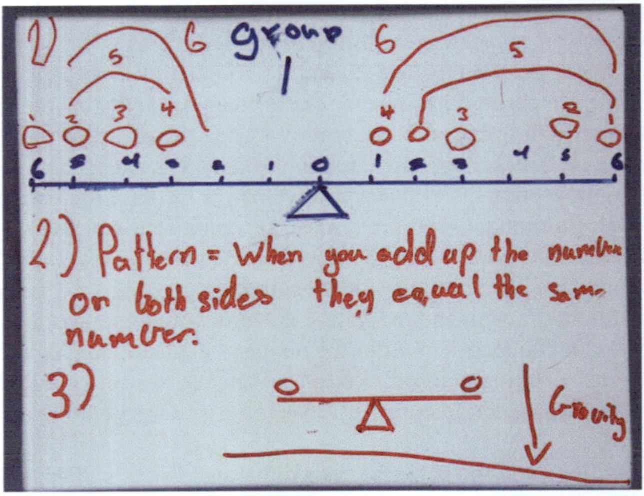

Fig. 5.11 Group drawing on a whiteboard of patterns of lever balance, showing abstracted representations of patterns of weights for balance

case served to structure the experimentation to some degree, making apparent the patterns of distance and weight through mathematical abstraction.

2. *Symbolic drawings were used to establish key ideas*

In a subsequent lever task, students were asked to provide advice to the owner of a donkey and cart, shown in a photograph where the donkey is raised in the air because of too heavy a load on the back of the cart. The drawings in the most productive cases were used to abstract the problem to lever principles with explicit reference to fulcrum and load. The affordance of the drawing in these cases is to help make apparent and force choices regarding the spatial and numerical features of the situation and to establish common meaning amongst students in the group. Figure 5.12 shows the abstraction involved in approaching a solution. The group had debated whether to draw an actual donkey or a symbolic representation.

3. *A key distinction is between the generative and consolidating roles for drawing*

One of the key distinctions we made in the analysis was that between generative and consolidating roles for drawing. In some cases students were challenged to draw in a task where the real exploration best took place with the physical exploration. For instance, students were challenged to work out the mechanisms and energy pathways for small toys such as wind-up cars, a jack-in-the box, a spring-loaded helicopter launcher, a balloon rocket or a mousetrap car. In many cases the students explored the toys to work out the mechanisms, but the drawing, with some exceptions, tended to be an after-the-event activity, and students did not have sufficient access to the hidden mechanisms to speculate on the spatial arrangement of cogs or springs that would give drawing its power. In other cases, however, for the astron-

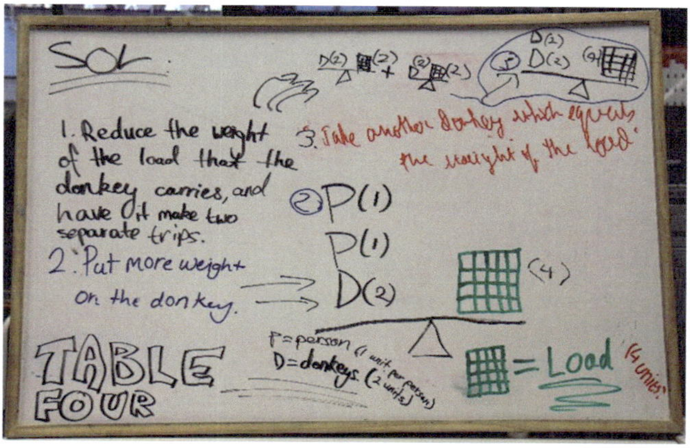

Fig. 5.12 Abstracted drawing of a donkey and a cart, making reference to lever principles

omy lessons, for example, the drawings were the site for collaborative problem-solving as students negotiated their understandings of the spatial arrangements that would explain night and day, or sun elevation, for instance (see below).

4. *Drawing was effective in framing/constraining student attention to relevant details of phenomena.*
5. *Drawing engaged students in focusing and maintaining attention on the task.*

In a task involving drawing and writing about a balloon-powered car, the act of drawing raised for the pair of students the question of the relative direction of the air coming from the balloon and the movement of the cart. The drawing required specificity and led the pair back to exploring with the car itself. It thus triggered attention to noticing of details, and the resolution was reflected in the final drawing. Similarly, in drawing a mousetrap car mechanism, a pair of students became intensely focused on the details of the mechanism layout, occasioned by a need to frame the drawing to illustrate this clearly. In this and other cases, the act of drawing led to deeper engagement with the phenomenon.

6. *Drawing acted as a common ground through which groups of students revealed their ideas and negotiated agreement about the visuospatial aspects of interpretations/explanations*
7. *Drawing can be powerful when coordinated with other representational modes*

Two boys were challenged to construct a drawing that might be shown to a 7-year-old to explain how it could be different times in London and Melbourne. One student drew and talked his partner through the specifics of his drawing. His partner responded with his own account, illustrating with pointing to features of a small earth globe, and then negotiated drawing his own account on the basis this would be clearer for a 7-year-old (Fig. 5.13).

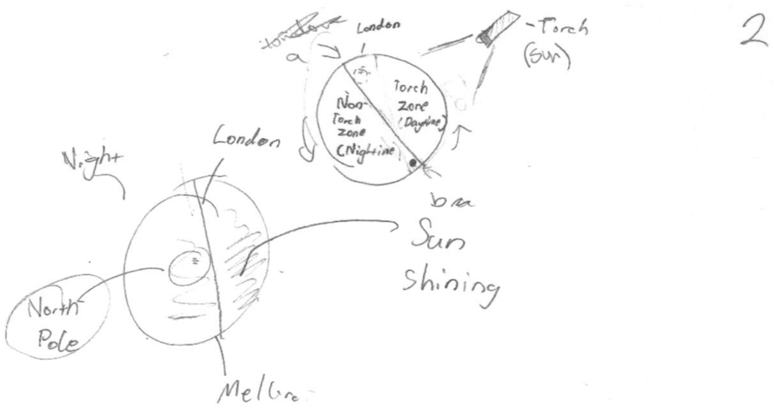

Fig. 5.13 Drawings to explain why London and Melbourne experience different times. The drawing on the right is the second of the two

8. *Drawing acted as a common ground through which groups of students revealed their ideas and negotiated agreement about the visuospatial aspects of interpretations/explanations*
9. *Drawing on whiteboards was advantageous in allowing preliminary thoughts to be rendered and refined and in allowing joint construction*

Figure 5.14 shows a whiteboard drawing designed to show how the sun can be at a higher angle in the sky in summer. This drawing was the result of considerable collaborative discussion between the two students, involving rubbing out and redesigning text, frequent exploration through a torch and model of the globe and discussion in which one student would illustrate a point using the drawing, and gesturing, and the other would exclaim 'yes, I've got what you mean!'. The drawing thus grew by degrees with each student contributing and refining their ideas of what about the geometry of the orbit and axis tilt was important. The whiteboard allowed this constant refinement and shared production and thus became an effective site for negotiation. We found generally that students were more ready to commit ideas on the whiteboard, because of this capacity to erase. The drawing in these cases often was used for, first, unrefined thoughts and was progressively modified through collaborative discussion and shared control.

10. *Drawing exposes visuospatial aspects of student conceptions that were accessible to teachers and provided an opportunity for negotiation of meaning*

During the lessons it was clear, as the teachers circulated round the class checking students' work and engaging with their ideas, that the drawings were a powerful focus for conceptual discussion through their specificity in visual, spatial and symbolic aspects. Teachers and students were able to focus their discussion through features of the drawings in ways that would not have been possible with text or talk alone.

These vignettes of student drawing illustrate the central role of representation construction in collaborative reasoning in science and show the central role of draw-

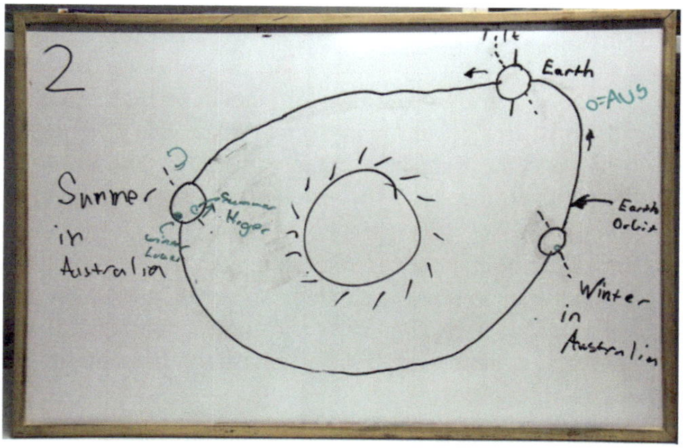

Fig. 5.14 Whiteboard drawing to explain why the sun is higher in summer than winter, in Australia

ing, alongside other modes of representation, in collaborative inquiry processes that focus on conceptual explanation. We argue that inquiry processes in science classrooms that include drawing as a central process engage students in approximating scientific discovery processes and support reasoning and learning in powerful ways. We further argue that these representation construction processes are central to scientific problem-solving within transdisciplinary contexts. These findings are consistent with a body of research that places representing and drawing as central to modelling processes in mathematics (Lehrer & Chazan, 1998) and engineering (Johri, Roth & Olds, 2013) as well as science. Thus, developing capability to represent and draw needs to be central to teaching and learning in each of the STEM disciplines, if students are to operate effectively in transdisciplinary contexts.

5.6 Further Development of the Representation Construction Approach

Since developing the representation construction inquiry approach, we have worked with many teachers and quite closely with a number of schools to refine and extend the approach and expand the range of topics for which we have generated resources. One school that has been particularly generative in this regard is Salsa College,[1] a metropolitan boy's school, where a group of dedicated teachers worked with Years 7 and 8 students over 3 years in collaboration with the research team. The experience and innovations of teachers Alice, Jaz and Kate[2] are described below under particular features of their approach.

[1] Pseudonym for the school

[2] Pseudonyms are given for all teacher names.

5.6.1 Student Record-Keeping

The teachers introduced learning journals for science, which were project books that were larger than A4 in size and were formatted so that when opened the left-hand page was lined and the right-hand page was blank. The use of these project books was new to the students who previously used fully lined A4-sized workbooks.

The project books facilitated the use of drawings in recording what they learned (see Fig. 5.15 for some examples). Drawings were often used in addressing the representational challenges (see Fig. 5.16 for some examples). The blank page encouraged visual forms of representations. The visual representations provided the teacher with ready insight into students' thinking:

> *Immediately by looking at their representations, I know, okay those boys have got it and those boys are on the right track but those haven't fully kind of understood. (Alice)*
>
> *But the books just having the blank page, I think sometimes, it's just all text that we kind of forget how much the use of those representations and diagrams can really help in Science, so it was a good reminder. (Alice)*

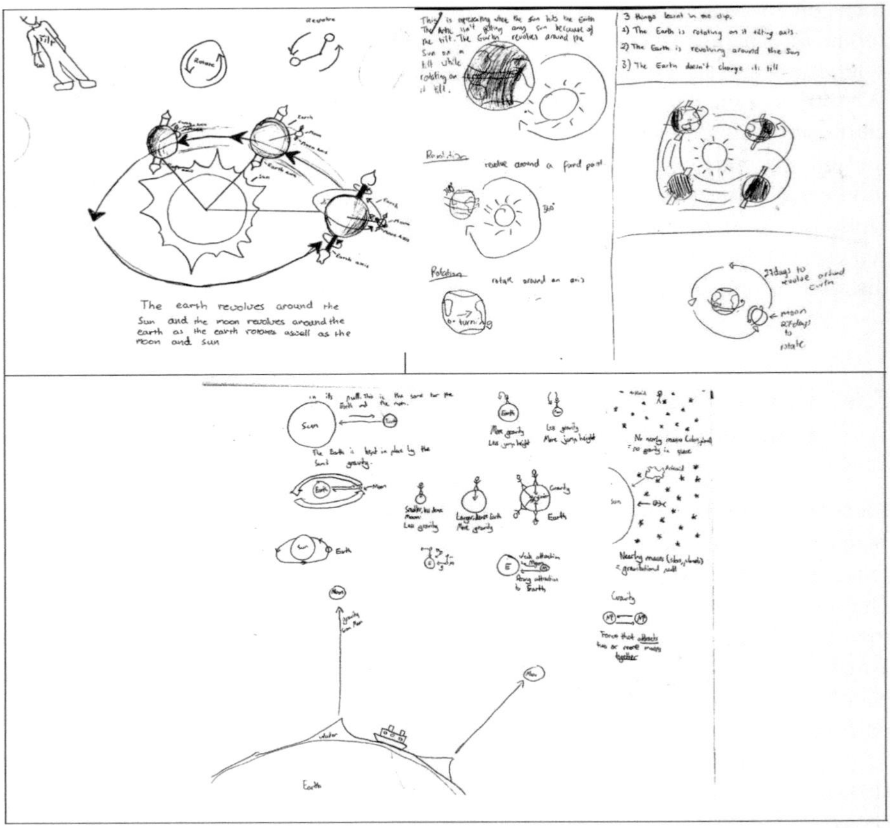

Fig. 5.15 Examples of students' entries into their learning journals

Representational Challenge: Represent in a drawing the height of the midday sun in winter from Melbourne

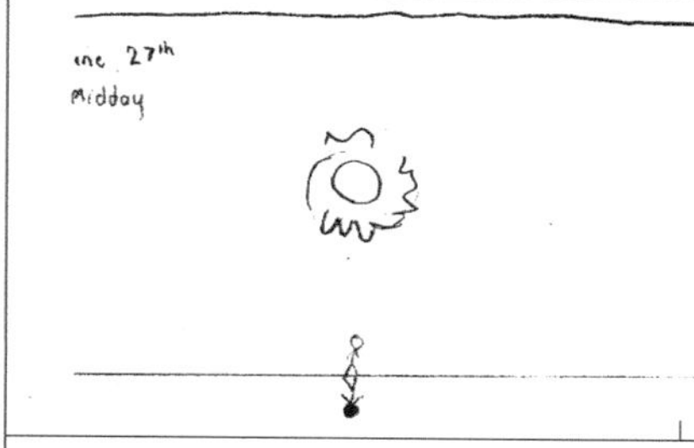

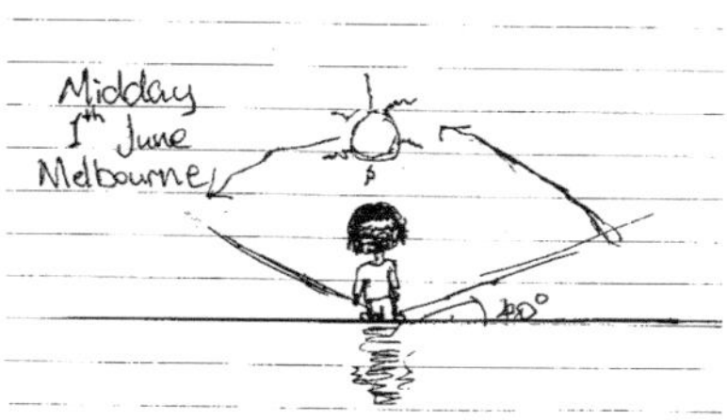

Representational Challenge: Why when we were watching the opening ceremony for the Olympics was it night in London, but morning here in Melbourne?

It is morning in Australia because the Sun is shining on the Earth bit it can't shine all of Earth like a shadow. It is because of the tilt; rotating, revolving around the Sun it rotate

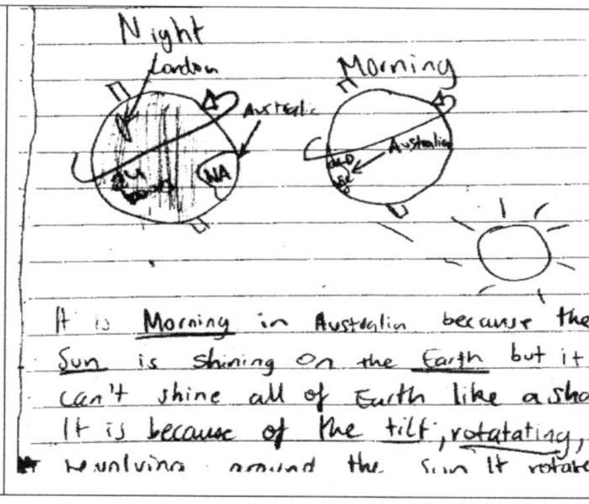

Fig. 5.16 Examples of students' responses to two representational challenges

The teachers found that students were more willing to use their journals to reflect on their learning:

> *...they seemed more willing to go back over their work and look back at their past stuff as well...And I don't think they do it very well if it's just written stuff and they had a sense of ownership over it which was good. (Kate)*

The entries the students' made in their learning journals were seen by the teachers as a vehicle for discussion:

> *And I think ...that while they're doing their representations you can have conversations with them and be active with them but it's not such a threat, it's not give me the correct response, it's more about why have you done it that way. (Alice)*
>
> *But I found that the discussions were a lot more sophisticated that they were having, around the topics than usually with the textbook. (Alice)*

5.6.2 Pretesting and Alternative Conceptions

A pretest for each topic was developed by the research team. Whilst the administration of pretests was not common practice at Salsa College, the teachers agreed to implement it. They initially viewed the pretest as part of the research rather than integral to the teaching sequence, but subsequently came to view it as an important part of the teaching approach: 'It just should be teaching practice; it should just be what we do [Kate]'. The prevalence of alternative conceptions was surprising for Jaz who commented, 'I didn't realize. I just thought once kids learn things that they keep a hold of it, but they don't'.

The teachers used the information gained from the pretests in their teaching as the illustrated by the following comments:

So I would say that, in that question [taken from the pre-test], what did we think and I'd get them to talk about it. And then at the end of the lesson, we'd say "Okay, so if we saw that question again, how would we be changing our answer to be more representative? ...we weren't pretending like they had this blank slate and they'd never seen astronomy before. They already had ideas, that we kind of – half the battle was challenging them, more so than teaching them new content. [Alice]

I did deal with the topics that they had the most trouble with. [Jaz]

...and the misconceptions we knew where the majority of the class were thinking so you could direct your teaching to that...it highlighted for me the numbers in the class who don't get it, don't get a concept. (Jaz)

5.7 Summative Assessment

The teachers at Salsa College had a long-standing practice of administrating pen-and-paper-based tests as a final summative task to the topics that were taught. This practice continued in the astronomy unit. However, a key insight the teachers gained from the students learning journals was the power of the multiple modes of representation that the students generated. This prompted a change to open-ended questions given on the final test, challenging students to construct representations and providing a space rather than the traditional lines for student to respond. Figure 5.17 shows the use of this expanded space for student responses to a test question asking, 'An astronomer investigating the motion of *Europa*, which is a moon, or natural satellite, of the planet *Jupiter*, found that it *revolved* as well as *rotated*. Use the space below to clearly explain what each of these motions mean'.

The teachers commented on the value of having these multiple representational responses to the test questions:

In their test answers if we gave them the space they would perhaps do a diagram to help with explanation or we might say use representation, they didn't just stick to the words. (Jaz)

And it valued those boys that do like to draw. (Alice)

The science team subsequently adopted this approach to assessment more widely:

And even with our year 8 exam last semester [outside of the Astronomy topic] like in our extended response more inquiry based we opened it up that they could represent that knowledge in multiple ways. (Alice)

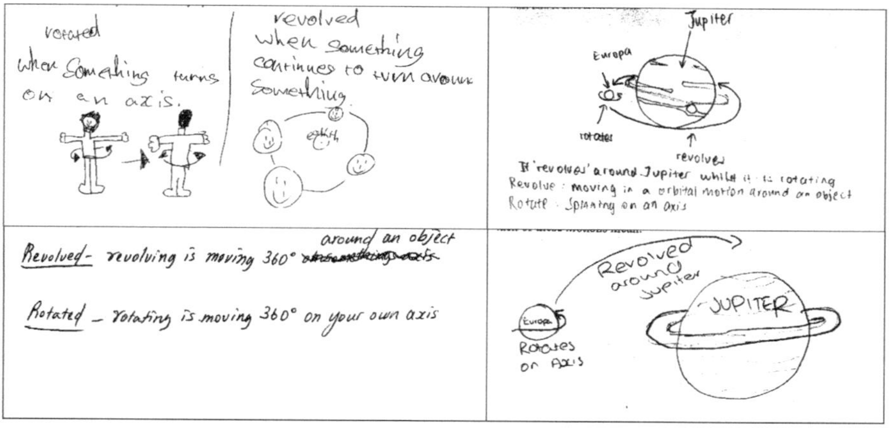

Fig. 5.17 Student responses to a test question

5.7.1 Teachers Focusing on Meta-Representational or Representational Competence

The Salsa teachers who have worked with us in developing and implementing representation construction over several years indicate that a key change to their teaching revolves around a more targeted need to develop students' meta-representational competence (diSessa, 2004) or representational competence (Kozma & Russell, 2005). Kozma and Russell (2005) point out that representational competence allows students to think, communicate and conceptualize about science concepts and includes the abilities to:

(a) Use representations for describing scientific concepts
(b) Construct and/or select a representation and explain its appropriateness for a specific purpose
(c) Use words to identify, describe and analyse features of representations
(d) Compare and contrast different representations and their information content
(e) Connect across different representations and explain the relationship between them
(f) Realize that representations correspond to phenomena but are distinct from them
(g) Use representations in discourse to support claims, draw inferences and make predictions

We have found that teachers who are new to implementing representation construction also point to representational competence as a new area of focus for their teaching. The following case refers to a Sydney metropolitan girl's school where Year 7 teachers were introduced to representation construction. They then trialled

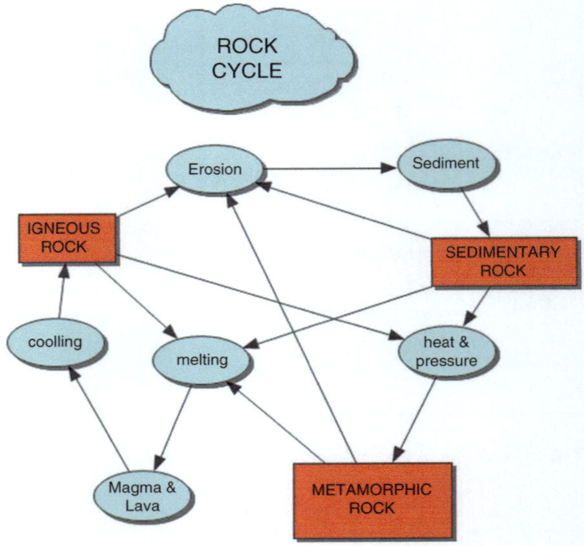

Fig. 5.18 Small group critique of diagrammatic forms of the rock cycle

the approach in the classroom sometimes taking classroom video to share with colleagues and researchers.

The following transcript is taken from a video clip one of the teachers shared with colleagues. The task involved small groups of students who were to critique a set of seven different rock cycle representations. For each representation the students were to answer the questions, 'What does it show well?' and 'What it does not show well?'. Figure 5.18 shows a particular rock cycle that is referred to in the following transcript of the teacher and students:

Teacher (T): *Looking at the cycle what can you tell me about it?*
Student (S)1: *It shows how everything is formed and connected.*
T: *When you say everything what do you mean?*
S1: *The types of rocks.*
S2: *And it is colour-coded too.*
T: *Does that help?*
S2: *Yes because if you follow the arrows you find what you are looking for.*
S1: *For example, both sedimentary and igneous rocks have similar processes that they can through heat and pressure form the metamorphic rocks* [pointing to the dark red arrows]*...it shows how they are connected to the metamorphic rock.*
S3...*it gives you options about where to go.*
S1: *The second example is sedimentary rocks can melt to form magma, which when it cools becomes igneous rocks; the igneous though can become a sedimentary rock once again through erosion [tracing the path with a pen].*
T: *So erosion is leading from that one* [pointing at igneous].
S1: *Connected to sediments to sedimentary...*
S2: *Its like a never ending cycle* [point out various cycle on the diagram].
T: *Does it show weathering?*
S1: *It shows erosion but doesn't show weathering.*
T: *So does this help explain the ideas?*

S2: *Looking at it first it was kind of confusing but once you had time to look at it and follow the arrows it makes a lot of sense.*

Following this activity a class discussion ensued as to the affordances and constraints of the rock cycle presentations. A key point expressed by the teacher was:

It is always important to acknowledge the fact that when you are making decisions about representing things compromises have to be made as to the level of detail wanted.

The next tasks for the students in one of the classes, as expressed by the teacher, were:

They made a poster of the rock cycle each from an initial critique and then peer assessed each others…this lead to a challenge to construct a stop-motion animation of a rock story; each pair were given a random sequence of rocks in pairs of students e.g. sedimentary to metamorphic.

In reflecting on any changes to their practice through the implementation of representation construction, the following views were expressed by the teachers:

I think getting them to try and represent something or I think particularly critiquing, comparing two different representations was useful when we did the rock cycle one, comparing two different ones.

Usually I would give them what I thought was the best diagram for what I was trying to explain…but getting them to compare two diagrams and pick out the best and draw their own version is better, because they've had to process that to put it into a diagram.

But the thing that I've changed the most is critiquing the representations…I think critical evaluation is probably the biggest change I've seen in myself… the kids would see this explicitly.

I think getting them to try and represent something or I think particularly critiquing, comparing two different representations was useful when we did the rock cycle one, comparing two different ones…getting them to compare two diagrams and pick out the best and draw their own version is better, because they've had to process that to put it into a diagram.

The students are now more critical of the representations that they find…before they would grab the first google image without critically analysing it for what it shows.

Because they had done it earlier – they were used to the language, used to critiquing and yes there are different ways to represent things…and then they would comment why are things like this represented in this textbook and not in another.

5.8 Conclusion

This chapter has introduced representation construction as a directed inquiry pedagogy approach that requires students to interpret and construct representations of scientific concepts, claims and processes. By representing some aspect of the world about them, students engage in the processes of knowledge construction of science as well as gaining scientific knowledge. The approach maps well with the creative processes in which scientists explore nature and construct new knowledge. The adoption of representation construction approaches addressed call for school science to better represent the epistemic practices by which knowledge is built in science.

Representation construction supports a more active view of knowledge than traditional structural approaches and encourages visual as well as the traditional text-based literacies. This is illustrated with the visual nature of the student-generated representations given as examples throughout this chapter and, in particular, the section related to drawing to learn in science. The examples of student-generated representations emphasize the manner in which students grapple with conceptual challenges in exploring, generating, evaluating and refining representations. Representation construction show promise in not only engaging students in inquiring into the world about them but supporting students to develop scientific literacies to a high level.

The representation construction approach places demands on the pedagogical skills of the teacher beyond those needed for transmissive approaches, for example, the skills to provide a representation-rich environment and opportunities for students to negotiate, integrate, refine and translate across representations. Teachers require good subject content knowledge that entails an understanding of the key representational resources underpinning science topics and an understanding of the role of representation in teaching and learning science. The approach requires of teachers a capability to run open discussions and develop the insights needed to guide the classroom tasks and conceptual negotiation.

The adoption of representation construction approaches does open up new directions and emphases for teachers to pursue in their teaching. For example:

- A change from students using their notebooks as repositories of distilled scientific knowledge provided by the teacher to use their notebooks as learning journals
- The affordances of the student-generated representations to provide insights into their thinking and formative tools that inform the teacher in addressing issues such as the prevalence of alternative conceptions
- A new emphasis in not only developing students' conceptual understanding of science but also developing students' meta-representational competence

Representation construction as a guided inquiry approach was born from extensive research in science classrooms. However, we feel that many of the ideas inherent with the approach have synergies with inquiry-based approaches in other disciplines. Certainly, the basic premise that representations are things that individuals use to understand the world as well as communicate meaning to other individuals applies to other disciplines to science such as the creative arts and, in particular, other STEM disciplines. For example, Dreher, Kuntze and Lerman (2016) point out that representations and their connections play a key role for experts in the creation of mathematical knowledge and for learners to build a conceptual knowledge in the mathematics classroom. Mathematical objects are abstract, and so experts as well as learners must use representations when dealing with them (Duval, 2006). Similar views are expressed by Johri, Roth and Olds (2013) for the discipline of engineering. These authors note in a special issue in the Journal of Engineering Education focused on 'Representations in Engineering Practice' that representations are central to engineering professional practice as well as learning about engineering design processes in the classroom.

Engaging with multimodes of representations in teaching and learning is important for each of the STEM disciplines. However, when considering the demands on the learner for integrated STEM education experiences, Honey, Pearson and Schweingruber (2014) indicate that, 'Students need to be competent with discipline-specific representations and be able to translate between discipline-specific representations thereby exhibiting what some scholars refer to as *representational fluency*' (p. 71). Representational fluency is synonymous with meta-representational competence. The role of the STEM teacher becomes one of not only introducing students to the individual disciplinary representations but also guiding them in constructing their own representations and developing their skills in representational fluency that allows them to move flexibly with and across disciplinary representations. It is therefore a worthy path for future research in representation construction to explore its efficacy in the teaching and learning within and across the STEM disciplines. Our current project is a Victorian Department of Education-funded project, Secondary STEM Catalysts: professional learning programme (2016–2018), which aims to build STEM engagement of Year 7 and 8 students in 30 government schools across the state of Victoria. Whilst we are early days in the project, teachers from all STEM disciplines initially find representation construction an appealing approach to pursue further in their teaching.

References

Ainsworth, S. (2006). DEFT: A conceptual framework for learning with multiple representations. *Learning and Instruction, 16*(3), 183–198.

Ainsworth, S. (2008). The educational value of multiple representations when learning complex scientific concepts. In J. K. Gilbert, M. Reiner, & M. Nakhlel (Eds.), *Visualization: Theory and practice in science education* (pp. 191–208). New York: Springer.

Ainsworth, S., Prain, V., & Tytler, R. (2011). Drawing to learn in science. *Science, 333*(26), 1096–1097.

Chi, M. (2009). Active-constructive-interactive: A conceptual framework for differentiating learning activities. *Topics in Cognitive Science, 1*, 73–105.

Chubb, I. (2014). Office of the Chief Scientist: *Science, Technology, Engineering and Mathematics: Australia's Future*. Australian Government, Canberra, Australia. Retrieved from http://www.chiefscientist.gov.au/wp- content/uploads/STEM_AustraliasFuture_Sept2014_Web.pdf

Cox, R. (1999). Representation construction, externalized cognition and individual differences. *Learning and Instruction, 9*, 343–363.

diSessa, A. (2004). Metarepresentation: Native competence and targets for instruction. *Cognition and Instruction, 22*(3), 293–331.

Dreher, A., Kuntze, S., & Lerman, S. (2016). Why use multiple representations in the mathematics classroom? Views of English and German preservice teachers. *International Journal of Science and Mathematics Education, 14*(2), 363–381. doi:10.1007/s10763-015-9633-6.

Duval, R. (2006). A cognitive analysis of problems of comprehension in a learning of mathematics. *Educational Studies in Mathematics, 61*, 103–131. doi:10.1007/s10649-006-0400-z.

Elkins, J. (2011). Visual practices across the University: A report. In O. Grau (Ed.), *Imagery in the 21st century*. Cambridge, MA: MIT Press.

Furtak, E., Seidel, T., Iverson, H., & Briggs, D. (2012). Experimental and quasi-experimental studies of inquiry-based science teaching: A meta-analysis. *Review of Educational Research, 82*(3), 300–329.

Gibson, J. (1977). The theory of affordances. In R. Shaw & J. Bransford (Eds.), *Perceiving, acting, and knowing: Toward an ecological psychology* (pp. 67–82). Hillsdale, NJ: Erlbaum.

Gilbert, J. K. (2005). *Visualization in science education*. New York: Springer.

Gooding, D. (2004). Visualization, inference and explanation in the sciences. In G. Malcolm (Ed.), *Studies in Multidisciplinarity* (Vol. 2, pp. 1–25). Elsevier.

Gooding, D. (2006). From phenomenology to field theory: Faraday's visual reasoning. *Perspectives on Science, 14*(1), 40–65.

Goodrum, D., Druhan, A., & Abbs, J. (2012). The status and quality of year 11 and 12 science in Australian schools. Canberra, Australia: Australian Academy of Science. Retrieved April 2016 from http://www.science.org.au/publications/research-reports-and-policy.html

Greeno, J. G., & Hall, R. P. (1997). Practicing representation: Learning with and about representational forms. *Phi Delta Kappan, 78*(5), 361–368.

Hackling, M., & Prain, V. (2005). *Primary connections: Stage 2 trial*. Canberra, Australia: Australian Academy of Science.

Honey, M., Pearson, G., & Schweingruber, H. (Eds.). (2014). *STEM Integration in K-12 education: Status, prospects and an Agenda for research*. Washington, DC: The National Academies Press.

Hubber, P., Tytler, R., Chittleborough, G., Campbell, C., & Jobling, W. (2012). *Evaluation of delivery of the switched on secondary science professional learning (SOSSPL) program (2011–2012)*. Waurn Ponds, Victoria: Deakin University.

Hubber, P., Tytler, R., & Haslam, F. (2010). Teaching and learning about force with a representational focus: Pedagogy and teacher change. *Research in Science Education, 40*(1), 5–28.

Johri, A., Roth, W., & Olds, B. (2013). The role of representations in engineering practices: Taking a turn towards inscriptions. *Journal of Engineering Education, 102*(1), 2–19. doi:10.1002/jee.20005.

Kapur, M. (2008). Productive failure. *Cognition and Instruction, 26*(3), 379–424. doi:10.1080/07370000802212669.

Kozma, R., & Russell, J. (2005). Students becoming chemists: Developing representational competence. In J. Gilbert (Ed.), *Visualization in Science Education* (pp. 121–145). Springer.

Latour, B. (1999). *Pandora's hope: Essays on the reality of science studies*. Cambridge, MA: Harvard University Press.

Lehrer, R., & Chazan, D. (1998). *Designing learning environments for developing understanding of geometry and space*. Mahwah, NJ: Lawrence Erlbaum Associates.

Lehrer, R., & Schauble, L. (2006a). Cultivating model-based reasoning in science education. In K. Sawyer (Ed.), *Cambridge handbook of the learning sciences* (pp. 371–388). Cambridge, MA: Cambridge University Press.

Lehrer, R., & Schauble, L. (2006b). Scientific thinking and science literacy. In W. Damon & R. Lerner (Eds.), *Handbook of child psychology* (6th ed., Vol. 4).

Lemke, J. L. (1990). *Talking science: Language, learning and values*. Greenwood Publishing Group.

Lemke, J. (2004). The literacies of science. In E. W. Saul (Ed.), *Crossing borders in literacy and science instruction: Perspectives on theory and practice* (pp. 33–47). Newark, NJ: International Reading Association/National Science Teachers Association.

Leviston, Z., Price, J., Malkin, S., & McCrea, R. (2014). *Fourth annual survey of Australian attitudes to climate change: Interim report*. CSIRO: Perth, Australia.

Linn, M., Lewis, C., Tsuchida, I., & Songer, N. (2000). Beyond fourth grade science: Why do U.S. and Japanese students diverge? *Educational Researcher, 29*(3), 4–14.

Moje, E. (2007). Developing socially just subject-matter instruction: A review of the literature on disciplinary literacy learning. *Review of Research in Education, 31*, 1–44.
Osborne, J. (2006). Towards a science education for all: The role of ideas, evidence and argument. In *Proceedings of the ACER research conference: Boosting science learning –What will it take?* (pp. 2–5). Retrieved from http://research.acer.edu.au/cgi/viewcontent.cgi?article=1000&context=research_conference_2006
Prain, V., & Tytler, R. (2012). Learning through constructing representations in science: A framework of representational construction affordances. *International Journal of Science Education, 34*(17), 2751–2773.
Schwab, J. J. (1962). *The teaching of science as enquiry*. Cambridge, MA: Harvard University Press.
Schwartz, D. L. (1995). The emergence of abstract representations in dyad problem solving. *The Journal of the Learning Sciences, 4*(3), 321–354.
Stieff, M. (2011). Improving representational competence using molecular simulations embedded in inquiry activities. *Journal of Research in Science Teaching, 48*(10), 1137–1158.
Stieff, M., & DeSutter, D. (2016, June). *Drawing from dynamic visualisations*. Paper presented at the International Conference of the Learning Sciences, Singapore.
Tytler, R., Ferguson, J., Aranda, G., Gorur, R., & Prain, V. (2016, June). *Drawing within experimental exploration as part of core epistemological and epistemic practices in science*. Paper presented at the International Conference of the Learning Sciences, Singapore.
Tytler, R., Haslam, F., Prain, V., & Hubber, P. (2009). An explicit representational focus for teaching and learning about animals in the environment. *Teaching Science, 55*(4), 21–27.
Tytler, R., Peterson, S., & Prain, V. (2006). Picturing evaporation: Learning science literacy through a particle representation. *Teaching Science, the Journal of the Australian Science Teachers Association, 52*(1), 12–17.
Tytler, R., Prain, V., Hubber, P., & Waldrip, B. (Eds.). (2013). *Constructing representations to learn science*. Rotterdam, The Netherlands: Sense Publishers.
Van Meter, P., Aleksic, M., Schwartz, A., & Garner, J. (2006). Learner-generated drawing as a strategy for learning from content area text. *Contemporary Educational Psychology, 31*, 142–166.
Van Meter, P., & Garner, J. (2005). The promise and practice of learner-generated drawing: Literature review and synthesis. *Educational Psychology Review, 17*(4), 285–325.
Vygotsky, L. (1981). *Thought and language* (Rev. and Ed. by A. Kozulin). Cambridge, MA: MIT Press.
Waldrip, B., Prain, V., & Carolan, J. (2010). Using multi-modal representations to improve learning in junior secondary science. *Research in Science Education, 40*(1), 65–80.
Zhang, Z., & Linn, M. (2008). Using drawings to support learning from dynamic visualizations. In *Proceedings of the 8th international conference on the learning sciences* (Vol. 3, pp. 161–162).

Chapter 6
Making STEM Curriculum Useful, Relevant, and Motivating for Students

Léonie Rennie, Grady Venville, and John Wallace

Abstract More than ever, we live in a connected, global community. In this chapter we argue for a STEM school education that helps students to explore and experience the kind of connectedness that reflects life outside of school. While many would agree that STEM curricula should be embedded in real-world, authentic contexts, much of the current policy and practice favours disciplinary approaches to knowledge narrowly focused on what is readily measurable or amenable to achievement testing. In contrast, the issues that affect students' lives outside of school are not unidisciplinary, neither are the solutions to problems that beset our world today. Here, we explore the contribution of an integrated approach to STEM education with the goal of increasing students' opportunities to engage in contextual, multidisciplinary issue-based learning.

6.1 Introduction

This chapter is based on the premise that we live in a connected, global community, and students should experience a school education that reflects this reality. In other words, the curriculum, particularly curriculum that is related to science, technology, engineering, and mathematics (the STEM subjects), must provide opportunities for students to explore and experience the kind of connectedness that reflects life outside of school. Most often, however, we find that schooling in these subjects does not encourage the embedding of the curriculum in real-world, authentic contexts, and this is especially so in secondary schools. Instead, current educational policy favours disciplinary approaches to knowledge that are translated into curriculum

L. Rennie (✉)
Curtin University, Perth, Australia
e-mail: l.rennie@curtin.edu.au

G. Venville
University of Western Australia, Perth, Australia

J. Wallace
University of Toronto, Toronto, ON, Canada

R. Jorgensen, K. Larkin (eds.), *STEM Education in the Junior Secondary*,
DOI 10.1007/978-981-10-5448-8_6

documents that espouse educational outcomes narrowly focused on what is readily measurable or amenable to achievement testing. Left out are the less cognitive-based outcomes that are so important in building the range of citizenship skills that we expect students to possess by the end of their school years.

The issues that affect students' (and their parents') lives outside of school are not neatly divided into traditional disciplines such as chemistry, physics, and geometry, and we can see this in almost any issue we care to examine. Global problems, such as changing weather patterns or sustainability of food and water resources, as well as local problems, such as building a road through a wetland or dealing with a fruit fly infestation, are not unidisciplinary and neither are their solutions. Multidisciplinary teams of specialists are required to examine the complexity of these issues and work out possible solutions. Of course, developing those solutions is also a complex matter because real-world problems are beset by political, economic, and ethical considerations. How can we prepare students to become involved in multidisciplinary issues relating to their world outside of school?

In this chapter, we explore how integrating STEM subjects as part of the curriculum can increase students' opportunities to engage in contextual, multidisciplinary issue-based learning. In our discussion, we use the term STEM as an acronym referring to one or more of the four component subjects. Because there is no clear or agreed definition of STEM integration or even STEM education (English, 2016), and there is no single subject called STEM, our discussion will refer to science (or sometimes technology or mathematics) as a starting point for integration among these subjects and with other subjects.[1] We begin by expanding our premise above, elaborating the consequences of the disparity between science in the real world and science in schools and suggest why integrating STEM subjects might make the curriculum more useful, relevant, and motivating for students. We overview the research base that informed our thinking about what is meant by curriculum integration and then elaborate points of tension between disciplinary and integrated approaches to curriculum in terms of the strategies that enhance and the barriers that hinder their implementation. We conclude with suggestions about how teachers of the junior secondary years might incorporate an integrated approach that can motivate students through the perceived usefulness and relevance of the learning outcomes.

[1] We acknowledge that this approach appears to privilege science above the other STEM subjects, and as English (2016) pointed out, science has the greatest representation in integrative approaches. In our experience, however, mathematics and technology were frequently involved, but not always referred to explicitly. Engineering was represented as the design part of technology, but engineering is rarely a named subject in the curriculum, so teachers referred to design and construction as technology.

6.2 The Disparity Between Real-World Science and Science in Schools: The Promise of an Integrated Curriculum

As noted in the first paragraph of this chapter, secondary school curricula are uni-disciplinary in the sense that disciplines, such as chemistry, biology, history, and mathematics, are readily recognised as separate subjects in school timetables; even general science at the junior secondary school level is often taught in terms of its component disciplines. Such separation is a convenient way to organise timetables and deliver lessons, but how true is it to the science in students' world outside of school? More than three decades ago, Capra (1982) wrote:

> We live today in a globally interconnected world, in which biological, psychological, social, and environmental phenomena are all interdependent. To describe this world appropriately we need … a new vision of reality, and a fundamental change in our thoughts, perceptions and values. (p. 16)

Capra was arguing for a holistic view of reality; he believed we had reached our limits of understanding if we looked at the world in a disconnected way, fragmented into individual subjects. Yet schools still have a curriculum that is far removed from the practice of science in the real world (Scott, 2008). Creating snippets of content for students to digest by stripping away the connectedness and context from real-world science also removes the excitement of knowledge building and its significance in solving the problems of the day. The result is a sterile subject in which many students have little interest and perceive peripheral relevance (Lyons & Quinn, 2010).

Perhaps it is not surprising that science education, particularly at the secondary level, has experienced decades of criticism for its inability to turn out students who are knowledgeable and interested in pursuing science (Stocklmayer, Rennie, & Gilbert, 2010). Research by Osborne and Collins (2001) found that science curricula at this level were often perceived to be difficult and content-driven, were poorly taught by transmissive methods, and so failed to engage many students. These authors reported "the sense that pupils were being frog-marched across the scientific landscape, from one feature to another, with no time to stand and stare, or absorb what it was that they had just learnt" (p. 450). Of course, science was not like this in every classroom. Osborne and Collins summarised: "From a more positive perspective, pupils saw the study of science as important and were engaged by topics where they could perceive an immediate relevance, practical work, material that was challenging and high-quality teaching" (p. 421). Nevertheless, it seems that even students who perceive science to be important do not necessarily want to participate in it themselves (Jenkins & Nelson, 2005).

Students have held this view of disconnection for some time, as Beane (1991) has described:

> To the students, the typical curriculum presents an endless array of facts and skills that are unconnected, fragmented, and disjointed. That they might be connected or lead towards some whole picture is a matter that must be taken on faith by young people or, more precisely, on the word of adult authority. (p. 9)

To Beane (1995), the solution was for teachers to work within a framework of curriculum integration, in which two things would happen:

> First, young people are encouraged to integrate learning experiences into their schemes of meaning so as to broaden and deepen their understanding of themselves and their world. Second, they are engaged in seeking, acquiring, and using knowledge in an organic – not an artificial – way. That is, knowledge is called forth in the context of problems, interests, issues, and concerns at hand. And since life itself does not know the boundaries or compartments of what we call the disciplines of knowledge, such a context uses knowledge in ways that are integrated. (p. 616)

Beane was an advocate of curriculum integration as a means of tailoring curriculum to the needs of students in the "middle years" of schooling (usually Grades 6–8 or 9). Despite a lengthy history, particularly in the USA, middle schools struggled to achieve their goal of improving adolescent education (Mac Iver & Ruby, n.d.), a key feature of which was an interdisciplinary curriculum, with, for example, "content focused on major concepts and unifying themes drawn from the areas of literature, social studies, mathematics, science, and fine arts" (Alexander & Williams, 1965, p. 221). The *Middle School Journal* began in 1970 (originally as the *Midwest Middle School Journal*) to support, in the words of one editor, those "middle school educators [who] are working diligently to fulfil our dream of a relevant education for this unique group of students" (Gaddis, 1973, p. 1). An integrated curriculum was part of that dream, but it wasn't easy to achieve, as Drake (1991) demonstrated in her description of the difficulties experienced by her team in implementing a holistic learning approach through integrated studies. Nevertheless, the efforts to focus on the middle years were often associated with efforts to integrate the curriculum as a means to increase its relevance, including in Australia, where the Schools Council (1993) recommended that curriculum should be worthwhile, integrated, and inclusive (see Russell, 2010, for a retrospective overview).

In the mid-1990s, as interest in middle schooling and integrated curriculum grew, we began a research programme to try to understand what curriculum integration was about and what its outcomes were. Our research programme is described in Rennie, Venville, and Wallace (2012a, pp. 7–11). In brief, we began with 12 case studies of programmes their creators described as integrated and synthesised the findings to describe different kinds of integration and document best practice among them. Five years later, 9 of those 12 schools were revisited to see what had become of the integrated programmes. This allowed the identification of factors that enabled or inhibited their longevity, and we validated these factors with six new case studies. We then turned our focus to examine the nature of student outcomes from integrated programmes, this time conducting a further eight case studies in Australia and Canada. The remainder of this chapter draws from this research base and our iterative efforts to reflect on and understand our findings in a more international context.

6.3 What Is Meant by an Integrated Curriculum?

While the etymology of the term "integration" is clear – it derives from the Latin word *integratus* meaning to make whole – its meaning in the context of curriculum is less so. In general terms, there is agreement that integration in curriculum means the bringing together previously separate areas of either content or skills (Kysilka, 1998). In specific terms, however, there is considerable diversity among curricula that are described by their creators as integrated, and many authors have suggested categories to describe the diversity of approaches. Most of these categorisations start from a curriculum where disciplines are taught quite separately and then describe degrees of overlap of the disciplines involved. An early categorisation was put forward by Meeth (1978), who defined four stages of integration beyond the single discipline approach: In cross-disciplinary programmes, one discipline is examined from the perspective of another, such as the science of music; in multidisciplinary programmes, two or more disciplines contribute to the study of a single problem or issue, but the disciplines retain their "identity", as in using aspects of physics and mathematics to design a solar cooker; in interdisciplinary programmes, the boundaries between the disciplines become blurred, and a more coherent, or integrated, approach is taken to study the matter at hand; finally, in a transdisciplinary programme, discipline boundaries are entirely dissolved. As Meeth pointed out, and Drake (1991) demonstrated in her team's development of an integrated curriculum, this is the most difficult form of integration because teachers need skills and knowledge from many areas. Nevertheless, transdisciplinary is the way solutions to major societal issues are approached. As an example, think of the complexity of the worldwide response to the recent threat of the spreading Ebola virus and the range of disciplinary teams needed to respond to it.

A number of authors have described integration based on the degree of discipline overlap; see Drake (1993) and Jacob (1989) for examples. Others have used more simple descriptions, such as Applebee, Adler, and Flihan's (2007) three-part categorisation beyond the unidisciplinary curriculum of correlated, shared, and reconstructed. In our own research, however, we became uncomfortable with the notion of the continuum these categorisations imply because they suggest that one end is better than the other end. As remarked elsewhere (Rennie et al., 2012a), "we found no inherent quality of 'betterness.' [in the programs reviewed] Instead, we believe that the effectiveness of each curriculum must be judged according to its purpose" (p. 5). Further, like Hargreaves, Earl, and Ryan (1996), we consider that a continuum simplifies the complexities of integrated approaches. We found it more fruitful in talking with teachers to use a categorisation of curriculum integration based on a description of its approach and purpose. Drawing on the variations we studied during our research programme, we settled on six kinds of approaches to integration in relation to STEM subjects. They are synchronised, thematic, project-based, cross-curricular, school-specialised, and community-focused programmes. These approaches are outlined in Table 6.1 with examples but are described more fully in Rennie et al. (2012a, pp. 20–33).

Table 6.1 Summary of approaches to curriculum integration

Approach to integration	Description of integrated approach with example
Synchronised	Teachers identify skills, knowledge, or understanding that are common to particular topics in two or more subjects and teach these topics separately but, in parallel, explicitly drawing cross-curricular links and reinforcing concepts.
	Example: For a Year 9 applied class, the science and geography teachers identified areas of curriculum overlap and collaborated to teach a 5-week energy topic in parallel, making clear the common concepts and using common assessment items.
Thematic	Teachers work collaboratively (perhaps in learning teams) to organise the curriculum around a local or global topic (e.g. insect infestation, the Olympics). They teach their subjects separately in a complementary way, making connections back to the theme. They may come together in a culminating event incorporating aspects of all learning areas.
	Example: All Five Year 8 teachers worked as a team to run integrated projects (e.g. environmental theme day) incorporating aspects of all learning areas.
Project based	The focus here is on a designated task requiring knowledge and skills from more than one subject area for its completion. Projects are often technology- or engineering-based requiring construction of some kind.
	Example: A Year 9 technology class was set a bridge-building project which incorporated knowledge of science, mathematics, engineering, design, and construction. The aesthetics of the bridge were judged by the English teacher.
Cross-curricular	These approaches are usually based around overarching skills or competencies such as literacy, numeracy, ICT skills, or perhaps social skills such cooperation or environmental responsibility. Integration occurs when these skills are the focus of more than one subject at the same time.
	Example: A school-wide literary focus aimed at assisting Indigenous students to learn English, which underpinned all subject areas and was supported by cross-curricular initiatives, such as horticulture relating to the school garden.
School specialised	When a school has a long-term focus on a specific area, such as aviation studies, teachers in the other subject areas may tailor their courses to have explicit links to this specialisation.
	Example: A coastal high school developed a marine studies specialisation. Teachers in each of the "core" subject areas taught one specified unit of marine studies in each of Years 8–10.
Community focused	Here, a significant community issue, such as maintenance of a local wetland, becomes the curriculum focus, and teachers orientate their subject teaching to help students understand the issue from different perspectives and seek potential ways of dealing with it.
	Example: Two Year 8 classes focused a term's work in science and social studies on access for the disabled in community venues and activities.

Based on Rennie et al. (2012a, Table 2.1)

Considered in terms of discipline overlap, we found that synchronised and thematic approaches were usually multidisciplinary because teaching within subjects prevailed, but in project-based programmes, the approach was more interdisciplinary, the subject boundaries became blurred, and the connections between subjects were more visible to students. Similarly, cross-curricular approaches tended to be interdisciplinary, blurring subject boundaries. School-specialised programmes were at least multidisciplinary but could be interdisciplinary, depending on how the curriculum approaches were implemented. Community-focused programmes were often interdisciplinary but had the potential to be transdisciplinary because meaning-making was focused on the issue, not on the content of the contributing subjects.

The integrated programmes identified in our research were differentially successful in terms of achieving their espoused objectives for several reasons. In sum, we found that integrating curriculum requires time, effort, and commitment, particularly at secondary levels of schooling, but we identified factors that underpinned success as well as the barriers that hindered it. Our overall findings were supported in a recent review of integration in science and technology curricula by Gresnigt, Taconis, van Keulen, Gravemeijer, and Baartman (2014). These authors synthesised other educators' models and proposed a "staircase" model of approaches to integrated curriculum with "steps" labelled as isolated, connected, nested, multidisciplinary, interdisciplinary, and transdisciplinary. Gresnigt et al. summarised their findings in a figure (Figure 2, p. 73), concluding that as one ascended the staircase, there were "higher chances at [achieving] 21st century skills, positive students' attitude towards the curriculum, teacher enthusiasm and commitment, more time spend [sic] on science and technology" (p. 73). Concomitantly, there was "more need for teacher commitment, professional development, teacher support, sustained facilities at school level (e.g. time, funding, schedule, room)" (p. 73). Although their review was limited to primary education, Gresnigt et al.'s findings were very consistent with our own. Integrating the curriculum is hard work.

6.4 Why Is Curriculum Integration So Difficult?

Our research, supported by other educators' analyses (such as Hall & Kidman, 2004; Pang & Good, 2000), suggested that all forms of curriculum integration, especially at the secondary level, challenge the culture and customs of schooling, the ways schools traditionally work. This challenge derives from an approach to curriculum that is flexible, multidisciplinary, and democratic, colliding with a schooling context that is rigid, disciplinary, and hierarchical. The barriers to integration include school traditions and policies that are tied up with organisational and administrative structures related to discipline-based timetabling and room allocation, teachers' training that is based in the disciplines, and assessment routines that privilege disciplinary knowledge over other educational outcomes. Further, parental expectations of "a good education" are often attuned to career choice, and further study and these expectations are linked with unidisciplinary approaches and high cognitive grades. We have discussed these difficulties at length elsewhere (see Rennie et al., 2012a, 2012b); here we look briefly at two key sources of tension that arise from integrated curriculum: the purpose of schooling and STEM curricula and the assessment of outcomes.

6.4.1 The Purpose of Schooling and STEM Curricula: The Knowledge Tension

There has long been a tension deriving from the need for high school STEM curriculum to prepare students for a career or further study and, at the same time, provide those students who do not pursue a STEM-related career with sufficient useful knowledge and skills to become thoughtful and responsible citizens (for illuminative discussions, see Fensham, 1997 and Millar, 1996). The current "solution" in Australia and some other countries is to teach general science, technology, and mathematics subjects at the junior secondary level and then more specialised subdisciplines at the senior level that are more likely to be prerequisite for tertiary study, such as physics and calculus. But do general subjects at junior secondary level provide students with sufficient background knowledge for citizenship, particularly those students who opt out of science as soon as possible? There is little evidence that they do (Fensham, 1997; Goodrum et al., 2001), yet this group represents about 80% of the student population.

During the final decades of the last century, there was increasing emphasis on scientific literacy as the overall purpose of science education (Bybee, 1997), championing a science education that would provide all students with those requisite knowledge and skills to become scientifically literate and responsible citizens. Sometimes the argument simply reduced to a cry of "scientific literacy for all" with the result that the term "scientific literacy" became little more than a slogan with mixed and minimal meaning. Roberts (2007) clarified the confusion by proposing two visions of science/scientific literacy. In Vision I, the "foundational" view looks "inwards at the canon of orthodox natural science, that is, at the products and processes of science itself" (p. 730) and builds a curriculum that focuses on science as discipline. In contrast, Roberts' (2007) Vision II of scientific literacy as "the character of situations with a scientific component, situations that students are likely to encounter as citizens" (p. 730), focuses on science in life outside of school, a more citizen-oriented science. In a subsequent analysis, Roberts and Bybee (2014) regarded Vision I as more appropriately describing science literacy and Vision II as scientific literacy. This distinction is consistent with our own view that school education should encourage students to connect with science in the community, and this is more likely to be achieved with an integrated curriculum.

The crux of the knowledge tension is this: Traditionally school science curriculum has been rooted in Vision I, and this situation has been perpetuated because of the belief that strong disciplinary knowledge is powerful knowledge. It is the knowledge that takes students into tertiary science and high status careers. Integrated, interdisciplinary knowledge, in contrast, is viewed as providing less powerful knowledge, because it is more subjective and qualitative. It is based in Vision II, a curriculum that:

> illuminates how science permeates and interacts with many areas of human endeavour and life situations. … This view is sometimes called science for citizenship, concentrating on matters of more obvious personal and social relevance to students than preparing to grasp more demanding science they might or might not study. (Roberts & Bybee, 2014, p. 546)

But is specialised disciplinary knowledge the kind of powerful knowledge required by junior secondary students? Our research suggests not. We found that integrated, interdisciplinary knowledge, that reflects real-world problems, issues, and concerns, can be powerful knowledge for these students. In one of our case studies, students in a middle school investigated a variety of issues related to the quality of water in a local lake (see Venville, Sheffield, Rennie, & Wallace, 2008). They learned a range of science concepts that were integrated across other subject areas to address the environmental as well as the community issues in the use and health of their local wetland. Further, linking science into those community issues introduced values, such as social and environmental responsibility, enabling students to think about how the problems and issues related to them personally. The students engaged in critique and debate and were able to evaluate and communicate ways that these problems and issues might be addressed. Note that, like Roberts' Vision II, learning about science in the real world does not negate the need for and use of disciplinary content knowledge; relevant content is required, but it enables adolescents' learning to be about life experiences in familiar, local contexts, as well as issues and problems in the larger, global world. In this, and in some other case studies, we found that a curriculum working with community-based, interdisciplinary topics provided the students with powerful STEM knowledge, as well as powerful knowledge to think about and act within their world.

6.4.2 Assessing the Outcomes of Integration

Another of the tensions associated with integrated programmes is the difficulty of assessing the outcomes. Traditional, discipline-based curricula are easy to assess because the expected outcomes relate to readily definable achievement of cognitive objectives. In contrast, studies of integrated programmes reveal that the student outcomes are much broader than can be measured on simple achievement tests. Reviews of the outcomes of integrated curriculum, such as Vars (1991), Hurley (2001), and Becker and Park (2011), have found some benefits of integrated courses over separate subject curriculum in performance on achievement tests, but this varied according to the nature of integration. Broadening the basis of assessment, however, revealed improved outcomes in higher-order thinking, problem-solving, creativity, motivation, and collaboration (Hargreaves et al., 2001). Using different "lenses" to examine learning in an integrated Year 9 class, where students were tasked with building a solar-powered boat, uncovered different but complementary outcomes; indeed we found that more than one frame of reference was needed to capture the variety of learning outcomes in that one integrated project (Rennie, Venville, & Wallace, 2011).

There are two issues underpinning the tension related to measuring the outcomes of integrated curriculum (Rennie et al., 2012a). The first is that the diversity among programmes described as integrated makes it almost impossible to compare outcomes among them or to compare the outcomes of a particular integrated curriculum with those of a subject-specific curriculum. Integrated programmes differ in terms of their purpose, their content, the way they are implemented, and the resources and administrative support teachers experience in their efforts to integrate. These contextual factors are significant determinants of the effectiveness of the programmes, and they cannot be ignored. In essence, this is an issue about definition; there is no single definition for integration, and there is no single definition of expected outcomes. They are as variable as the programmes themselves. This introduces the second issue: How can these outcomes be measured? Traditional approaches to measurement tend to be narrowly focused and content-based, particularly in situations where high-stakes testing is involved. Integrated curricula have a range of intended outcomes relating to skills in problem-solving, knowledge transfer, attitude, motivation, collaboration, and cooperation, and we still have much to learn about how to measure outcomes that are beyond cognitive gains. Comparison of outcomes between integrated programmes, or between integrated and subject-based programmes, is fraught because assessment methods are unlikely to be aligned to the context, the purpose, or the intended outcomes of each. In their meta-analysis of the effects on learning of integrative approaches to STEM subjects, Becker and Park (Becker & Park, 2011) found only 28 studies over a two-decade period that contained quantitative measures comparing students' achievement. These authors lamented the lack of empirical data provided in other studies, yet we suggest that empirical data are not always appropriate to provide, given the diversity of programmes and outcomes. Developing high-quality assessment tools to measure the diverse outcomes of integrated curricula remains a challenge for the education community.

6.5 Factors That Facilitate Integration

Our research programme gave us sufficient insight into the problems and possibilities for curriculum integration to identify the attributes of successful integrated programmes (Rennie et al., 2012b). We found that a small and stable learning environment, effective leadership, teacher-team activities linked to the classroom, time for in-school planning, a flexible timetable, and the ability to make community links were paramount in determining the success of attempts to integrate at the junior secondary level. These attributes are not mutually exclusive; they overlap and are mutually supportive. The more prevalent they were, the higher the chance of a sustainable and successful integration programme. We overview these attributes in the following section.

6.5.1 Small and Stable Learning Environment

The likelihood of a successful innovation was much greater if the teachers involved in the integrated programme had worked together before and particularly if they "shared" a class or classes in their usual teaching. It was important that these teachers had similar ideas about teaching and similar views of the purpose of integration. Being co-located in their work area was also helpful as it facilitated communication among the teachers involved. Stable environments where teachers could reteach the integrated topic in subsequent years enabled teachers to hone their skills, tweak the topic activities, and deliver increasingly effective programmes.

6.5.2 Leadership

Although a single teacher with broad-based pedagogical content knowledge could integrate part of the curriculum within their class, or two teachers could work cooperatively and successfully together, all participating teachers needed to have their efforts supported by the school administrators. Even if the administrators were not teaching themselves, their support was needed to provide encouragement and resources, including meeting spaces and flexibility in curriculum delivery and assessment. Larger groups of teachers needed someone to lead the effort to integrate; someone who had oversight of the curriculum, an ability to assist with planning, could provide support and focus and smooth the way when difficulties arose.

6.5.3 Teacher-Team Activities Linked to the Classroom

We found programmes involving more than one teacher were more effective when teachers worked closely together in their classroom activities, by sharing instructional materials and assessment tasks, for example. Teachers in integrated programmes are often team taught, bringing their classes together for particular activities or events, so the students worked together as a large group. Where sharing did not occur, communication and co-planning tended to lapse, and teachers found it easier to go their own way, and the integration effort gradually foundered.

6.5.4 In-school Planning Time

As mentioned earlier, integrating curriculum requires time and effort, and some of that time must be during school hours. Lack of planning time increased teachers' out-of-school workload, creating considerable stress. Effective programmes had explicit arrangements and spaces where teachers could meet and work together to plan and monitor the progress of their integration efforts. Further, when time was formally set aside for teachers to plan and collaborate, it signalled that their efforts were both important and valued by the school administration.

6.5.5 Flexible Timetable

In-school planning time is closely related to a flexible timetable, both controlled by the school's administrative team. To pursue their goals, there needed to be organisational flexibility for teaching teams to make the necessary pedagogical decisions about bringing classes together, arranging in-school visits at times that suited visitors, or engaging in activities that took students outside of school. Off-site activities often required travel, so flexible timetables allowed longer class-time allocations. Long periods also facilitated the construction work in project-based integrated programmes as less time was wasted in frequent setting out and packing up.

6.5.6 Community Links

Unlike the previous five attributes that are associated with organisational structures within the school, community links specifically relate to taking curriculum content and classroom activities outside of the school and/or bringing community people or practices inside the school. We found two levels of interaction with the community. At one level, students engaged with community members or organisations as a means of obtaining information to use in their school activities. For example, a visitor might come to school to discuss insect control measures in local wetlands. The second, deeper level of interaction occurred when students became involved in actions within the community. This was more than information exchange; it was active participation in community issues, such as helping to replant a community area being rehabilitated.

6.6 Integration and Junior Secondary Schooling

Throughout this chapter reference has been made to middle schools, or the junior secondary level of schooling. We consider that integrating at least part of the curriculum is eminently suited to students in these grade levels (Wallace, Venville, & Rennie, 2010), who fit between the flexibility of primary schooling and the rigidity of senior secondary schooling. Much teaching in primary schools is based around themes, where teachers may focus for a whole or half term on some overarching concept dealt with in an interdisciplinary way. For example, one primary school's website has themes for its Years 5 and 6 curriculum as "How do we remember World War 2?" "Why do people go to the theatre?" "How did the Roman invasion change Britain?" "How has the Shang dynasty shaped modern China?" "What makes Britain 'great'?" "Will the legacy of the Olympics be positive for Brazil?" (http://www.hawkingeprimaryschool.co.uk/curriculum/learningthemes/). At the primary level, curriculum integration is relatively easy because teachers have specific training to teach basic skills across subjects and the flexibility of timetabling to teach in themes, where the boundaries between the various subject areas become quite blurred. In senior high school, in countries where there are end-of-schooling exams for tertiary entrance, unidisciplinary approaches dominate the curriculum. For those students not bound for tertiary studies, there are mixed approaches, sometimes combining vocational subjects, but rarely dealing with subjects like science and mathematics in a serious way. In junior high school, although timetables are usually set to the curriculum subjects, and students move to different rooms for different subjects with different teachers, there can be sufficient flexibility to integrate at least some parts of the curriculum, as shown by the innovative teachers in our research programme, as well as elsewhere.

In the junior high school, there is opportunity to focus STEM curriculum on the "big ideas" that have personal and social relevance in the outside world. Big ideas consist of overarching themes, problems, or issues that stand "above, across, and beyond" disciplinary concepts. They can be global in nature, addressing important and persistent international problems such as world poverty, human migration, climate change, arms escalation, or overpopulation (for more examples, see Wilson's (1998) notion of consilience). Big ideas can also be more local, personal, or place based, such as the study of a local wetland, the problem of garbage disposal, the merits of wind power, or the study of food banks. Another way of thinking about big ideas is in terms of crosscutting concepts such as those proposed in the US National Science Education Standards (NRC, 1996), for example, "scale, proportion, and quantity". Crosscutting concepts are "themes or concepts that bridge the engineering, physical, life and Earth/space sciences; in this sense they represent knowledge about science or science as a way of knowing" (Duschl, 2012, p. 7). While big ideas require robust technical knowledge, invariably they also incorporate the social, cultural, and political and as such are used to help students become better informed and better connected and better able to take appropriate action on the problems of their world.

We have described six ways that integration might occur (see Table 6.1), and all involved integration with the STEM subjects. Although not always successful, due to the absence of some of the enabling factors discussed above, all had potential to engage students in relevant, engaging study that resonated with their interests. In one of our reflections on integrated curriculum (Wallace et al., 2010), we used Habermas' (1971) work to analyse the findings through the perspective of curriculum interests; the technical interest, concerned with rules and regularities; the practical interest, concerned with relating and communicating, in terms of both personal meaning-making and the solving of practical problems; and the critical interest, concerned with political action. Two cases were analysed, and, drawing on the earlier work of Lloyd and Wallace (2004), a framework to tackle a big idea in an integrated way was suggested, including teaching strategies (see p. 202). Essentially, the framework revolves around five elements. First, select a topic that is related to a big idea in the curriculum and identify the relevant skills students need and the issues that are of immediate concern to them, in the context of bigger local and possibly global issues. Second, elicit students' current understandings to illuminate the gaps and therefore the directions to pursue. It is important to acknowledge students' viewpoints to ensure that they can feel personally involved in ways that have meaning for them. Third, determine how this topic will assist students to understand "how the world works" (the technical interest) and what pedagogical strategies can achieve this. Fourth, consider what kinds of guided inquiry will assist students to make personal sense, through solving practical problems (the practical interest). Fifth, and finally, is the critical interest. Often this will involve responsible and thoughtful action related to the local community. The last three of these points may not be present equally in the chosen topic; it depends on the nature of the topic or the big idea. Essentially, the three interests might be seen as anchor points to help teachers ensure that they can give students a more holistic approach to their understanding of the topic.

6.7 Discussion

We began this chapter by noting that we live in a connected, global community and students should experience a school education that reflects this reality. We pointed out that the school curriculum tends to be unidisciplinary, whereas the problems and issues that students will face in the world outside of school require multidisciplinary approaches to their solution. We drew attention to some of the negative consequences of this mismatch, including students' lack of interest and commitment to pursing science and other STEM subjects, and put forward an argument for an integrated curriculum as a means of offering a more holistic approach that is more likely to appeal to students, particularly at the junior secondary level. We quoted from Beane (1991, 1995), whose purpose was to address the needs of these students in ways that made their curriculum relevant and engaging, and believed an integrated curriculum could do this.

We turned then to the meaning of curriculum integration in schools and reviewed the kinds of integration encountered in our research programme and also the factors that determined the effectiveness of their implementation. Although the curriculum approaches observed varied in terms of their purpose, how they were organised, how many teachers or classes were involved, and the extent to which subject overlap or boundary blurring occurred, we saw no inherent patterns that indicated which programmes might be better or worse. Instead we argued that the worth of an integrated programme depends on how well it achieves its purpose and objectives, which in turn depends on how well it was able to exploit the facilities and resources of the school and the pedagogical and organisational skills of the teachers and administrators involved.

Next we noted the importance of making explicit connections between school knowledge and community issues. An important outcome of working in the community was the greater freedom students had to control what they did and what they learned. Our research revealed that although students' learning outcomes from community links were dependent on how students engaged, science concepts were likely to be integrated across other subjects as well as issues in their local environment. In addition, linking science into community issues introduced values, such as social and environmental responsibility, in association with the relevant science concepts. This enabled students to think about how the problems and issues relate to them personally, to engage in critique and debate, and to evaluate and communicate ways that these problems and issues can be addressed (Venville et al., 2008). Finally we recounted a framework for tackling a big idea or topic that could form the basis of developing an integrated curriculum at the junior secondary level.

We acknowledge here that other educators have been similarly concerned to address the needs of students in ways that prolonged their engagement in science and other STEM subjects. Based on his extensive review of science education (and a career devoted to its betterment), Aikenhead (2006) argued for a humanistic approach to science education as a means of making science learning more attuned to everyday life and promoting the engagement of more students. Following their comparative review of formal and informal science education, Stocklmayer et al. (2010) proposed a greater crossover between the two to increase the effectiveness of science education and suggested how this might be done by teasing out the factors that encourage learning (Table 5, p. 25). Although these authors' factors and Aikenhead's (2006) description of the characteristics of a humanist perspective in science education (Table 1.1, p. 3) seem very different at first glance, perusal of the content of these two tables makes it obvious that both endorse the pursuit of an integrated curriculum, a curriculum that deals not only with the content of the disciplines but with the real-world contexts in which they can be applied. Importantly, both involve significantly greater connection with the community than is currently the case in most schools.

We summarise our argument by drawing attention to what we call "a worldly perspective" on curriculum integration. We first wrote about this perspective as we reflected on the first stage of our research (Venville, Wallace, Rennie, & Malone, 2002). Our subsequent work and reflections strengthened our view. It comes from the essential features that were common in the programmes, especially the effective integrated programmes we reviewed. We found that they all combined both disciplinary and integrated knowledge; disciplinary knowledge was never forgotten; it just wasn't the central concern, and students' interests and understandings were central. We found that integrated approaches generally dealt with local issues and concerns, which led to broad-based knowledge and skills which could be transferable to global topics and problems. In other words, disciplinary and integrated approaches to understanding issues and topics can be complementary, rather than incompatible. We think of it in terms of balance and connection: balance between disciplinary knowledge and integrated knowledge and connection between local types of knowledge and global types of knowledge. We found that a holistic, worldly approach to STEM curriculum allowed the disciplines and real-world approaches to problem-solving to co-exist in a balanced and connected way. Further, the greater the balance, and the greater the connection, the more power the curriculum has for the students. We proposed that a curriculum:

> in which both disciplinary and integrated approaches to solving science-related problems co-exist in a balanced way, provides a powerful model for STEM curricula because it enables science learning to go beyond cognitive, conceptual outcomes by including the social processes and real-world contexts that enable students to become effective citizens. Further, we propose that such a curriculum will demonstrate connection between local issues and global concerns. In other words, we are suggesting that that STEM curricula provide a mix of disciplinary and integrated knowledge, set in carefully chosen local problems that can be applied to more global issues. The nature of that mix, finding the point of balance and the degree of connection, is dependent on the particular educational context, and will vary from school to school and from place to place. (Rennie et al., 2012b, p. 140)

In 1949, Tyler asked, as one of four basic questions about curriculum, "What educational purposes should the school seek to attain?" (Tyler, 1949, p. 1). Internationally, we are still struggling to find answers, as illustrated by the variety of curricula evident in various countries and the diversity of views existing about their suitability. The bottom line is about what will be of most benefit to students. Is it a curriculum that provides strong disciplinary knowledge or a cross-disciplinary approach that is more attuned to the world in which they live? Interestingly, the recent national moves in Australia (Education Council, 2015), in the USA (National academy of Engineering and National Research Council, 2014), and in Europe (European Commission, 2015) to promote STEM subjects are now showing specific support for an integrated STEM approach, although all acknowledge that the road towards integration is not a smooth one.

Our research suggested that integrating at least parts of the curriculum in STEM subjects is a fruitful way to build a curriculum that students find useful, relevant, and motivating. We know that teaching and learning are complex, multifaceted processes, and as researchers, we acknowledge that integration is not easy and that there is still a long way to go in supporting teachers in terms of how best to develop their content knowledge across fields and their pedagogical knowledge and assessment practices that relate to integration. Nevertheless, the integrated approach provides a powerful curriculum model because it enables science learning to go beyond conceptual content and involve the social processes and contexts that empower students to become effective citizens.

References

Aikenhead, G. (2006). *Science education for everyday life* (Evidence-based practice). New York: Teachers College Press.

Alexander, W. M., & Williams, E. L. (1965). Schools for the middle years. *Educational Leadership, 23*(3), 217–223.

Applebee, A. N., Adler, M., & Flihan, S. (2007). Interdisciplinary curricula in middle and high school classrooms: Case studies of approaches to curriculum and instruction. *American Educational Research Journal, 44*(4), 1007–1039.

Beane, J. A. (1991). The middle school: The natural home of integrated curriculum. *Educational Leadership, 49*(2), 9–13.

Beane, J. A. (1995). Curriculum integration and the disciplines of knowledge. *Phi Delta Kappan, 76*(8), 616–622.

Becker, K., & Park, K. (2011). Effects of integrative approaches among science, technology, engineering and mathematics (STEM) subjects on students' learning: A preliminary meta-analysis. *Journal of STEM Education, 12*(5&6), 23–37.

Bybee, R. W. (1997). *Achieving scientific literacy*. Portsmouth, NH: Heinemann.

Capra, F. (1982). *The turning point: Science, society and the rising culture*. New York: Simon and Schuster.

Drake, S. M. (1991). How our team dissolved the boundaries. *Educational Leadership, 49*(2), 20–22.

Drake, S. M. (1993). *Planning integrated curriculum: The call to adventure*. Alexandria, VA: Association for Supervision and Curriculum Development.

Duschl, R. A. (2012). The second dimension: Crosscutting concepts. Understanding a framework for K-12 science education. *Science Scope, 35*(6), 6–11.

Education Council. (2015). *National STEM school education strategy: A comprehensive plan for science, technology, engineering and mathematics education in Australia*. Retrieved from www.educationcouncil.edu.au

English, L. (2016). STEM education K-12: Perspective on integration. *International Journal of STEM Education, 3,* 3. Retrieved from http://stemeducationjournal.springeropen.com/articles/10.1186/s40594-016-0036-1

European Commission. (2015). *Science education for responsible citizenship* (Report to the European Commission of the Expert Group on Science Education). Luxembourg, Europe: European Union.

Fensham, P. J. (1997). School science and its problems with scientific literacy. In R. Levinson & J. Thomas (Eds.), *Science today: Problem or crisis?* (pp. 119–236). London: Routledge.

Gaddis, H. D. (1973). President's Corner. *Middle School Journal, 4*(1), 1.

Goodrum, D., Hackling, M., & Rennie, L. (2001). The status and quality of teaching and learning of science in Australian schools: A research report. Canberra, Australia: Department of Education, Training and Youth Affairs.

Gresnigt, R., Taconis, R., van Keulen, H., Gravemeijer, K., & Baartman, L. (2014). Promoting science and technology in primary education: A review of integrated curricula. *Studies in Science Education, 50*(1), 47–84. doi:10.1080/03057267.2013.887694.

Habermas, J. (1971). *Knowledge and human interests* (J. Shapiro, Trans.). Boston: Beacon Press.

Hall, C., & Kidman, J. (2004). Teaching and learning: Mapping the contextual influences. *International Education Journal, 5*(3), 331–343.

Hargreaves, A., Earl, L., & Ryan, J. (1996). *Schooling for change: Reinventing education for early adolescents.* London: Falmer.

Hargreaves, A., Earl, L., Moore, S., & Manning, S. (2001). *Learning to change: Teaching beyond subjects and standards.* San Francisco: Jossey-Bass.

Hurley, M. M. (2001). Reviewing integrated science and mathematics: The search for evidence and definitions from new perspectives. *School Science and Mathematics, 101*(5), 259–268.

Jacobs, H. H. (1989). *Interdisciplinary curriculum: Design and implementation*. Alexandria, VA: Association for Supervision and Curriculum Development.

Jenkins, E. W., & Nelson, N. W. (2005). Important but not for me: Students' attitudes towards secondary school science in England. *Research in Science & Technological Education, 23*(1), 41–57. doi:10.1080/02635140500068435.

Kysilka, M. L. (1998). Understanding integrated curriculum. *The Curriculum Journal, 9*(2), 197–209.

Lloyd, D., & Wallace, J. (2004). Imaging the future of science education: The case for making futures studies explicit in student learning. *Studies in Science Education, 39*, 139–177.

Lyons, T., & Quinn, F. (2010). *Choosing science: Understanding the declines in senior high school science enrolments*. Armidale: National Centre of Science, ICT and Mathematics Education for Rural and Regional Australia (SiMERR Australia), University of New England. Retrieved February 15, 2010, from www.une.edu.au/simerr

Mac Iver, D., & Ruby, A. (n.d.). *The emergence of Middle Schools, growth and maturation of the Middle School Movement*. Retrieved from http://education.stateuniversity.com/pages/2229/Middle-Schools.html

Meeth, R. L. (1978). Interdisciplinary studies: A matter of definition. *Change, 10*(7), 10.

Millar, R. (1996). Towards a science curriculum for public understanding of science. *School Science Review, 77*(280), 7–18.

National Academy of Engineering and National Research Council. (2014). *STEM integration in K-12 education: Status, prospects, and an agenda for research*. Washington, DC: The National Academies Press.

National Research Council. (1996). *National science education standards*. Washington, DC: National Academy Press.

Osborne, J., & Collins, S. (2001). Pupils' views of the role and value of the science curriculum: A focus-group study. *International Journal of Science Education, 23*(5), 441–467. doi:10.1080/09500690010006518.

Pang, J. S., & Good, R. (2000). A review of the integration of science and mathematics: Implications for further research. *School Science and Mathematics, 100*(2), 73–82.

Rennie, L. J., Venville, G., & Wallace, J. (2011). Learning science in an integrated classroom: Finding balance through theoretical triangulation. *Journal of Curriculum Studies, 43*(2), 139–162.

Rennie, L. J., Venville, G., & Wallace, J. (2012a). *Knowledge that counts in a global community: Exploring the contribution of integrated curriculum*. London: Routledge.

Rennie, L. J., Venville, G., & Wallace, J. (Eds.). (2012b). *Integrating science, technology, engineering, and mathematics: Issues, reflections and ways forward*. New York: Routledge.

Roberts, D. A. (2007). Scientific literacy/science literacy. In S. K. Abell & N. G. Lederman (Eds.), *Handbook of research on science education* (pp. 729–780). Mahwah, NJ: Lawrence Erlbaum.

Roberts, D. A., & Bybee, R. W. (2014). Scientific literacy, science literacy, and science education. In N. G. Lederman & S. K. Abell (Eds.), *Handbook of research on science education* (Vol. 2, pp. 545–558). New York: Routledge.

Russell, V. J. (2010). Perspectives on schooling in the middle years. In P. Peterson, E. Baker, & B. McGaw (Eds.), *International encyclopedia of education* (3rd ed., pp. 233–241). Amsterdam: Elsevier Ltd.

Schools Council, National Board of Employment, Education and Training. (1993). *The compulsory years in the middle: Schooling for young adolescents*. Canberra, Australia: Australian Government Publishing Service.

Scott, D. (2008). *Critical essays on major curriculum theorists*. London: Routledge.

Stocklmayer, S. M., Rennie, L. J., & Gilbert, J. K. (2010). The roles of the formal and informal sectors in the provision of effective science education. *Studies in Science Education, 46*, 1–44.

Tyler, R. W. (1949). *Basic principles of curriculum and instruction*. Chicago: The University of Chicago Press.

Vars, G. F. (1991). Integrated curriculum in historical perspective. *Educational Leadership, 49*, 14–15.

Venville, G., Sheffield, R., Rennie, L. J., & Wallace, J. (2008). The writing on the wall: Classroom context, curriculum implementation, and student learning in integrated, community-based projects. *Journal of Research in Science Teaching, 45*(8), 857–880.

Venville, G., Wallace, J., Rennie, L., & Malone, J. (2002). Curriculum integration: Eroding the high ground of science as a school subject? *Studies in Science Education, 37*, 43–84.

Wallace, J., Venville, G., & Rennie, L. (2010). Integrated curriculum. In D. Pendergast & N. Bahr (Eds.), *Teaching middle years: Rethinking curriculum, pedagogy and assessment* (2nd ed., pp. 188–204). Sydney, NSW: Allen & Unwin.

Wilson, E. O. (1998). *Consilience: The unity of knowledge*. New York: Vintage.

Chapter 7
Innovations in Teacher Preparation for STEM: The Value of the Theory-Practice Nexus

Robyn Jorgensen and Kim Alden

This paper discusses an innovative partnership between schools and universities where the state department has sought to produce high-quality graduates who are school-ready. The intent of the program is to develop centers of excellence in targeted areas. This paper discusses the development and outcomes of the STEM Center of Excellence. Using an action research approach, the team has undertaken continuous monitoring of the program to develop a quality program that best meets the needs of the learners and the profession. Follow-up interviews have been undertaken with employing principals to assess the quality of the graduates and the value-adding of the program.

In 2009, the Australian government announced, through its Coalition of Australian Governments (COAG), funding support for new approaches to teacher education. As the key employers of graduate teachers, it was seen that there was a stronger need for education departments to have input into the preservice teacher education programs. As a result, the various states were provided funds to enable them to implement changes through partnering with the various providers of teacher education. In 2011, the Queensland government implemented their initiative which was to create five centers of excellence in preservice teacher education across the state in targeted areas. These targeted areas were identified areas of need – teaching in remote communities, working in low SES communities, working with students with disabilities, teaching in rural communities, and the teaching of STEM. We note, that only one of the centers focused on a curriculum area. This paper focuses on the Center of Excellence in STEM teacher education. Initially this center was based at the campuses of Benowa State High School and Benowa State School

R. Jorgensen (✉)
University of Canberra, Bruce, Aust Capital Terr, Australia
e-mail: robyn.jorgensen@canberra.edu.au

K. Alden
STEM Teacher Education Centre of Excellence, Griffith University, Nathan, Australia
e-mail: kalde3@eq.edu.au

R. Jorgensen, K. Larkin (eds.), *STEM Education in the Junior Secondary*,
DOI 10.1007/978-981-10-5448-8_7

(primary) but currently involves Benowa SHS, Merrimac SHS, and Helensvale SHS – all secondary schools. The partner university is Griffith University. In its early iteration, the center was known by the host school's name, but as the project expanded in 2015, it now includes more schools, and the name has been changed to STEM Teacher Education Centre of Excellence (TECE) to reflect the inclusion of other schools.

Within the broad context of preservice teacher education and the effectiveness of university-based programs, there are a number of reviews of teacher education currently being undertaken. In part, these reviews are founded in governments' concerns about the quality of teacher education. In Queensland, the state where this center is located, the 2013 Commission of Audit report provided guidelines for Queensland Department of Education, Training and Employment (DETE). At a federal level, a review was undertaken in the early years of the project, and much of the recommendations arising from that review have been implemented. Further to these more general reviews, the Queensland Teachers Union (2014) more specifically targeted the attrition of teachers last year since there were concerns with the regard to the high attrition rates of graduating teachers and their preparation for entering the profession and their ongoing support once in the field.

The centers of excellence focus primarily on the attraction, recruitment, and development of high-quality teaching graduates and secondarily on developing the capability of the existing workforce. The purpose of the national partnership is multiple, and these are listed as attracting the best entrants to teaching, preparing teachers and school leaders for their roles in the school environment, placing teachers and school leaders to minimize skill shortages and enhance retention, developing teachers and school leaders to enhance their skills and knowledge throughout their careers, retaining and rewarding quality teachers and school leaders for the value they bring to the classroom, and collecting and maintaining teacher workforce data (Education Queenland, 2012). The STEM TECE commenced operations in 2012 (a year later than planned by the department) and has funding extended through until the end of 2014. From 2015, the center has been funded by the Queensland Department of Education (DET) (which is the new iteration of DETE). It has been further extended via funding provided by the state government. It is hoped that more funding will be forthcoming as the model has been yielding very successful outcomes.

The STEM TECE has adopted aspects of the Stanford model of teacher education (Graduate School of Education, 2013) in particular those where there is the intent to link theory and practice (Allen, 2009) and has many characteristics of the Melbourne University Model (McLean Davies et al., 2012), particularly in relation to the use of data to inform practice. The program has also sought to make strong links between theory and practice as this is known to be a positive approach to initial teacher education that has profound benefits for teacher preparation (Smith & Hodson, 2010). It is also recognized that the tripartite relationship between the school, the university, and the preservice teachers makes for considerable learning possibilities for all partners (Hughes, 2006); however, the primary focus has been on the learning and development of the initial teacher preparation. The process adopted in the program helps to fill some of the spaces that Gravini (2008) has noted that exist between the various partners in initial teacher preparation.

While constrained with issues pertaining to the governmentality of the various authorities that oversee higher education teacher education and the registration of graduating teachers, the STEM TECE model has allowed for 2 days each week to be undertaken in the school context while university studies are undertaken and then an intense period in schools where there are no university study commitments. The amount of days teaching in a preservice degree is controlled by the Queensland College of Teachers (QCoT), and preservice teachers must meet this requirement to gain registration to teach in that state and for registration as a teacher within Australia. By having school-based days running in parallel with university studies has allowed the preservice teachers to experience a strong theory-practice nexus, something that has not been possible in current programs where the university studies are completed prior to school-based experiences. The structure of the school days has been under constant revision in attempting to find the best balance between managing workloads for the preservice teachers (who must continue their on-campus commitments alongside the extra 2 days each week in school) and maintaining the quality and integrity of the overall program.

The program is a one-year, end-on graduate entry program. Selection to the program is unashamedly elitist since the Queensland Department of Education is seeking to ensure the selection of the best STEM graduates into government schools. To this end, entrants must have a strong score in their academic studies and undergo a suitability interview prior to acceptance into the program. While the students predominantly undertake studies in mathematics and science due to their original studies/professions, there are some graduates who enter the program with computing backgrounds. Many of the students have been practicing engineers. So, while STEM is the focus of the program, the students are predominantly mathematics, science, and computing graduates, some of whom were engineers. The STEM TECE has provided not only work-ready training throughout the program but also has guaranteed permanent employment to those preservice teachers who have been deemed to be outstanding in the exit interviews. In the current employment context, this guarantee has been highly sought after by the preservice teachers. While the numbers are relatively low for a graduate program, the entrance requirements limit the intake, and the work demands have meant that a number of preservice teachers have been counseled out of the program since 2013 making for the smaller cohort.

The school context is unique as the preservice teachers are part of a secondary program, and hence their practicum is usually conducted in secondary schools only. But with special dispensation and support from the teacher registration authority, the preservice teachers have also worked in the upper primary years, but this has ceased and the program is now only offered in the secondary school sector. This early experience has given the preservice teachers a feel for both primary and secondary education. Within the current STEM context, the value of mathematics and science in the primary school is paramount, so new strategies for curriculum leadership and design in the primary years have been solicited. One of these strategies is the need for specialist STEM teachers in the primary years (Campbell & Malkus,

2014). Preservice teachers who have been involved in this program exit as secondary teachers but with the unique experience of having worked across both sectors of schooling so are ideally situated for being specialist teachers in the primary STEM area. There is also a strong emphasis on the junior years of secondary school. Many of the preservice teachers have no background in mathematics yet as a STEM teacher, and with a shortage of teachers in the junior mathematics classrooms, the participants have been encouraged to take mathematics curriculum courses as part of their coursework so that they will be prepared to teach in these years.

This chapter reports on the initial establishment of the program – that is, the first 2 years of operation – where the authors worked in partnership to develop the model. It is noted that the model has continued to evolve since that time based on the action research model used to make improvements. Graduates from the first year were assessed by their employers midway through their first year as teachers in schools. All graduates employed by the state were assessed significantly more favorably by their employers compared with the normal graduate employed in schools. Furthermore, interviews with employing principals were also undertaken as part of this paper, and very favorable reports were tendered by the employers. Collectively these data suggest that the program is producing graduates of a high quality who are ready to take positions in government schools.

7.1 Method

As this is an innovation in teacher education with considerable government investment, there has been continuous monitoring of the program. We have adopted an action research model (Kemmis, 1999) where, based on feedback from the various stakeholders, revisions to the program have been incorporated into the overall design of the program. Feedback/data have been in the form of mid- and end-of-semester evaluations, where the students complete both Likert and open-ended questions, and an exit interview in which they map their highs and lows on a graph and then speak with the university coordinator to explain their graph (see Fig. 7.1 for an example). Both of these techniques have yielded data that has been reflexively used to modify and improve the program based on student input but also to provide data as to the success and limitations of the program. The Department of Education has also conducted interviews with employing principals to gauge the effectiveness of the center in terms of work readiness of the new employees. In this paper we draw on aspects of these two forms of university/school data collection across the two cohorts of preservice teachers to document their experiences of the theory-practice nexus.

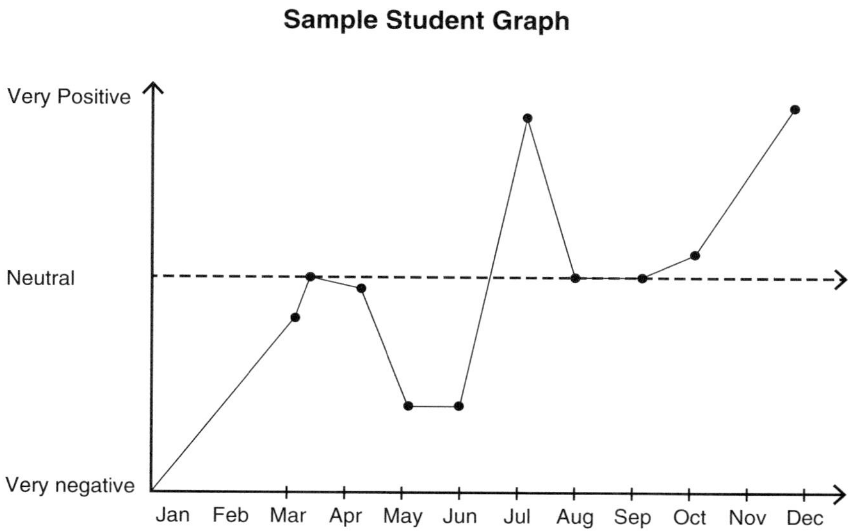

Fig. 7.1 Sample student graph

7.2 The Student Cohorts

The student intake has varied over the 2 years, based on incorporation of feedback from the stakeholders. In 2012, no midyear intake was allowed, but in 2013, we found that there were two key issues we needed to address. First was that a number of preservice teachers were failing to cope with the demands of the course and were counseled out ($n = 4$) and that there were a number of preservice teachers who had missed the initial intake dates and were keen to come into the program during the middle of the year. We trialed the midyear intake in 2013 and were very pleased that this candidate was the first to secure permanent employment. In 2014, we have revised the possibilities for counseling out and admitting preservice teachers into the program midway through the year as this has proved to be very successful.

The preservice teachers come from varied backgrounds including engineering, pharmacy, dentistry, allied health areas, and computers/ICT. To ensure compliancy with teacher registration, all preservice teachers must have two teaching areas. Preservice teachers accepted into the STEM TECE have teaching areas in senior mathematics, and/or physics, and/or chemistry as these are high need areas in schools. Most preservice teachers will teach in these areas, but some also have backgrounds in ICTs as well as other teaching areas (business, geography, and accounting). Two preservice teachers have been French speakers, one of whom secured employment in the French Immersion Program at a high school in his third year of teaching. Another student has been teaching Chinese and became employed in Chinese immersion programs where he taught STEM in Chinese. All preservice teachers are encouraged to take a mathematics methods course regardless of their teaching areas. This is in recognition that as science graduates, it is highly likely they will be allocated classes in mathematics even if they do not have a strong mathematics background. As such, the graduates of the STEM TECE are well prepared for their potential employment in the STEM area (Table 7.1).

7.3 The Theory-Practice Nexus

To establish points of differentiation that have marked this program as a program of excellence different from what is currently offered in many preservice programs, a number of key foci have informed our work. These have been based on current emphasis in educational practice at the level of schooling. First, it is recognized that government schools cater for very diverse students, so STEM teachers need to be prepared for teaching in these contexts. A differentiated curriculum (Lawrence-Brown, 2004) and integrated teaching/theory experience are emphasized in the program. Second, evidence-based teaching (Hattie, 2005) is central to quality pedagogy, so preservice teachers are expected to work from data to inform their teaching practices and, progressively throughout their program, come to develop practices to collect data and use that data to inform their practice. These are significant moves in current teaching practice.

Where university studies have tried to build links between theory and practice, these are most often at the rhetorical level since the preservice teachers do enter the school context when they have completed their theoretical components of their university-based courses. Such a process frequently separates the theory (university) component from the practical (school-based) component. In contrast, the STEM TECE program has had preservice teachers in classrooms in their orientation week, to gain a sense of the reality of classroom life and teaching in the weeks to come. As such, there have been three main aspects of the insertion into classrooms – in the initial weeks of the course, preservice teachers were placed in schools and classrooms, initially to observe. Once they have had some grounding from their university studies, they are asked to take small groups, short lessons building toward full lessons, and then multiple lessons over a day. By the time they have commenced block practicum, they are teaching up to three lessons per day – either in the secondary and/or primary schools. This lead-in experience not only creates a strong sense of belonging to the classrooms in which they work but also a sense of confidence prior to commencing the block practicum. Through the lead-in days, preservice teachers have been able to incorporate their learnings from the university courses into their practical experiences in the classrooms and to also reflect on their practical experiences via their learnings from their university coursework. This has provided for a strong integration of theory and practice that has built a robust knowledge for the block practicum experience. Two examples of a semester's program have been included in Appendix 1 to provide an account of the integration of structure of the program – one from early in the program and a more current one.

Table 7.1 Student cohorts

	2012	2013	2014	2015	2016
Commenced February	17 (including 3 part-timers in their first year)	13 (includes 3 part-timers due to graduate in 2013 and another in her first year)	11	9	25
Exited during Sem 1	0	2 (one could not financially afford time commitment of 2 days per week, the other dropped out of teaching altogether)	0	2	8
Completed Sem 1	17	11	11	7	17
Exited before Sem 2	1 (of her own accord due to work and other commitments)	4 (preservice teachers counseled out of the program	1		
Midyear intake	Not applicable	1	3	0	0
Commenced Sem 2	16 (including 3 part-timers)	8	13	7 Includes 2 part-timers	17
Completed Sem 2	16 (including 3 part-timers)	8	tba	7	17
Offered permanency	11 of 13 completing GDE (plus 3 part-timers continue into 2013 = 16)	6 of 7 completing GDE (plus 1 part-timer continuing = 8)	tba	5 That is 100% – 2 part-timers continue into 2016	17 Have already been made offers September 2016

Preservice teachers have seen the value of being immersed in the school setting from very early on their course. The preservice teachers were required to have their clearances to work with children prior to commencing the course. This has been somewhat contentious as can limit the enrollment of preservice teachers. Those who enroll late are unable to commence at the start of a year as they would not have clearance. They commence orientation week with 2 days of orientation at the university, and then the final 3 days of the week are spent in the school. So, by the time they start their formal studies in week 1 of the university year, they have already had a "taste" of schools and teaching, albeit, only observation.

> Jules: …there was quite a lot of excitement and stress with the STEM TECE program… At the beginning, everything was new, a lot of juggling, very hard to get your head around this thing, different organisations do different things, people… But the good thing was, that we had a chance to do observation and, um, even though after a while it was quite boring, this was good in the sense that this was a moment in time when you could draw this connection

between a few things, start to understand at university and what's happening in the classroom. And because we were not teaching, we were just, I wouldn't say passive, but we were not interacting much with the children, um, I really liked it actually. I was very happy to see how things were done, getting some notes, and I think this was a really good time.

From the preservice teachers' perspective, the immersion into the field very early in their course has been highly valued as it has given them a strong sense of the world of schooling and STEM education. The lead-in days prior to the practicum in which they integrate their university learnings with their practical learnings have been a consistent positive message in the feedback cycles. The following comments offered by preservice teachers highlight the value-adding of the days in schools alongside their on-campus studies. In his comments, Bill articulates how the practical experience in schools enhanced his university learnings in ways that were very practical and grounded:

Bill[1]: I think what we were taught at the university, you get quite a rounded point of view of research or evidence based teaching, however, there is certainly plenty of things that we don't get taught at the university that we learn in the classroom, particularly the practical things taught about your particular contact area, about keeping the tempo going, they talk about scaffolding and development, we know all the theories but putting it in practice can be another level altogether. So ultimately, this program supplements university in a positive way.

On another level, Selina talks about the practical issues of getting to know the school and the preservice teachers so that when the block practicum commences, the preservice teachers are very comfortable with their role and place in the schools:

Selina: We don't really realize, you know, what 4 years out of high school but you can relate to the children but you forget what school is like, you forget what it is like to be in a classroom, you forget what teachers do and all of that kind of stuff so it was good to come back in and kind of just settle in the groove before you do some sent out prac like the one to two lead in days that the others teachers get, that would be very stressful but we were already in there and we already knew the kids so we could just ease our way into it a little bit easier and make everything a lot more enjoyable.

At a very personal level, Ashley articulates her sense of being very happy as part of the school and being able to apply her learnings from the university into the school context, and the value of the mentors in the development of her personal knowledge of teaching:

Ashley: So, started off the year January, February, March – fairly happy, ready… Oh, very happy, ready to start my new career, ready to start a new degree, my new way of life. Um, so, it was very positive and, um, I was very willing to take on advice from mentors and from university courses. So, every day was a victory, it was "I've got something new, everything's great. I can apply it, I can use it". And I felt comfortable doing that. And I felt, obviously, like I was growing. If this was [gestures] up here, I could see I could excel even further, so…

While this aspect of the theory/practice nexus is somewhat predictable, it is further evidence of the successful preparation of the graduates to be prepared for their ongoing learning as teachers.

---

[1] Pseudonyms are used in this paper as a requirement of University Ethics.

At the same time, while undertaking their school-based work, preservice teachers have attended weekly seminars that have built understandings about the nuanced workplace expectations of the government employer. These seminars have been targeted for the employment in Department of Education, Training and Employment (DETE) schools so that graduates will be work-ready for DETE schools. The seminars have been presented by key personnel within DETE. Many of these seminars are very topical and are related to the overall preservice experience, and the presenters had been invited to the university to present to the full cohort of preservice teachers. The contemporary nature of these seminars has been integral to the work readiness of the graduates. Many of those presenting the workshops are staff at either one of the schools and who were seen to be key players in the region/state in the area of the nominated topic. Through these seminars, preservice teachers have gained knowledge of operating systems within DETE and expectations of a teacher employed within the system. Being located within the school setting has enabled preservice teachers access to such knowledge pertinent to their current work in STEM TECE but also as potential employees of the state.

7.3.1 Praxis Assessments

In concert with the practical placements in classrooms, university assessments have been tailored to meet the underpinning principles of the program (differentiation and data-driven pedagogy). This practical assessment has further reinforced the principles of the program but has also enabled preservice teachers to work with real students, to generate real data that will then be used in their teaching. This very hands-on assessment is often impossible to undertake in programs where the theory components of coursework are completed prior to the practicum components. In one assessment piece, preservice teachers worked with two students from their classes and had to identify an appropriate assessment item on a particular mathematical concept (e.g., fractions, 3D shapes), test the students, and then develop interventions to teach to where the student was working. They then reassessed the children at the completion of the assessment to gauge progress and evaluate their own teaching and the value of the tests selected.

7.4 Extra Work, but Worth the Effort

A consistent theme across all evaluations and across all cohorts has been the recognition that the program has required a lot of extra effort from the participants. While this has been a recurring theme, the preservice teachers have also recognized that the extra effort has paid important dividends to their learnings as teachers. In some responses, the early immersion into the practicum meant that preservice teachers confronted misconceptions they held about teaching and could work with these views. One student found that her dream of being a teacher was not going to be as easy as she thought and that early on in the program she realized she was going to have to work hard to be successful.

> Gaurlien: … the beginning of teaching and I expected to walk into the classroom and just be amazing at it because I'd wanted to do it my whole entire life. So… But that wasn't the case. And so that was a bit hard. And it was a struggle to learn…

By the end of their courses, students have consistently reported that they felt much more informed about their work as teachers to the point where there was a clear difference between them and their peers not involved in the STEM TECE. The preservice teachers remarked that they felt they had gained significantly more knowledge and practical wisdom than their peers. By the midway part of their program, many of the preservice teachers were acutely aware of many aspects of theory and practice, and the link between the two. They reported that this made equal conversations quite difficult when they were involved in interactions with these peers. They felt significantly more informed than their non-STEM TECE peers.

> Bill: … those people who aspire to be great teachers and to hit the ground running and to be given the opportunity to think from different perspectives on the teaching profession and look at from all of their different perspectives and then to enter that profession, having that knowledge. I think that there are a lot of benefits and it is certainly well worth the extra effort involved. I think that whilst we have learned that the extra hours we do in front of the class teaching pays off, and you become more confident now having all of those hours under one belt, so that is also another benefit. Again if you wanted to aspire to be an okay teacher then certainly the normal pathway would suit a lot of other people as well. I don't see this as being suited to everybody.

In the comment by Mark (below), he is very forthright in believing the program has given him the edge on his teaching:

> Mark: They give you that much better as a teacher experience wise then you could in the normal [program]. I have friends in the normal one and we just seem to be head and shoulders over them in all teaching ways as we are confident and we know how we're doing and all that sort of thing, how to assess and it has also been told from like our relief teachers have said that it didn't even look like I was a student teacher. She was sitting up the back and said that it didn't even feel like a student teachers class, it felt like you were the real teacher, so I am sure that happens with other individuals in the normal course but I think that everyone here would have had that same experience.

Selina similarly commented that she recognized the extra work coming into the school context gave her an advantage over her peers, making her a much better teacher.

> Selina: I would say that the start is a bit tough and all of your friends at uni are having two days off and you are coming in here so it is tough trying to juggle uni assessment, which I found the hardest with coming but the level I am at now. I know I wouldn't have been if I had just done the standard [program]. Coming to school on those two days made the difference.

In 2013, we recognized that late enrolling students were unable to access the program at the start of the year, effectively locking them out of the entire program. The program commences in orientation week, with preservice teachers in schools in that week, requiring police clearance to work with children. As these permissions take some weeks to obtain, it is the case that preservice teachers who decided to leave their applications to later in the enrollment period did not have police clearance and so could not participate in the first weeks of the in-school experiences, effectively excluding themselves from participation in the program. Similarly, it was recognized that some preservice teachers struggled with balancing the extra time of the STEM TECE, their on-campus work, and their lives beyond the program. As a result, in 2013, a midyear strategy was implemented. First, an exit strategy was implemented for some preservice teachers who were not coping with the demands of the intense program and who were then able to transition into the standard (non-STEM TECE) program. Second, recognizing that some preservice teachers may want to opt into the STEM TECE program, it has been opened up for high-performing preservice teachers. Such preservice teachers are required to meet the initial entry requirements, are interviewed and assessed on the basis of referee reports and practicum assessments that comment directly on their professional development as a teacher, and have performed strongly in their first semester coursework and practicum. Laura is one of the preservice teachers who opted into the program mid-2013. In her comments below, she is articulate about the difference between the two options available to the graduate entry preservice teachers.

> Laura: Because, coming from my other prac, I do feel like the expectations are very high in this program – which is good – um, if you feel like you can live up to them.

There is a cautionary note in her concluding comment with regard to the high demands of the course. We note that Laura was the first of the preservice teachers to be offered permanent employment for 2014 – before she finished the program.

7.5 Employer Assessments

While we are able to confidently claim that the preservice teachers have felt that they gained substantially from participating in this type of program, it is a good test of teacher preparedness to consult with the employers of the graduates. Two forms of data have been included here. First, DETE consulted with the employing principals of the STEM TECE (and other centers) as to the preparedness of the graduates from the center. They were asked to compare the STEM TECE graduates with other graduates from standard programs across a number of dimensions. These data confirmed that the principals rated the preservice teachers as better prepared than the usual graduate. These scores were aggregated across all centers, but some of the data identified the STEM preservice teachers, and these were very positive. To further elucidate on these quantitative assessments, five principals were interviewed in relation to their STEM TECE employees. The strength of targeting high-achieving preservice teachers with strong content knowledge and then the value-adding through the STEM TECE program are aptly summed up by the principal comment:

Southside[2]: We teach to the middle and we don't have the capacity to teach to the top. If I take a look at Kristy on the other hand, she's a very smart operator, very bright young lady. She academically is way superior to what I'm getting out of just education courses. She analytically understands the physics involved. Her mathematics is extraordinary so she doesn't have to double take herself. She can think clearly, she can put things on the board in articulate manner, she's able to be engaging, because she actually has outside external information to base her illustrations on. So I look at that and I go this is exactly what the teaching of maths and science should be about.

What can be seen in this comment is the strength of the selection process in terms of the knowledge and dispositions of the graduates. Moreover, the principal is forthright in his claim that the STEM TECE program is producing graduates of a much higher caliber than the mainstream programs. It is also a common theme in the responses that the graduates are not only developing strong pedagogic content knowledge but they have many other attributes that were highly desirable including teacher readiness, behavior management, and confidence. In their first semester as a teacher, principals commented that the graduates were well ahead of their peers who had undertaken the usual teacher preparation courses.

Golden Palms: And across the board [in her teaching], so she would be where we would hope someone would be by their third semester of teaching not their first semester of teaching. So much so that in their third year here we usually ask them to take on a role as a course coordinator.

Similarly another principal commented on the preparation of the graduates to be able to walk into the school and be part of the team.

Southside: So what do I see that's good about Kristy from what I've seen within your program is that she comes out really confident about how she can handle kids in class. So we saw that growth while she was doing teaching with us, and then we were able to actually see her come into the class and really feel at home quickly. And that's a bit to do with what we've done here, but it's also to do with the fact that she's left your program with a deal of confidence in her own ability to manage kids. So as I say her teaching's exemplary, and I gather that's come from your course. The strategies she uses are fantastic. She has a wealth of knowledge, she has a great deal of enthusiasm,

This was further reinforced by a principal who had opened a new school where there were no existing resources or history for the graduate student to rely on, but the graduate was not only able to work within that context but took a leadership role with the new teachers. The context in this school at that time was that the final year of primary school (year 7) had transitioned to the secondary school, so many of the teachers in that sector were also transitioning into the secondary school context. The graduate "stepped up to the mark" and took a leadership role with these teachers in terms of discipline knowledge.

Ruralside: I mean, for us as well, she came into a brand new school where there was no existing curriculum for her to pick up, and obviously she had a team of teachers with her, but she copes quite well given that, that she had to start from scratch, and was part of, she took all Year 7 last year maths and science, and she really, she was probably fairly integral

---

[2]Pseudonyms are used to ensure anonymity of the schools as per the University Ethics guidelines.

> in the science, given that our other Year 7 teachers were all primary, ex-primary and didn't have the science background that she had, so she sort of stepped up and had a bit of leadership around that.

Similarly, the principal at Golden Palms also acknowledged how the STEM TECE graduates were ahead of their peers who had not undertaken this program:

> Golden Palms: She moved into this team with a lot to add from the first day and a confidence and a willingness to do that whereas the typical beginning teacher comes with an apprenticeship mindset and they work alongside the people that have been on their course longer.

In a similar vein, the principal at Northside was able to comment on his graduate and also compare her against other graduates he had employed since from other programs.

> Northside: Yeah, she had a high level of confidence coming in. I mean, for a person at her stage of her career. Because I know we've had some other beginning teachers since then, and yeah, the issues that I know we've dealt with with them, they probably weren't evident with Tess, so she came with a level of capability, ready to go, and that was really evident.

A final comment that aptly sums up the views of the principals can be seen in the following comment where the principal recognizes that the program is preparing the preservice teachers much better than other programs:

> Ruralside: I think just, yeah, that little bit of maturity around the classroom management stuff and dealing with kids, they've obviously spent more time in the classrooms. Some of the ones you get, particularly who are doing the graduate diploma 1 year, sort of thing, and heaven forbid the Teach for Australia programs that are coming soon. They're certainly ahead of them as far as what they're like in that first term in a school

Nearly all principals agreed that the selection of the entrants into the program was paramount and summed in the comment – "…so the selection of her as a candidate is really important. You just can't take anyone." So while there are some reservations about the program being elitist in its selection of preservice teachers, it also pays to attract and develop high-quality preservice teachers into being STEM teachers.

7.6 Challenges to the Rollout of Innovations

The intense nature of the program where preservice teachers must attend schools for 2 days each week while their non-STEM TECE peers have these days "off" means that the STEM TECE preservice teachers are under enormous workload pressures. They have to complete their university requirements under the same time frames as their non-STEM TECE peers while also undertaking their off-campus requirements. It is undeniable that the extra days in schools to allow for the tight theory-practice nexus have created a considerable workload for the preservice teachers. The preservice teachers have limited time available to undertake part-time work, which, in many cases, created financial challenges to families. It also creates a limit on the time available for parenting, thus creating greater demands on the partner in terms of parenting arrangements. In the comments below, the preservice teachers raise their concerns about the extra workload and the impact on their lives both in and out of the program. There is a peak period when preservice teachers are commencing teaching full lessons, planning for those lessons, and being expected to participate in the wider school activities while still undertaking their university components of the program.

> Bill: … I thought that I knew what it was like and that I had a handle on the work load and stress and everything else associated with the program, but I was wrong. We were warned from the very first day, I thought that we were even warned back at the open days that it is very challenging year. So that year started off with the first week or two being a honeymoon period and then things started piling up, the work started to increase quite rapidly, assignments, the two days that we spent here at Benowa were two days not spent working on uni stuff so all of that accumulated to a very busy working week.
>
> Nate:… [the] university had not changed anything pertaining to us, all of our assessment, all the multitude of our assessment and work load was the same as everything else and the only differences was that we were down two days.
>
> Rochelle: The university program … Just tried to get that part over and done with so I could get back to worrying about this [teaching]. Um, and I guess probably there was a bit of a lower point here maybe, when started teaching full on and all the uni stuff as well… But this semester I think I coped a little bit better.

The most critical period for the preservice teachers was immediately before commencing the block days in schools. This was also the period when they were completing their on-campus assessments and examinations, so they were very stressed with the extra workloads. In refining the program, the days in school immediately before the block period have now been made flexible so that preservice teachers are able to do more days prior to this period so that they have time. Timetables have also been changed so that the week immediately before the block period is now a school-free week so that the preservice teachers are freed of their school commitments (and lesson plans and teaching).

> Clinton: I was approaching the course in a positive frame of mind, I think that everyone is upbeat and supportive and then you have to face reality and that is what happens over here. I think that for me it was quite a steep learning curve to get back into academic life and to cope with it and Benowa and to cope with work … to cope with that as well. …in April and we had assignments and exams coming up so that was it and I think that personally what happened here was that [the] academic side was finished I could start to enjoy the Benowa side which was much nicer without the uni side. Especially when we got into the five-week block situation, that was much nicer so you could focus.
>
> Clinton: I think that one of the things with this program is that you are quite time poor at certain sections when you are doing the uni bit

The preservice teachers have also acknowledged that the stress of the program created doubts about whether or not to continue in this program and/or return to the standard program. But once they were through the intense period noted and had completed the entire program, there was a resounding acknowledgment that the pressure was worth it personally and professionally.

Michael: All up I think that this program has had its ups and downs but overall it is what has made it that much better, 'cause I thought about dropping out here, but I'm glad that I didn't because it would have been a big mistake. Just because of the work load basically and I just thought that it was maybe just a little bit too much.

Selina: I know it's life, but just from a science kind of background having to do all of those essays killed me and killed a lot of people here, it just really did.

Perhaps one of the largest challenges to the STEM TECE has been the buy-in by some of the university staff. The original Memorandum of Understanding (MOU) established between the center and the university sought to have 50% of assessments to be school based. In the idealized world of reform, this was a benchmark that would create a real sense of a theory-practice nexus. In Semester 2 of the first year, two courses for the semester had been changed, and only by the fourth semester (and with intervention from the senior executive of the university to direct curriculum staff to create assessments in accordance with the MOU) were all curriculum courses offering alternative assessments for STEM TECE preservice teachers. In some cases, these assessments have become the assessment for the entire cohort of preservice teachers whether or not they are in the STEM TECE program. Understandably in a work-intensive higher education sector and without the certainty of the program being ongoing, the extra workload for university staff can cause some resistance. It needs to be acknowledged that lecturing staff have only parts of their cohorts undertaking the STEM TECE program, so they have had to have assessments for the mainstream preservice teachers as well as alternative assessments for the STEM TECE preservice teachers. This has created an extra workload in an already intensified workload, so not only are they required to devise a further assessment, but the ongoing references to two assessment items during on-campus teaching have further exacerbated teaching loads.

There has also been a significant shift for school staff to provide strong mentoring and curriculum leadership in their interactions with the preservice teachers. All mentor teachers have completed a mentoring program so as to provide strong guidance for the preservice teachers. While there have been some of the usual personality mismatches, the mentoring arrangements have been largely successful. However, there have been some cases where this was not the case. In these cases, mentors were removed from the program if they were not ensuring high-quality experiences for the preservice teachers.

Despite the initial teething problems, as is to be expected with the rollout of any new reform, the program has met with considerable successes. Staff at both sites (school and university) have recognized the value-adding of the integrated model of teacher education. Since undertaking this formative research, the program has continued past the initial stage of funding and is now funded through another source (Department of Education), so it does appear that the STEM TECE is now established and embedded into the practices of the partner sites.

7.7 Conclusion

To conclude, we draw on some of the final comments offered by the preservice teachers in terms of their overall experiences of participating in this innovation. While as teacher educators, we value the theory-practice nexus and see this as one of the clear strengths of the program, it is, however, not without problems. As the preservice teachers have clearly articulated, and as we have recognized in counseling preservice teachers out of the program, we need to be cognizant of the high demands that this creates for preservice teachers. When their peers exit Griffith University with the same award, there needs to be some value for the extra work undertaken by the participants in the program. While one of the benefits is the offer of permanent employment in an environment that is currently constrained in terms of employment offers, it is often difficult for the preservice teachers to see this goal when they are deeply immersed in teaching practicum, assignment deadlines, and balancing the other parts of their lives. It is easy for them to lose sight of this benefit. However, when the immediacy of the demands of the program has been completed, preservice teachers have graduated, with employment offers, and have had space to reflect on their experiences, we see that there is scope for deeper reflection on the value-adding made possible through the strong theory-practice nexus that underpins this STEM innovation. As can be seen from the final comments, the preservice teachers saw considerable value in their program, despite the intense workload.

> Clinton: On reflection, I think that it is a brilliant course.
>
> Mona: It is intense but at the end if it I have become a better teacher than I would have been if I hadn't done it. The experience I have got is amazing. This program has bettered me as a teacher as I have resources, heaps of feedback, I have some really close friends that help me.
>
> Nate: …I feel that I have grown amazingly compared to where I started and I have more of an idea of what a teacher is and what a teacher does, the roles of a teacher.
>
> Selina: I am really glad that I did this program as I would have nowhere near of gotten my marks if I was in normal [program] as it has just helped develop myself and all the others so much that I honestly didn't think that I would get here but I have so I am really happy with that.
>
> Tom: Very happy with the program. Can't think of any faults with it really….It's hard – definitely hard – and intense. But, um, I think, it was very lucky that we're all part of it. It's very enjoyable.
>
> Mona: I have learnt so much on how to be a better teacher so I have definitely grown as teacher

Each of the participating principals expressed similar comments, each of the following coming from different principals regarding different graduates:

> Her results are already better than the average teacher in her cohort.
>
> So as I say her teaching's exemplary, and I gather that's come from your course. The strategies she uses are fantastic.
>
> She definitely was high quality, she was enthusiastic, she was ready to teach.
>
> Yeah, the thing that I guess stands out is preparation; you know, he is always prepared, has a plan of action and has got a system that he works through to achieve those goals, so I've never walked in there and he's not been planned, ready to go

In the time post of this research, the program has been sustained, and ongoing funding has been provided by DET to ensure that the model continues to produce high-quality STEM teachers for the public system. The program has also been progressively refined and developed based on reflexivity between the practices and the feedback provided by the various stakeholders. At the time of publishing this work, the program is now in its 6th year of operation and has been spread over a further three schools in the region. This extension of the program has helped to build the sustainability of the program as more teachers and schools are now familiar with the model. It is hoped that the model will continue into the future as it is clear from the feedback shared in this paper that there are profound benefits for all participants – preservice teachers, mentor teachers, and schools in general.

Of particular importance in this program is the preparation of the preservice teachers for junior secondary STEM teaching. In this program, the graduates enter their initial teacher training with the anticipation that if they are successful in both their academic and teaching studies, they are likely to gain secure employment with the Department of Education in an urban area. This is a difficult outcome to achieve in the normal teacher education program. The government is keen to attract high-quality teachers into public schooling, and the STEM TECE is seeking to have the best teachers in government schools. To this end they have continued to fund the program past its initial lifespan as the outcomes have been very positive. The teachers graduating from this program are very work-ready when they exit their teacher education program and are more than prepared to teach in government schools, particularly the junior secondary STEM years where there are often other teachers who are teaching outside their discipline area. A program such as this helps to ensure that qualified STEM teachers are being prepared through initial teacher education to work in this area of schooling. As the program expands into other schools, it is becoming more of the norm for quality STEM teacher education, the winner of which are the teachers and students.

2014 Benowa Teacher Education Center of Excellence Schedule:

Semester 1 **Version: February 6, 2013**

Practicum days ☐ School holidays ■

This program may be subject to change.

Preservice teachers are expected to undertake the equivalent of 1 day per week as classroom-based teaching experience. The second day (the equivalent of 1 day per week) relates to STEM TECE practice-based assessments that draw together the theory/practice nexus focus of the program. The relationship between evidence-driven practice and pedagogical practices that cater for diversity of learners and learning is central to these "second-day" experiences (Source: STEM TECE Operational Plan August 2011, available on Oneportal)

DATES	UNI WEEK	SCHOOL WEEK	DAYS/WK AT STEM TECE	PROGRAM
24-28 FEB	**0**	**5**	**2(WPE) Thurs Fri**	**Orientation THURSDAY and FRIDAY – full detail provided separately** Includes Code of Conduct, Child Protection, Ethical Behaviours , issue of laptops, tours of schools, meeting with mentor teachers ,time with mentors in classrooms and a Welcome Celebration 3pm Friday
3-7 MAR	**1**	**6**	**2 (WPE) Tue Wed**	**Tuesday: All day shadowing high school mentor** **Wednesday – All morning until 1.40 shadowing primary school mentor** ***Wednesday Seminar*** **Topic :** Debrief and planning – why are we here & what do we want to achieve individually/collectively? **Facilitator:** Kim Alden, Head of Mentoring and Lin Esders ,Organiser, QTU
10-14 MAR	**2**	**7**	**2 (WPE) Tue Wed**	**Tuesday - All day shadowing secondary mentor** **Wednesday – All morning until 1.40 shadowing primary school mentor** ***Wednesday Seminar*** **Topic :** Lesson Planning – the links between curriculum, assessments and planning **Facilitator:** Professor Robyn Jorgensen, Griffith University
17-21 MAR	**3**	**8**	**2 (WPE) Tue Wed**	**Tuesday - All day shadowing secondary school mentor** ***All Day Wednesday Professional Development*** **Topic :** The Symphony of Teaching and Learning (day 1 of 3) – using digital technologies in your classroom **Facilitator:** Lissa Hodson, Education Queensland
24-28 MAR	**4**	**9**	**2 Tue Wed**	**Tuesday** Working with mentors' classes as per timetable and completing clinical practice university assessments ***Wednesday Seminar*** **Topic :** Data to Performance – How to use Evidence to Guide your Planning and Teaching **Facilitator:** Glenn Chippendale DP Benowa SHS
31-4 APR	**5**	**10**	**2 Tue Wed**	**Tuesday :** Working with mentors' classes as per timetable and completing clinical practice university assessments ***Wednesday Seminar*** **Topic :** Outwit, Outplay and Outlast – Behaviour Management Strategies **Facilitator:** Tony Maher, Responsible Thinking Room Coordinator, Benowa SHS
7-11 APRIL	**6**	**SCHOOL**		**EASTER SCHOOL VACATION**
APR	**7**	**HOLIDAYS**		
21-25	**UNI BR**	**1**		***University Assessment Finalisation and Exam preparation*** ***A minimum of two practicum lead in days are to be scheduled over these three weeks in consultation with mentors and with the Head of Mentoring.***
	K			***During these two days PSTs will work with their mentors' classes as per their timetable.***

28–2 MAY	**8**	**2**		
5-9 MAY	**9**	**3**		
12-16 MAY	**10**	**4**	**5**	**Block Practicum commences** **8.25 – 8.30 Monday:** Check in for the week and weekly focus with Head of Mentoring in V05 Working with mentors' classes as per timetable – full timetable to be commenced ASAP ***Wednesday Seminar*** *– topic to be determined by perceived student need, or on student request*
19-23 MAY	**11**	**5**	**5**	**8.25 – 8.30 Monday :** Check in for the week and weekly focus with Head of Mentoring in V05 Working with mentors' classes as per timetable ***Wednesday Seminar*** **Topic :** Working Effectively with Students with Special Needs **Facilitator:** Lieve Rimbaut, Head of Special Education Services (HOSES) Benowa SHS **Interim reports by end of week** –Dr David Geelan to sign off Friday May 23
26-30 MAY	**12**	**6**	**5**	**Monday :** Working with mentors' classes as per timetable ***All Day Wednesday Professional Development*** **Topic :** The Symphony of Teaching and Learning (day 2 of 3) – using digital technologies in your classroom **Facilitator:** Lissa Hodson, Education Queensland
2-6 JUN	**13**	**7**	**5**	**Monday :** Working with mentors' classes as per timetable ***Wednesday Seminar*** **Topic :** Differentiation for beginners – catering for diversity in your classroom **Facilitator:** Brendan Zischke, Differentiation Coordinator Benowa SHS
9-13 JUNE	**14**	**8**	**5**	**Tuesday :** Working with mentors' classes as per timetable (Monday is a Public Holiday) ***Wednesday Seminar*** *- topic to be determined by perceived student need, or on student request*
16-20 JUN	**15**	**9**	**5**	**8.25 – 8.30 Monday :** Check in for the week and weekly focus with Head of Mentoring in V05 Working with mentors' classes as per timetable ***Wednesday Seminar*** **Topic :** The DETE Employment Process **Facilitator :** Kim Alden, Head of Mentoring **Final reports written** – Dr David Geelan to sign off Friday June 20 **Friday - debrief and planning for Semester 2, End Semester Celebration 3pm Friday June 20**
23-27 JUN	**16**	**10**		
				WINTER SCHOOL VACATION AND END SEMESTER ONE

Semester 1 2016 Overview				Version as at 8/02/16 May be subject to change.
DET WEEK	UNI WEEK	Orientation at Benowa - all	Wed Feb 24	All PSTs, as per separate program. Will include Code of Conduct and Child Protection training
5	0	Orientation at Helensvale and Benowa	Thurs Feb 25	As per separate program, for Semester 1 Helensvale PSTs only at Helensvale, and Semester 1 PSTs only at Benowa
		Orientation at Merrimac	Fri Feb 26	As per separate program, for Semester 1 Merrimac PSTs only at Merrimac
6	1	Preservice Teachers participate • 1 day per week at Benowa Room V05 participating in workshops and seminars to support professional development, generally 8am to 4pm • 1 day per week at First Practicum school (most commonly Wednesdays)	Tues March 1	Seminar : 8.30 am Why are we here and what do we want to achieve individually/collectively – Kim Alden : 9.30 am How to make Science (or anything else) Interesting - Jeremy Newton-John Merrimac SHS Break for Morning Tea : 11.30 am Well-being for teachers, Brendan Zischke, _ Benowa SHS 12.30 Lunch : 1.00pm A tour of DET platforms – Oneschool, Oneportal, the LP and the intranet –Kim Alden : 2pm Learning from Experience – TECE graduates talk about being a 'tekkie'. Making the most of the 'second day' in schools and classrooms – Kim Alden
			Second day	PSTs at practicum school shadowing mentor
7	2		Tues March 8	Seminar : 8.00 am Check in for the week and weekly focus topic with Kim : 8.30 am Digital Technologies for Teaching – Timm Hayer, Teacher & Microsoft Innovative Educator, Tallebudgera State School Morning tea/lunch : 12pm Dan Meyer's '3 Acts' of Problem Solving - Darren Rackemann , 2014 Recipient of QLD's Science Innovation Champion Award, Program Director of Startup Apprentice http://www.startupapprentice.com.au/about-us/, ex Varsity College school leader Short break : 1.30 pm Working Effectively with Students with Special Needs –Lieve Rimbaut, Head Of Special Education Services (HOSES), Benowa SHS
			Second day	PSTs at practicum school shadowing mentor
8	3		Tues March 15	Seminar : 8am Check in for the week and weekly focus topic with Kim : 8.30 Microsoft's 21st Century Learning Design Framework – Darren Rackemann Morning Tea :10 am Essential Skills for Classroom Management– Andrew McMahon, HOD Positive

				Behaviour, Merrimac SHS Lunch : 12.30 pm Teaching mathematics to 21st century learners.- Ramon Doon, Maths HOD , Windaroo Valley SHS,
			Second day	PSTs at practicum school shadowing mentor
9			Tues March 22	Seminar : 8am Check in for the week and weekly focus topic with Kim (Lesson Planning – Kim and Harry) : 9am Diamonds in the Rough - What's special about teaching Junior Secondary? – Glenn Chippendale DP BSHS gchip1@eq.edu.au . This session will involve the bulk of our time today and we will break for morning tea and lunch as we are ready : 2.30 pm Making the most of the Block Practicum – Kim Alden
			Second day	PSTs at practicum school shadowing mentor NB For GC Students enrolled in 7032EDN (Junior Science) Harry Kanasa has organised a field trip for Wed March 23 to the Eco Centre at Nathan Campus. Mt Gravatt students also enrolled in this course are invited to join in. All other preservice teachers will be shadowing their mentor at their practicum school.
Hols	Hols			
Hols	5			
1	6	Prac lead in day *		University assessment finalisation and exam preparation. A minimum of two practicum lead in days are to be scheduled over these four weeks in consultation with your mentor and with the Head of Mentoring
2	7	Prac lead in day*		
3	8			
4	9			
5	10	Block Practicum at School 1	Mon May 9	Block Practicum commences - Working with mentors' classes as per timetable
6	11			Working with mentors' classes as per timetable
7	12			Working with mentors' classes as per timetable - Interim reports due
8	13		**June 2**	Working with mentors' classes as per timetable **Thursday After School Workshop at Benowa : The DET Employment Process – Kim Alden**
9	14			Working with mentors' classes as per timetable
10	15		Fri June 17	Final reports due by end of week –sign off Friday
11	16			Opportunity to catch up on any missed practicum days this week
Hols	End			

References

Allen, J. M. (2009). Valuing theory over practice: How beginning teachers re-orient their practice in the transition from the university to the workplace. *Teaching and Teacher Education, 25*(5), 647–654.

Campbell, P., & Malkus, N. N. (2014). The mathematical knowledge and beliefs of elementary mathematics specialist coaches. *ZDM, 46*(2), 213–225.

Education Queenland. (2012). *Improving Teaching Quality National Partnership*. Retrieved 5/12/2013, from http://education.qld.gov.au/nationalpartnerships/imp-teach-quality.html
Graduate School of Education. (2013). *Stanford Teacher Education Program*. Retrieved 1/12/2013, from https://gse-step.stanford.edu/
Gravini, M. N. (2008). Academics and practitioners: Partners in generating knowledge or citizens of two different worlds. *Teaching and Teacher Education, 24*(3), 649–569.
Hattie, J. (2005). What is the nature of evidence that makes a difference to learning? *Using data to support learning conference*. Melbourne, Australia: Australian Council for Educational Research.
Hughes, J. A. (2006). Bridging the theory-practice divide: A creative approach to effective teacher preparation. *Journal of Scholarhsip and Teaching, 6*(1), 110–117.
Kemmis, S. (1999). Action research. In J. P. Keeves & G. Lakomski (Eds.), *Issues in educational research* (pp. 150–159). Kidlington Oxford, UK: Pergamon.
Lawrence-Brown, D. (2004). Differentiatied instruction: Inclusive strategies for standards-based learning that benefit the whole class. *American Secondary Education, 32*(3), 34–62.
McLean-Davies, L., Anderson, M., Deans, J., Dinham, S., Griffin, P., Mameniart, B., et al. (2012). Masterly preparation: Embedding clinical practice in a graduate preservice teacher education program. *Journal of Education for Teaching, 39*(1), 93–106.
Ministerial Inquiry. (2014). http://www.studentsfirst.gov.au/news/submissions-teacher-education-ministerial-advisory-group-now-open .
Queensland Teachers Union. (2014). http://www.qct.edu.au/
Smith, K., & Hodson, E. (2010). Theorising practice in initial teacher education. *Journal of Education for Teaching, 36*(3), 259–275.

Chapter 8
Successful Students – STEM Program: Teacher Learning Through a Multifaceted Vision for STEM Education

Linda Hobbs, John Cripps Clark, and Barry Plant

Abstract The current STEM education agenda is driven by the belief that STEM skills are crucial to innovation and development in our contemporary, technological, knowledge-based, competitive global economy (Office of the Chief Scientist, Science, technology, engineering and mathematics: Australia's future. Australian Government, Canberra, 2014; Australia's STEM workforce: science, technology, engineering and mathematics. Australian Government, Canberra, 2016). This chapter articulates a comprehensive, multifaceted and coherent *STEM vision* that addresses the subtle and complex challenge of preparing "twenty-first-century" citizens within the constraints of a traditional school system and curriculum. For STEM education to be incorporated effectively and sustainably in schools, a STEM vision needs to be inclusive of school-specific needs. In this chapter, we report on our preliminary insights from a teacher professional development programme operating in ten schools in Victoria, Australia, designed to develop year 7 and 8 science, technology and mathematics teachers' capacity to teach STEM. Evaluative data from the first year of this three-year programme is presented to illustrate the variety of classroom activities that can arise from a comprehensive STEM vision. The research is showing that a STEM vision needs to be more than discrete STEM-related activities slotted into an already bulging curriculum to be sustainable.

8.1 Introduction

While science, technology, engineering and mathematics are disciplines in their own right, their synergies championed under the "STEM" banner are igniting a flurry of political, professional and business discussions, with significant implications for education. The current STEM education agenda is driven by the belief that STEM skills are crucial to innovation and development in our contemporary

---

L. Hobbs (✉) • J.C. Clark • B. Plant
Deakin University, Burwood, Australia
e-mail: l.hobbs@deakin.edu.au

R. Jorgensen, K. Larkin (eds.), *STEM Education in the Junior Secondary*,
DOI 10.1007/978-981-10-5448-8_8

economy. Because STEM is being positioned so centrally to a country's competitiveness, it is influencing funding in industry, education and research. A utilitarian conceptualisation of education, while not new, is often promoted through this drive to prepare students for a STEM-dominated future in which three-quarters of jobs are forecast to need STEM skills and capabilities (Office of the Chief Scientist, 2014, 2016).

In Australia, concerns have been voiced about both performance and participation of students in STEM-related subjects through all sectors and whether they are fully prepared for the modern workplace (Australian Industry Group, 2013). In international comparisons over the last decade, Australian school students performed better than the OECD average, but achievement in science is stagnating and declining in mathematics (Thompson, De Bortoli, & Buckley, 2013). Since 1992, many STEM-related subjects at senior secondary level have shown declines in participation, particularly in more demanding subjects (Office of the Chief Scientist, 2014). State and federal education authorities have reacted by initiating a range of policy changes culminating in the National STEM School Education Strategy (Education Council, 2015) that aims to raise student STEM participation and achievement through increasing student aspirations; improve teacher capacity and quality; support within school systems; create partnerships with tertiary providers, business and industry; and build an evidence base. These aims resonate with initiatives in other parts of the world, such as the European Community where attempts have been made to raise student STEM awareness, establish industry and school links and build up STEM teaching skills (Scientix, 2014).

The challenge for educators is to translate an ill-defined, politically charged and narrowly utilitarian policy agenda of securing a future workforce, into a valid and coherent curriculum. The objective of this chapter is to articulate a comprehensive, multifaceted and coherent STEM vision that addresses the subtle and complex challenge of preparing "twenty-first-century" citizens within the constraints of a traditional school system and curriculum. We will argue that, for STEM education to be incorporated effectively and sustainably in schools, a STEM vision needs to be more than individual activities or projects dropped into an overcrowded curriculum: what is needed is a vision that is inclusive and interdisciplinary in nature and specific to school needs.

This chapter explores the question: How can a multifaceted vision of STEM education in a teacher professional learning programme sustainably and effectively meet the specific needs of schools? We report on insights from a teacher professional development programme, *Successful Students – STEM Program*, operating in ten schools in Victoria, Australia, designed to develop year 7 and 8 science, technology and mathematics teachers' capacity to teach STEM. In order to situate this STEM initiative into the broader story of STEM, the following section examines the rise of STEM as a policy agenda and how this has been translated in recent years into an imperative for schools. The structure and focus of SS-STEM Program are then outlined, with particular attention to the components of the STEM vision framework. To illustrate how schools are using this framework, a series of case studies show how the teachers' efforts to plan and implement STEM initiatives are developing over the course of the programme. This chapter is written after the second of four professional learning sequences and therefore presents the change process in action.

8.2 The Emergence of STEM Education in Australia

Worldwide, there has been an explosion of policy, programmes and pronouncements on STEM education in the past decade, with an endless stream of reports and conferences. This drive emerges largely from a business imperative (Bybee, 2013) and is based on concerns that the emerging workforce appears unprepared to meet the challenges of the future. In Australia, international competitiveness has been triggered by changes to employment and lifestyle arising from the digital revolution, dubbed the third industrial revolution (Economist, 2012). The disciplines of science, technology, engineering and mathematics are heralded as the drivers of this change, as well as the means by which we can maintain our competitiveness (Office of the Chief Scientist, 2016). These disciplines, through the construct of "STEM", have become the mechanism by which a business imperative has secured a place within the current political innovation agenda.

Underpinning this STEM agenda is the rhetoric around jobs and growth. STEM employment is predicted to grow 50% faster than other jobs (Australian Government, 2015a) with significant increases in particular in professional, scientific and technical services and health care (Education Council, 2015). Irrespective of job or industry, STEM capacity supports innovation and productivity; it is estimated that "changing 1 per cent of Australia's workforce into STEM-related roles would add $57.4 billion to GDP" (PWC, 2015).

This issue of growth is then projected onto education. Concerns have been expressed about students' declining rates of participation and engagement in STEM subjects in the post-compulsory year levels of secondary school, particularly in mathematics, and falling performance compared to the best performing countries in international mathematics and science testing (Education Council, 2015). Within the Australian population, there is concern that groups of students, such as girls, Aboriginal and Torres Strait Islanders, and students with low social and economic background and from rural and remote communities are being excluded from full participation in STEM education, which has implications for their employment and participation in the economy (Professionals Australia, 2015).

The problem is also characterised as leakage from the STEM education pipeline, where loss of interest and engagement occurs at critical junctures such as the transition from primary to secondary, when STEM subjects become electives, and the transition to tertiary study (Tytler, Osborne, Williams, Tytler, & Cripps Clark, 2008). The problem has a number of interlocking elements: increasing disenchantment with STEM study in secondary school, leading to lower post-compulsory participation, leading to STEM-qualified shortages in both industry and education (Tytler, 2007). It is worth noting, however, that while there do exist shortages in some

STEM areas, based on 2011 census data, there has been 18% increase (from 2006 to 2011) in "STEM-qualified" individuals not in employment: 33% of scientists, 40% of mathematicians and 34% of IT professionals not in the labour force (Office of the Chief Scientist, 2016).

Another dimension to the problem is a reported decline in the "STEM skills" of the upcoming workforce. A number of industry groups (such as Australian Industry Group, 2013) state that the "soft skills" such as communication, teamwork, critical thinking, creativity and problem-solving are not adequately taught at schools. Creativity, problem-solving and entrepreneurial skills are promoted as crucial to participating in the emerging economy (Australian Government, 2015b).

Arising from these observations and concerns about current and projected workforce needs are a number of policies and initiatives. The most recent is the *National STEM School Education Strategy* by the Education Council in 2015, which proposed a set of actions to guide the distribution of resources by the state and federal governments over the next 10 years. The document is inclusive in its framing of STEM education pertaining to subject-specific and cross-disciplinary teaching. As well as delivering "core subject knowledge", STEM education is positioned as supporting the development of "skills of collaboration, critical thinking, creativity and problem solving". The goals of the strategy are aimed towards, firstly, the general population by developing a fundamental level of "STEM literacy" and for "young people to become more STEM capable", and, secondly, for preparation of the new STEM elite by developing "higher levels of STEM capability" and building STEM aspirations (p. 5).

A number of initiatives are emerging at the state and national levels. At the state level, the Victorian State Government has initiated a number of programmes to improve STEM education. This chapter reports on one of these STEM initiatives, the *Successful Students – STEM Program*. This initiative involves teachers of STEM, by which we mean teachers of one or more of the science (S), digital technology (T), design and technology (E) and mathematics (M) learning areas. Engineering is not a discipline within the Victorian Curriculum, but the "engineering principles and systems" sub-strand of "technologies contexts" do provide a platform for teachers to engage with the engineering design process.

8.3 The *Successful Students – STEM* Program

The *Successful Students – STEM* (SS-STEM) Program (2015–2017) is one of the 11 initiatives of the *Skilling the Bay* programme in Geelong, which is funded by the Victorian State Government and administered through the Gordon Institute of Technical and Further Education (TAFE) (http://www.successfulstudents-stem.org.au/).

Skilling the Bay was established in 2011 in response to the changing economic climate of Geelong where a number of major manufacturing industries have closed down, giving way to a new, more "knowledge-based" economy. *Skilling the Bay* funding of $11 million is key to supporting this transition by focusing on skills development, workforce participation and education. The SS-STEM Program is one of the four programmes meeting *Skilling the Bay's* goal of improving educational attainment and participation rates in STEM subjects. The programme is being developed and implemented by a team of researchers (SS-STEM team) from the School of Education at Deakin University. The programme funding supports teacher professional development, academic leadership and administration, partner school programmes (such as excursions), a STEM into Industry programme, and teacher participation in the STEM Education Conference.

The expected outcomes are:

- Increased sophistication of teachers' incorporation of STEM practices into their teaching
- Increased student awareness and aspirations
- Improvement in amount and quality of student participation in STEM activities and studies
- Improvement in students' confidence in subjects like science and mathematics
- Increased incorporation of the STEM practices into the school programmes

The programme involves ten partner schools from the Geelong region, focusing explicitly on year 7 and 8. Three teachers from each of the ten partner schools committed to professional development for two and a half years. These teachers could be mathematics, science or digital or design technology teachers or teachers in positions of leadership who can support the change process of the SS-STEM teachers and generally within the school. Teachers undergo four intensive professional learning (PL) sequences. Each sequence involves two intensive days focusing on building teachers' knowledge of STEM practices and pedagogies. They then plan and implement a STEM initiative in their school. The teachers return after 8–12 weeks and report on their initiatives to the other project schools on a third reporting and planning day.

Schools decided their own focus for improving STEM, such as subject-specific innovations (e.g. focusing on mathematics or science only), innovations requiring integration of subjects (e.g. different models of developing activities that involve teaching in science and maths) or innovation across a suite of subjects that promoted particular STEM pedagogies (such as design-based learning, across maths and science). The intention is that, as the programme progresses, the teachers focus on not only their own development but also act as change agents in their school to lead sustainable STEM innovation.

In addition, a Deakin Project Officer works with schools to support their developing practice. There are a number of other programmes and initiatives within the SS-STEM Program. A "Secondary STEM Teacher Network" was established to extend support to project and non-project schools in the Geelong region. Network meetings are hosted by BioLab (a state government-funded science centre) in Geelong, at least once a school term. A "Student Ambassador Program" involves students from Deakin University's science, engineering and information technology faculties to act as ambassadors for STEM and their chosen career and to assist

Table 8.1 The SS-STEM professional learning: key questions and main activities

Professional learning event and "focus"	Key questions	Key activities
Initiation and planning day 2015 term 2, ½ day	Introduction to programmes and schools	Initial planning by schools
Sequence 1. PL1 "Pedagogies and contemporary STEM practices" 2015, term 3, 2 + 1 days	What does effective teaching and learning look like? What pedagogies are needed for an innovative STEM curriculum? What cooperative approaches to teacher learning will lead to improved practice? (supports, purposes)	*Days 1 and 2* Descriptions and activities for the STEM pedagogies School planning *Day 3* School reports School planning Support structures
Sequence 2. PL2 "Assessment, upscaling and leading change" 2016, term 1 + 2, 2 + 1 days	How can a school develop a vision for STEM education? How do you frame STEM in your school? What does STEM teaching and learning look like? What model/s of curriculum development and teacher collaboration are needed for STEM education? How is teacher learning of STEM education provided for? How will you engage with industry and community?	*Day 1 and 2* Mapping intentions against a "STEM vision" framework Activities relating to the STEM pedagogies Illustrations of assessment practices *Day 3* School reports School planning Targeting and sustaining change
Sequence 3. PL3 "Sustaining change" 2016, term 3 + 4, 2 + 1 days	What is needed to sustain change? How can teacher and student learning be captured as convincing evidence of sustainable change? How can STEM teaching and learning be documented and shared?	*Day 1 and 2* Representational challenge *Focusing and sustaining change* Design challenge *Approach to research* Problem-solving *Communicating change* *Day 3* School reports School planning Assessment and evaluation

(continued)

Table 8.1 (continued)

Professional learning event and "focus"	Key questions	Key activities
Sequence 4. PL4 "Embedding practice and generating evidence of change" 2017, Terms 1 + 2, 2 + 1 day	What is needed to sustain and embed change? What does sustaining change look like? How can teacher and student learning be captured as convincing evidence of sustainable change? How can STEM teaching and learning be documented and shared?	*Day 1 and 2* Planning for teacher action research Contemporary science and linking with industry Creativity frameworks *Day 3* Showcase and celebration day

teachers with implementing their STEM initiatives and projects. A "STEM into Industry program" assists schools to use industry links within their units; the Australian Industry Group is assisting with establishing relationships between schools and Geelong-based companies involving twenty-first-century technologies. The "Deakin STEM Education Conference" was convened as a national conference in 2016 (http://stemedcon.deakin.edu.au) to, in part, provide a forum for teachers from the SS-STEM partner schools to showcase their innovations and for all teachers (especially in the Geelong region) to see possibilities for STEM education.

The PL sequences and the ongoing support are key to supporting each school's approach to STEM innovation. The four PL sequences focus on different aspects of the change process. A STEM vision framework was not formalised until the second professional learning sequence, although teachers were exposed to all elements in earlier professional learning days. The key questions and main activities of each of the PL sequences are outlined in Table 8.1.

8.4 A STEM Vision for Teachers and Schools

The programme is longitudinal and responsive to the developing needs of the teachers involved. In order to consolidate the objectives and activities of the SS-STEM Program, we felt the need to articulate a comprehensive, multifaceted and coherent *STEM vision* that addresses the subtle and complex challenge of preparing "twenty-first-century" citizens within the constraints of a traditional school system and curriculum. The process of developing the STEM vision has been iterative and developed through a process of crystallisation (Ellingson, 2009) where the SS-STEM team has built a deeper understanding of what STEM education is, how it can be effectively implemented in schools and how teachers need to be supported. The STEM vision framework has been presented to a number of audiences and, based on feedback, continues to be refined. The latest version is represented in Fig. 8.1.

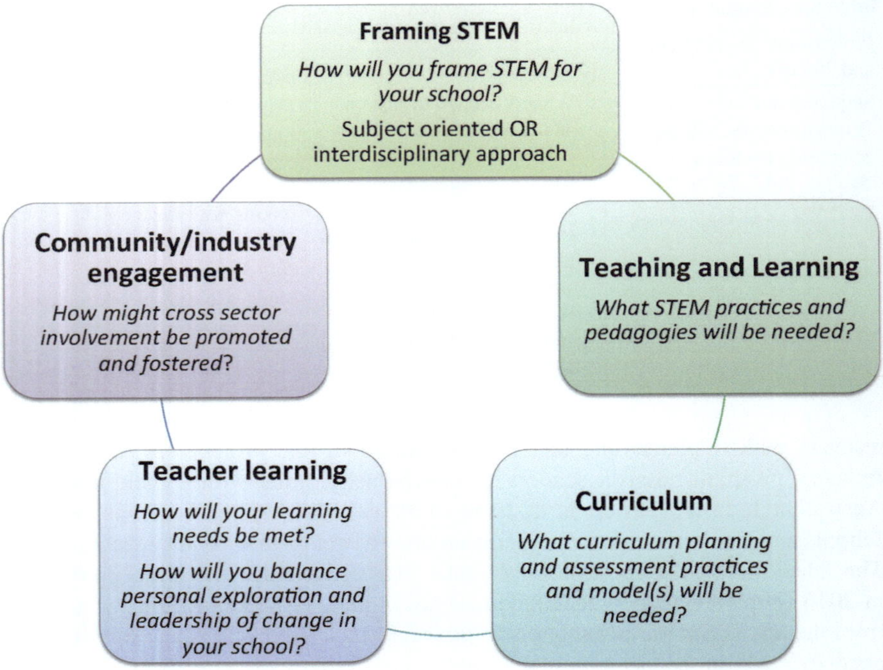

Fig. 8.1 The Successful Students – STEM Program "STEM vision"

The STEM vision framework was designed to enable teachers to appreciate that STEM innovation is not simply a matter of introducing disconnected STEM activities but must engage with the STEM agenda through a targeted and deliberate framing of STEM, be underpinned with the language of STEM, draw on a range of pedagogies that enable students to engage seriously and deeply with the STEM practices, draw on curriculum and assessment practices that align with the STEM practices, incorporate teacher learning opportunities that supports the change processes of not only the targeted teachers but also the wider school community and include meaningful links with industry and community that provide opportunity for students to see (and possibly participate in) STEM as it is practised.

Our multifaceted vision for STEM education has the following elements:

1. *Framing STEM*: STEM education is conceptualised as inclusive and interdisciplinary.
2. *Teaching and learning*: A common set of STEM practices underpin planning and pedagogy.
3. *Curriculum*: Curriculum is locally developed using multiple models of discipline integration, assessment practices and teacher collaboration.
4. *Teacher learning*: The professional learning programme is both intensive and ongoing and supported by an expanded teacher network.

5. *Community-industry engagement*: Community and industry are engaged to support and provide meaningful contexts, authenticity and depth to the teaching programmes.

The five elements of the STEM vision are described below, drawing on current theory, pedagogy and commentary on STEM education. To support the teachers' planning and learning, a STEM vision document (Appendix 1) incorporating these five elements was used to prompt teacher reflection on current and existing practices. The teachers completed the template on day 2 of PL2, and these have been analysed to support case study development for this chapter.

8.4.1 Framing STEM

Increasingly the STEM community is looking to integration of the STEM disciplines in real-world design problems as a way of engaging students in imaginative and collaborative problem-solving and reasoning (Bybee, 2015). STEM practice should be seen as inclusive of the knowledge-generating practices of the individual disciplines, as well as what is common across the disciplines. Vasquez (2015) describes STEM not as curriculum but "as an approach to learning that removes the traditional barriers separating the four disciplines and integrates them into real-world, rigorous, relevant learning experiences for students" (p. 11). By presenting real-world problems that require solutions from across the four disciplines, the barriers between the disciplines can be broken down.

Figure 8.2 depicts a scale of the different models for how STEM can be framed for education. At one end, and in line with Vasquez's definition, STEM can be represented as a meta-discipline that relates to only the overlaps between the disciplines (amalgamated model) and refers to the generic or "soft" skills that are common to all four disciplines. At the other end, STEM is inclusive between the interconnections and the individual practices of each discipline (holistic model), recognising that science and mathematics learning can similarly represent the discursive practices of the STEM disciplines. In the middle, STEM is framed as relating only to when the disciplines work together, thus excluding the work of the individual disciplines (interconnected model).

Schools are structured as predominantly subject specific; however, there is a move towards some degree of interdisciplinarity (Bybee, 2015). In order to be inclusive of the current teachers in schools, STEM was presented as being relevant to the mathematics, science, design and technology and digital technology learning areas of the Victorian curriculum, as well as offering a means of bridging these disciplines. Therefore, an amalgamated model framed our approach to STEM.

Teachers were asked to consider how they intended to frame STEM at their school and what processes and mechanisms would be needed to promote and sustain their framing of STEM. This framing of STEM has implications for how teachers then respond to other parts of the framework.

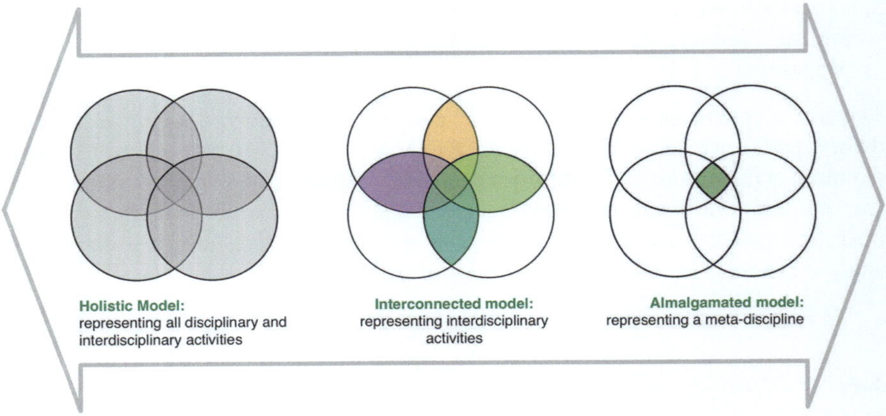

Fig. 8.2 Different models of STEM in education

8.4.2 Teaching and Learning

STEM practices relate to the disciplinary practices recognisable across the four disciplines. Table 8.2 summarises four interconnecting skills or proficiencies that are common to the STEM disciplines, derived from Clarke (2015), and describes the related STEM teaching and learning practices that can support their development.

These STEM practices can be aligned with learning tasks and inform curriculum planning around STEM in order to justify and direct curriculum innovation or provide alternative foci for assessment. Essentially, these practices are what make a programme "STEM".

A school needs to have a clear vision for what STEM practices will be promoted through their STEM programmes. These STEM practices need to be aligned with the pedagogy used, with specific attention to the discursive practices – the talking, writing and doing – of the disciplines involved.

Some pedagogies relevant to STEM education are listed below. These pedagogies formed the basis of the professional development offered through the SS-STEM Program.

Inquiry Through Representations Guided inquiry pedagogies in science and mathematics have been shown to engage low SES students in active learning and improve learning outcomes. These approaches align school STEM curricula with the knowledge-building practices of science and mathematics and exemplify the use of the discursive, representational tools and artefacts, such as drawing and modelling, animations and a range of digital tools and resources that now pervade STEM professional and research practice.

Focus on Student Problem-Solving/Reasoning Reasoning, including socio-scientific reasoning, which acknowledges multiple sources of evidence in decision-making (Morin et al., 2015), is central to guided inquiry in science through student

Table 8.2 STEM practices and teaching and learning practices

Interconnecting STEM skills/proficiencies	STEM teaching and learning practices
Flexible reasoning skills	Problem-solve
	Create
	Generate own questions
	Inquire
Effective and adaptable use of artefacts	Use conceptual, digital, physical tools
	Explore and investigate artefacts
	Use a range of modern tools
	Use artefacts of the discipline in a flexible way
	Apply constructed artefacts to new contexts
Proficiency in professional/technical discourse	Understand and engage with the disciplinary representations
	Know the language
	Share and communicate
	Work in teams
Understanding of the nature of evidence in different settings	Collect real data in a variety of situations
	Use evidence to validate a solution to a problem or justify a decision
	Make judgements about the accuracy and reliability of information

construction of representations. In mathematics students are engaged in problem-solving, modelling and reasoning through generalising, abstracting and justifying.

Design-/Challenge-Based Approaches Design- or challenge-based activities are commonly associated with integrated approaches to STEM; however, they also have a strong tradition of producing powerful learning in science and mathematics. The design process is a means by which the design and technology learning area can be integrated into mathematics and science, potentially breaking down disciplinary barriers through engaging in real-world problem-solving. Effective design challenges are linked to both assessment and curriculum.

Digital Technologies Digital technologies can be incorporated in two ways: the use of inquiry-, design- and challenge-based approaches to teach the digital technology learning area and the use of digital resources to support contemporary innovative approaches to mathematics and science that reflect disciplinary practices. The use of digital technology helps to:

- Develop students' critical thinking to evaluate digital resources, tools and algorithms
- Develop and explain computational processes
- Solve problems
- Think logically, algorithmically and recursively
- Develop creative thinking through designing digital interfaces to communicate information

8.4.3 Curriculum

In Australia there is no subject called STEM. As teachers need to meet the state reporting requirements, which are usually subject based, the traditional structure of the siloed curriculum can be a barrier for schools engaging with STEM. However, a STEM education can still meet the school's reporting requirements and curriculum demands, but teachers and leaders often require strong argument and evidence to be convinced that integration of STEM subjects can still meet the curriculum requirements.

Teachers individually and collaboratively develop curriculum in their schools in ways that reflect the priorities and cultures of the school. Teachers work in a variety of ways, by working as individual teachers or as teams of teachers, either subject-based or as interdisciplinary teams. There is a need for the schools to develop strategies for teachers to collaborate, especially when subject integration is the aim.

Because there is no mandate or curriculum framework for "STEM" so far in Australia, teacher collaboration strategies will vary across schools, depending on how they choose to frame STEM. In response to the needs of the partner schools, the SS-STEM team elaborated on Dugger's (2010) categorisation of teacher collaboration and curriculum models, as shown in Fig. 8.3.

The first four are based on Dugger's scheme, and the fifth was introduced to represent another model exhibited by some of the SS-STEM teachers. In keeping with an inclusive framing of STEM, we do not limit STEM to any particular model of teacher collaboration.

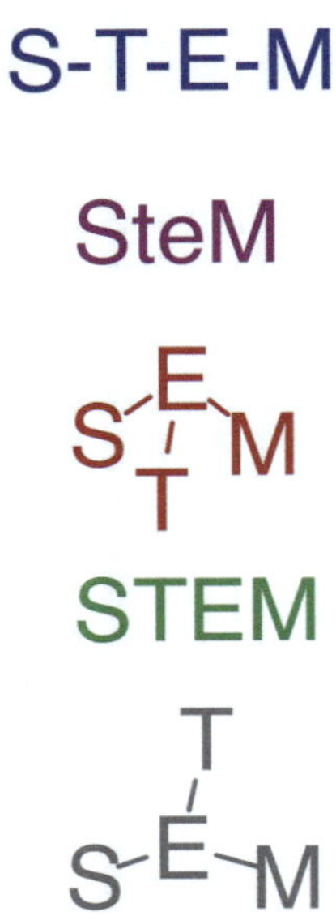

1. Teach each discipline separately
   In science classes, there is a renewed focus on using representations to enhance concept development. In mathematics, teachers could use complex problem solving to challenge their students.

2. Teach all four but more emphasis on one or two
   A teacher integrates mathematics and science through a challenge based unit of work where students design a vehicle.

3. Integrate one into the other 3 being taught separately
   The engineering processes of team work, identify and investigate a problem, design a solution, and testing and evaluation is added into some science and mathematics units, but there are limited links across the science and mathematics subjects.

4. Total integration of all by a teacher
   Science teacher integrating, T, E and M into science.
   A school introduces a new STEM elective focusing on designing digital solutions to real world problems

5. Divide a STEM curriculum into the separate subjects
   Technology, science and maths teachers design a combined unit and each teacher teaches different components of the unit in their separate subject, and with clear contributions from science, maths and technology subjects in solving a common problem.

Fig. 8.3 Teacher collaboration and integration models used by schools in 2016 of the *Successful Students – STEM Program*

8.4.4 Teacher Learning

The teacher is a powerful conduit for student engagement. There is an opportunity for the leading teachers and school leadership to work closely with teachers to build their capacity for delivery STEM curriculum. It is important to understand the learning intentions of both the students and the teachers in these initiatives and to provide teachers with opportunities to enhance their own learning by trialling new practices and reflecting on their practice. Teacher learning may be enhanced through teacher network meetings (across schools); professional development that has intensive and ongoing components; school-based teacher collaboration, provided through ongoing and systemised teacher working groups (such as professional learning teams); and real opportunities for teachers to develop school curriculum that incorporates the STEM experiences in authentic and valued ways.

Teacher learning is a multifaceted process and can be likened to crossing a boundary between a familiar and an unfamiliar domain, for example, a familiar traditional subject-oriented approach teaching to an unfamiliar STEM approach. Encountering such a "boundary" raises the possibility for learning. Akkerman and Bakker (2011) pose four potential learning mechanisms that can occur at the boundary between the familiar and unfamiliar: identification, coordination, reflection and transformation. These mechanisms were incorporated within the STEM vision in the following way:

1. *Identify areas of concern/need for learning*: This occurs where teachers can consider their current practice and establish goals for their learning and goals for STEM education in their school.
2. *Secure resources, knowledge and people that can support the learning*: Teachers need to gain access to resources and knowledge of content, strategies and STEM practices that can be linked to their school curriculum. They also need support from specialists in the area to assist in the development of theory-informed practice, both in learning about new practices and when back in schools while implementing and embedding these new practices.
3. *Reflect on and critically analyse practice*: Teachers need to be able to trial new approaches, reassess outcomes and revise beliefs about learning and core purposes of STEM. Reflective practice should be embedded in the programmes, with each of the programmes involving evaluation of the effect on their students, with opportunities for teachers to report on new curriculum and teaching and learning approaches.
4. *Transform practice and identity*: Transformation can occur at multiple levels: the teachers' practice and identity in relation to STEM, teachers' ability to support and build capacity of other teachers in the school and through transformation of STEM education within the school

Sustaining STEM reform in a school will depend on where transformation is focused, whether it is with the teacher, the teaching team at the subject or year level, the entire discipline groups and leadership direction from the school executive and through school policy and planning documents. The use of evidence to evaluate STEM programmes in terms of effecting quality teaching and learning is essential to the learning process and in leading change. One of the tools we have used to collect evidence and provide feedback to teachers is the *Components of Effective STEM Teaching and Learning* (Appendix 2).

The *Components of Effective STEM Teaching and Learning* is a teacher development tool that describes effective teaching and learning in mathematics, science and technology. The components were originally developed for science teaching and learning as part of the Science in Schools (SIS) project (Tytler & Waldrip, 2001) and then redeveloped to include mathematics in the Improving Middle Years Mathematics and Science (IMYMS) project. The IMYMS components have been redeveloped for STEM to support a range of Deakin STEM-related initiatives (Successful Students – STEM and evaluation of the University of Sydney STEM Teacher Enrichment Academy). They incorporate the latest research relating to developing and using representations, the design and engineering elements in STEM and learning technologies. The tool is designed for respondents to place their practice on a scale of 1 to 5 for a series of components and subcomponents. Descriptors are placed at 1, 3 and 5, with 5 being the most sophisticated level of practice for that component or subcomponent. The respondent also records the degree of importance accorded to each component.

Teachers and curriculum planners can use this framework to encourage a consistent use of language around STEM pedagogies and for teachers to evaluate their current, intended and post-initiative STEM teaching practices. Teachers can use the tool to identify components they wish to focus on in their learning, either individually or as a teaching team or school.

8.4.5 Community-Industry Engagement

Linking science and mathematics with industry, the community and families is an effective way to emphasise the relevance of science and mathematics in all facets of human activity and in particular to acknowledge the social and cultural aspects of the disciplines. Community-industry engagement assists teachers and students to make connections between ideas within a discipline, with other disciplines and with the digital, analogue and real world. Programmes that illustrate how such links can be made can involve one-off industry talks or through in-depth exploration of contextualised issues or problems.

A range of science- and mathematics-related industries, companies and research organisations are available in many areas, and through partnership, these potentially offer powerful resources for schools. However, real-world practices, and their underpinning concepts and processes, can be quite complex and not necessarily easily translated in the classroom. Translation of the contemporary STEM practices into the classroom by teachers may require intense support to understand the relevant science or mathematical concepts, processes and representations. It is necessary to select elements of the industry practices that might be engaging for students and develop the teaching strategies that maximise the student learning.

Industry and the community can be linked with curriculum in schools in a variety of ways:

1. *Engagements*: Industry representatives to offer an immersion experience at the beginning/middle/end of a unit in order to provide a context and purpose for learning the unit content. For example, an engineer talks to students about their job during the immersion phase of a bridge-building unit.
2. *Elaborations*: In-depth immersion in a real-world problem generated by or related to industry that exposes students to industrial processes and problem-solving and career opportunities. For example, *Rip Curl* provides materials for a materials technology programme where students do tests with neoprene to design a wetsuit.
3. *Contexts*: Contextualised issues or problems can provide multiple links to industry, industry practices and subject-related outcomes. Such links help students to understand the problem, develop possible solutions and project themselves into these contexts, for example, a unit on bees that explores the scientific, mathematical, economic and social implications of bee parasitism.

Local industries, companies and education centres can provide meaningful contexts for exploring the latest digital technologies and STEM applications. Schools need to be supported in thinking through how industry links can inform the school curriculum.

8.5 Methodology

As a government-funded programme administered by the local TAFE college of the Gordon Institute, the SS-STEM Program is subject to evaluation against a set of key performance indicators (KPIs). Evidence generated to respond to the KPIs includes student surveys and focus group interviews, teacher questionnaires and interviews, curriculum documents, reports from teachers, teaching documents and observations.

For this chapter, we selected data from the first 12 months of the initiative to illustrate the variety of activities that can arise from a comprehensive STEM vision. Drawing from the data listed in Table 8.3, we showcase the work of two partner schools.

The SS-STEM team support teachers at their schools and during the professional learning intensives. During these times, the team recorded observations of teacher reflection, planning and difficulties and classroom activities as they unfolded. These

Table 8.3 Data used to generate representation of the partner school STEM initiatives

Tool	Data description
Partnership negotiation tool	Summary of partnership arrangements between Deakin University and the school[a]
Questionnaire	Components of Effective STEM Teaching and Learning[b], a teacher self-assessment tool
Observations	Written descriptions by researcher of communications with teachers, involvement in activities at the schools
Presentations	Teacher PowerPoint presentations on projects from PL1 and PL2
Planning documents	Teacher planning of STEM activities
Professional learning artefacts	Templates and activities provided by STEM team to represent teacher thinking and planning: concept map, STEM vision template
Focus group interview	Interview with teaching team at each school

[a]From Hobbs et al. (2015)
[b]Modified from Tytler and Waldrip (2001); see Appendix 2

observations inform the case studies by providing on-the-ground evidence of the enacted STEM activities and teachers' responses to them.

The STEM vision document (see Appendix 1) was introduced to prompt teachers' thinking beyond their own practice and the units that they were planning. This document was an important planning document for teachers but also provided important information about the current and future direction of STEM at each school.

The presentations and planning documents collected from the reporting days describe the STEM initiatives either as intentions for forthcoming school terms or as reflection on enacted activities or units. For enacted units, teachers evaluated the impact on student engagement and learning, although this process was often not rigorous in PL1. For PL2, we gave teachers evaluation questions for students to use post-initiative. These presentations (especially when reporting on enacted initiatives) and curriculum documents have been essential in constructing a list of the projects, which is used below to illustrate the diversity of activities emerging from the project.

A focus group interview was conducted during the PL2 sequence in order to ascertain how the STEM vision is developing at the schools, identify what further assistance is needed in implementing the project and collect data (such as curriculum documents) from teachers that would inform documentation of the school's project-related activities. Teachers were presented with data already gathered from the school, such as the STEM vision document, and data from the initial student survey (not included in the analysis leading to the case studies). These interviews were approximately 60 minutes long and audio-recorded and usually involved one to three of the project teachers and school leaders. The interviews provided additional data relating to the STEM activities being developed and how teachers were working as a team.

8.6 Schools Developing Their Own STEM Vision

In this section the projects emerging across the ten schools are collated to show diverse teacher collaboration, subject focus, student activity and STEM practices. Then, two partner schools have been selected as case studies to illustrate how the STEM vision was interpreted and applied by the teachers. Each case study emphasises one or more elements of the STEM vision framework and illustrates different teacher collaboration models.

8.6.1 STEM Initiatives

Table 8.4 draws on the presentations from the reporting days in PL1 and PL2 to capture the variety of teacher collaboration models, year levels and curriculum content, student activities and STEM practices emphasised.

Schools A, C and D did not attend or did not present at PL2 day 3, so only the PL1 project is represented. The other schools showed progress towards becoming more confident in planning and implementing STEM activities. This confidence was evident in how they refined, extended, diversified and applied STEM pedagogies to new contexts.

Schools F and G presented on the same project at both reporting days but reported on refinements to assessment practices, implementation, level of support provided by the teacher (usually more support was needed) and changes to the materials used (such as for construction purposes).

Other schools either applied the pedagogies trialled in PL1 from 1 year level to another (School B) or to a different topic in the subject (Schools E, H, I and J).

For some schools, changes from 1 year to another meant that additional teachers needed to be inducted into the programme. This was the case at Schools A and I (as new SS-STEM project teachers), B (due to expanding the scope – from year 7 to year 8), E (new teachers to year 8 science) and F and G (teachers of other year 8 classes).

Where the PL2 project differed to PL1 project, teachers developed the new activities to be variation on the same STEM pedagogy, such as new design challenges (School J), new representational challenges (School E) or new complex problem-solving (Schools B and H). School I changed their approach from a design challenge in PL1 to a series of problem-solving tasks in PL2.

Table 8.4 SS-STEM Projects per school in professional learning sequence 1 (PL1) and 2 (PL2)

School	Teacher collaboration	Year level and curriculum content	In these projects, students…	Related STEM teaching and learning practices
A	STEM	PL1 and PL2: year 8 technology (design processes and simple machines)	Investigate types and uses of simple machines	Problem-solving Create Use tools
	Tech teacher complements science programme		Design a machine that lifts 250 g weight	Explore artefacts Share and communicate Work in teams Collect and make judgements about real data Use evidence to validate solutions
B	SteM Maths teacher complements science programme	PL1: year 7 mathematics (angles, generating and using data), science (gravity and ramp slope)	Investigate ramps for people in wheel chairs: assay ramps in the community, test effects of angles on ramps and design a school ramp	Problem-solving Create Inquire Explore artefacts Know the language Share and communicate
	SteM Maths teacher complements science programme	PL2: year 8 mathematics (area, volumes, measurement and managing data), science (transpiration and evaporation)	Investigate *Are our garden water storage tanks large enough?*, including analysis rainfall data, garden bed water loses and size of water tanks in relation to collection roofs	Work in teams Collect and make judgements about real data Use evidence to validate solutions
C	S-T-E-M Tech teacher makes connections to STEM practices	PL1 and PL2: year 7 technology (developing coding skills)	Complete online coding skill-building activities, leading to control of robotic Lego devices	Create Use tools Explore artefacts Apply learning to new concepts Know the language Work in teams

D	SteM Math/science teacher coordinates maths and science activities	PL1 and PL2: year 8 mathematics (generating and using data), science (experimental design)	Design, test, evaluate and retest a "Barbie bungee" in order to build up problem-solving and inquiry skills	Problem-solving Inquire Apply learning to new concepts Know the language Work in teams Collect and make judgements about real data Use evidence to validate solutions
E	S-T-E-M Science teacher makes connections to STEM practices	PL1: year 8 science (plant and animal cells)	Represent the function of plant and animal cells as two challenges: (1) different objects to represent the different organelles and (2) bucket of cells	Inquire Use tools Explore artefacts Use discipline objects flexibly Engage in representations Know the language Share and communicate Work in teams Collect and make judgements about real data
	S-T-E-M Science teacher makes connections to STEM practices	PL2: year 8 science (particle theory of matter)	Use representations to (1) physically show temperature effects during chemical changes and (2) design a communication that shows relationships between reactions, equations and the particle theory	
F	S E M T Math/science teacher complements tech programme	PL1 and PL2: year 7 and 8 science (human senses), mathematics (measurement and variables), technology (robotics and coding)	Develop and apply programming skills when solving problems in science, mathematics and technology using Lego EV3 Robotics	Problem-solving Create Use tools Apply learning to new concepts Know the language Work in teams

(continued)

Table 8.4 (continued)

School	Teacher collaboration	Year level and curriculum content	In these projects, students…	Related STEM teaching and learning practices
G	T / S–E–M Coordinated approach to design challenge by science, maths and tech	PL1 and PL2: year 8 mathematics (scale, circles and measurement), science (properties of materials), technology (design and construction)	Investigate, design, create and evaluate a vehicle that will travel farthest down a ramp and represent learning in a portfolio that was assessed in the three subjects	Problem-solving Create Use tools Explore artefacts Use discipline objects flexibly Apply learning to new concepts Know the language Share and communicate Work in teams Collect and make judgements about real data
H	SteM Sciences and maths teachers making connections to science and engineering	PL1: year 7/8 mathematics (generating and using data), science (experimental design)	Design and construct a number of small projects – parachutes, cranes, bridge building – in order to build up problem-solving and inquiry skills	Problem-solving Create Use tools Apply learning to new concepts Know the language Share and communicate Work in teams
	SteM Science and maths teachers making connections between science and maths	PL2: year 7/8 mathematics (indices, powers and exponential change), science (acids and bases, radioactivity and cells)	Link three diverse topics from real life (pH levels, radioactivity and the Zika virus). In mini-workshops, students share and communicate as peer tutors	Collect and make judgements about real data Use evidence to validate solutions

I	SteM Math/science teacher approach complements tech programme	PL1: year 8 science (forces), technology (design process)	Design and construct a small self-propelled vehicle to carry water over a set distance, after completing immersion experiences and team building activities	Problem-solving Create Inquire Use tools Apply learning to new concepts Know the language Share and communicate Work in teams Collect and make judgements about real data
	SteM Math/science teacher making connections between science and maths	PL2: year 8 science (investigating), mathematics (basic operations, data and statistics)	Inquire into links between mathematics and science through a series of paper-based or hands-on "curiosity problems" relevant to students	Use evidence to validate solutions
J	SteM Science teachers connect science and engineering	PL1: year 8 science (simple machines, forces and energy transfer)	Design a Rube Goldberg machine focusing on energy	Problem-solving Create Use tools Explore artefacts Know the language
	SteM Science teachers connect science and engineering	PL2: year 8 science (forces)	Design, construct, evaluate and communicate to others about three design challenges: balsa bridge, a mouse trap car and a water rocket	Share and communicate Work in teams Collect and make judgements about real data

8.6.2 Case Studies

School B is one of the three secondary schools for girls in the Geelong region. In PL1, a team of mathematics teachers developed a year 7 mathematics inquiry sequence involving complex problem-solving in real-world contexts. In PL2, the teachers applied this inquiry model to year 8. The case study describes how the teachers built on an existing culture of innovation within the mathematics department, but recruiting other year 8 teachers to the inquiry model has met with initial resistance and the steps they needed to take to convince the teachers of the benefits of employing such approaches in year 7 and 8 maths.

School G is a co-educational government secondary school (years 7–12) in a growing part of Geelong, with a steady student population of 800–900 since 2009. During PL1, science, mathematics and technology teachers developed, trialled and reported on an interdisciplinary project for some year 8 classes. In PL2, teachers reported on their plans for a modified second iteration of the project to be implemented in PL3. The case study describes the project and aligns it with the STEM practices and how STEM is becoming prioritised at the school.

8.6.3 School B: Building STEM into Mathematics

School B joined the SS-STEM Program in order to attend to a continuing problem of students entering the school at year 7 with weak mathematics background, low student aspirations towards STEM careers and a decline in senior mathematics and science enrolments with many students not seeing the point of doing STEM subjects. In order to address these issues, a teaching team focusing principally on mathematics was chosen to participate in the SS-STEM Program. The teachers' framing of STEM was based largely on a need to improve students' engagement and application of mathematical ideas, so was largely subject oriented. The school endeavours to have the same teacher teaching mathematics and science at years 7 and 8, allowing for some coordination of the teaching of science and mathematics; therefore, the two activities generated by the team have some scientific concepts embedded in them to assist in understanding the nature of the problem being explored mathematically, as will be explained below.

Prior to SS-STEM, the school had a history of using novel approaches to mathematics teaching, for example, students design a chair as part of their year 9 mathematics class. Being inspired also by the problem-solving activities completed as part of the first PL intensive, it was a natural progression for the teachers to continue to develop their approach to open-ended problem-solving. From then on, collaboration and planning by the teachers focused on developing a common approach to teaching years 7 and 8 mathematics based on regular use of investigations involving complex problem-solving in real-world contexts where the focus is on big ideas and core mathematical principles instead of topics and essential learning and proficiency.

The intention is to have at least one problem-solving investigation each year from years 7–9.

Two investigations have been developed based around "big questions" (see Table 8.5): a year 7 investigation of *What would be the best ramp for wheelchair access to the deck in our Garden*? with a mathematical focus was on measurement, modelling and investigating the relationships between mathematical objects and the science ideas of gravity and ramp slope and a year 8 investigation of *Are our new tanks big enough for our garden*? focusing on area, volume, measurement and managing data in mathematics and transpiration and evaporation in science.

The year 7 investigation was designed to engage the human needs of students through posing a problem that had a real purpose. The school was intending to build a ramp for the newly established school garden, so the investigation enabled students to recognise disability ramps in the community and get a physical sense of elevation versus effort by testing ramps of different lengths and pushing wheelbarrows up ramps. The ultimate aim was to have students design an appropriate ramp to scale.

On reflection, teachers felt that the unit was successful at a number of levels. At a personal level, the investigation had the effect of raising some students' awareness of ramp use in their community and the need to make regulations about ramps:

> "The students started to make connections out into the community. One girl recently said that she had noticed a person in a wheelchair, and had felt empathy with the person, so it made me realise that the students saw it as an authentic learning activity."
>
> "They loved getting the wheelchair out."
>
> "They were surprised how many ramps were in the school building."
>
> "Many of the students found out through their investigations that there are regulations covering ramps, and some of school ramps failed these regulations. They also found out why long ramps have horizontal sections mid-way."

Mathematically, students engaged in the problem-solving process in rich ways, leading to a contextualised understanding of the mathematics involved:

> "Even the recording of effort made the girls think deeper, they started to inquire and solve how to best represent it graphically."
>
> "There was even powerful learning occurring through the students having to make decisions about how to record information, how to describe differences between ramps, and even how to represent the different ramps they investigated on paper."
>
> "One girl came up to me and asked – if she knew the angle and the length, is there a formula to work out the height? This and journal evidence indicates that many girls had started to make deep connections between the key maths ideas."

The teachers learned how to maximise the learning opportunities through such investigations, in particular, the need to confine the investigation "to little over two weeks, otherwise the students lose focus". The teachers had not initially intended to build assessment into this activity, but a fortuitous use of a student journal in the first iteration has since informed the development of an assessment opportunity:

> "Initially we saw the project as a deep learning activity, and had not built standards based assessments into the unit, but just asked the students to keep a learning journal. Reviewing the journals made us realise that the students had clearly met the standards, and it would have been very easy to construct a suitable rubric."

Table 8.5 Learning sequences for B College year 7 and 8 mathematics investigations

Phase	Year 7 ramp activity (Term 3 2015)	Year 8 tank activity (Term 2 2016)
Immersion		
Understanding the problem	Investigate different disabled ramps in the broader community by walking around town, including a local hospital and the school itself	Investigate the school vegetable garden watering system. Students asked to observe how the beds are watered from rainwater tanks as well as determine how these tanks are replenished
	Are all ramps the same?	
Guiding tasks	Measure various ramps and produce scale drawings of ramps	Science workshops on transpiration and evaporation
	Investigate different ramps using the dynamic trolleys in the science-inclination affected effort	Groups of students allocated to individual garden beds, and the big question is introduced: *Are our new tanks big enough for our garden*?
	Test how the ramp elevation affected speed of descent	Brainstorming to determine a range of mini-inquiries to be undertaken
Mini-inquiries		
Asking questions	Students undertake a series of mini-inquiries	Students generate inquiry questions that are used to direct classroom learning such as:
	Compare different ramp lengths in the garden and then rank the effort needed to get a wheelbarrow up each ramp	How much water do our garden beds receive and use?
	Hands-on activities give students a physical sense of elevation versus effort	What is the area of the collection point, and how much rain is collected?
	Explore the ramps around the school using a wheelchair, to investigate ease of being pushed up or down, and the difficulty of wheeling oneself up or down	What is the total area of the garden beds?
		How much water is used by our garden beds?
		Additional mini-workshops:
		What 1 mm of rain really meant (depth of rain recorded and collection area led to volume of water collected)
		Transpiration rates linked to daily temperature readings
The big question		
Analysis and conclusion	The big question is introduced – "What would be the best ramp for wheelchair access to the deck in our Garden?" Students design a ramp to scale	Groups of students work on the findings arising out of the mini-inquiries and then report findings to the rest of the class

The success of the year 7 investigation has been essential in identifying how the year 8 and 9 mathematics programme might also benefit from problem-solving investigations. Involvement in the SS-STEM Program not only validated existing efforts by these teachers to develop complex investigations but also raised the profile of STEM within the school through the efforts of the "STEM group":

> "Before the [SS-STEM] project started there was already a small number of STEM based tasks in year seven but none in year eight. When the STEM project came along it quickly validated what we were doing. It gave it a bit of credibility, we are about to run with it and refine it, and it expanded very quickly. I think we are now very clear on what it is, which is the hardest part. We can say to our colleagues, this is STEM, we want you to try it."
>
> "It is in year 7, year 8, and now going into year 9 mathematics (and science), partly because of a curriculum restructure, and also because we have a STEM group together

What did we learn from B College?

- **Differentiate to make it work:** Different entry points and endpoints were needed for the diverse learning needs in the student cohort to be met and for all students to experience success.
- **Assessment:** Assessment is not needed for every learning activity, as long as there are clear learning intentions. Such learning intentions can focus on students' ability to apply mathematical concepts to real-world problems.
- **Subject-focused stem activities:** STEM activities can be rich-learning experiences even when principally focusing on one of the STEM subjects. Even though some degree of science was included in B College's units, the core focus was mathematics. STEM was framed as a set of STEM practices that provided a language for solving real-world problems, rather than as an integration of curriculum areas.
- **Recruiting other teachers:** When expanding the unit beyond the immediate team, teachers needed to reassure their nervous fellow mathematics teachers. They did this by using arguments such as the following: "The unit only takes just over two weeks out of the semester's normal program, only 1/20th of mathematics time". "The unit incorporates a couple of the expected learning outcomes of the Victorian Curriculum". "Our colleagues needed to see the level of engagement of the girls as they tackled the Big Question".
- **Empowerment of teachers:** Teachers said that they felt empowered by their participation in the programme and used their experience of presenting to the other STEM schools on the PL days to prepare for presenting to their colleagues back at school.
- **Need to document:** Convincing more of their colleagues to attempt these STEM engagement units, unit programmes need to be well documented and use a common structure that would become familiar to the other teachers, such as using the common procedural descriptions of "Immersion task, mini-investigations and guiding activities, the Big Question (inquiry or design challenge), and final student presentations".

8.6.4 G College: Using STEM to Bridge Technology, Science and Mathematics

Two main concerns led this school to participate in the programmes: firstly, declining numbers of students are enrolled in post-compulsory STEM subjects, and, secondly, capable girls were opting into the life sciences and avoiding the physical sciences. Linked to the first concern were observations about disengaged students:

> Disengaged students are often those who present as more 'hands on'. The reality is that 'hands on' jobs are increasingly reliant upon STEM skills. We need to do more as a school and as a system to connect these students with these subjects.

The school leadership felt that students needed to be explicitly shown to these connections across subjects, using a coordinated STEM project, aimed at year 8, before students start to make choices about study and career pathways:

> By teaching the components of the project in different subject areas, as a combined project, it flagged to the students that there were connections between these STEM subjects. This relates to working practice in engineering.

A cross-discipline team was established and given professional learning time and support. A design-based strategy was chosen because it was applicable across all three disciplines. Technology, science and math teachers planned a combined year 8 unit, which was taught to three classes. The unit sought to expose students to industrial processes and problem-solving. Each teacher taught a different component of the unit in their separate subjects, with clear contributions from science, mathematics and technology subjects in solving the common problem: the *Rolling Vehicle Challenge*. The unit also had the objectives of stimulating an understanding of STEM and the career opportunities of STEM-related subjects and allowing students to make connections between mathematics, science and technology. The students worked in groups on a design challenge drawn from the Victorian Curriculum: Design and Technologies. An excursion to Deakin University CADET engineering teaching and research facility culminated in a celebration day where the student teams presented their solutions to the *Rolling Vehicle Challenge* to guests (internal and external) and then competed in a run-off. See Table 8.6 for the weekly programme. Assessment was based on performance of their vehicle in the run-off and a presentation of their journal, covering their research and their design process and findings. Timing and coordination of classes remained a challenge, and a survey showed that students sometimes failed to make connections between mathematics and science activities.

Teachers taught in their individual disciplines, but there was a conscious cross-curricular coordination of learning activities, and a common STEM language was used which was made explicit to the students and reinforced by the use of a reflective journal across all three disciplines.

Table 8.7 shows how the teachers aligned the vehicle challenge project to the STEM practices (extract from teacher planning).

Table 8.6 G College vehicle challenge interdisciplinary weekly programme

	Week 1	Week 2	Week 3	Week 4	Week 5	Week 6
Science	Material research	Why friction/force	Visit to CADET engineering building	Practical: forces	Property evaluation/ justification	Presentations (DU project officer present)
		Mass types of properties		Speed		
Mathematics	Scales and drawings	Circle properties	% of materials in F1 cars	Speed	Work on presentation	Excursion and celebration day
				Distance time		
				(Decimals)		
Technology	Lego Prototypes	Concept drawings	Creation and construction of Design	Construction (DU project officer present)	Construction/testing (DU project officer present)	
	Reflection	Reflection	Reflection	Reflection	Reflection	

Table 8.7 Grovedale programme aligned to the STEM practices (from planning document)

STEM practice	Programme alignment
Flexible reasoning skills	Problem-solving – The task requires students to solve the problem of designing a vehicle that will travel farthest down a ramp within their grouped teams
	Creativity – Students are encouraged to be as creative as possible with respect to the design and creation of their vehicle within the constraints of the project
	Generating own questions – Throughout the accompanying mathematics, science and technology curriculum, students will not only answer inquiry questions but also generate their own questions for investigation
	Inquiry – All of the accompanying project curriculum in mathematics, science and technology introduced students to an inquiry or focus question relative to that session
Effective and adaptable use of artefacts	Conceptual, digital and physical tools – Students access and use such tools to aid their understanding, knowledge and completion of their vehicle: design tools in technology for vehicle prototyping, design and construction and scientific and mathematical equipment for experiments and measurement are all utilised in accompaniment with those conceptual tools
	Exploring and investigating artefacts – In science, students are asked to explore and investigate the properties of a variety of materials that would be considered for the use in their vehicles
	Using a range of modern tools, digital tools – Many modern tools are to be included ranging from Google sketch-up for design briefs, scientific data loggers (temperature analysis), thermal imaging camera for temperature distribution on vehicles and 3–0 printers to design and produce elements of the vehicles, such as the wheels
	Being able to use objects of the discipline in a flexible way – This is always an aim of all curriculum facets within the project
	Application to new contexts – Considering this project is cross curricular in design, students are asked to transfer their knowledge gained from different subject areas (mathematics, science and technology) into the final design and construction of their vehicle
Proficiency in professional/technical discourse	Understanding and engaging with the disciplinary representations – Students produce a portfolio/journal of their learning which forms a representation of all elements of their group learnings throughout the project
	Knowing the language – As part of the student group journals, language specific to STEM will be collated as a glossary and expected to be utilised in their reflections and investigations throughout the project
	Sharing and communicating – In groups, students identify their own personal role within the group to ensure accountability to each group member. Students are expected to present at the "celebration day"
	Working in teams – A high focus is placed on students' ability to work productively in teams. Student groupings were reviewed by staff to ensure they remain productive and as equitable as possible

(continued)

Table 8.7 (continued)

STEM practice	Programme alignment
Understanding of nature of evidence in different settings	Collective real data in a variety of situations – Data logging tasks are used in science. Measurement data is generated and investigated in mathematics. Students undertake data investigations via prototype modelling in technology
	Using evidence to validate a solution to a problem or justify a decision – Given that the inquiry task for students was designing and creating a vehicle, student's final design was a product of solution validation and problem troubleshooting/justification. This evidence was not only obvious with their final car design but also via their journals and in their celebration day presentations
	Making judgements about the accuracy and reliability of information – Again this is part of the student journals

In the second iteration, the project is to be rolled out to the entire year 8 cohort. This requires preparing other teachers in how to implement the project, prompting the SS-STEM teachers to document the project. Also, because the project is a substantial change to the school's subject-oriented curriculum, an interdisciplinary approach to curriculum was legitimised by including STEM objectives in the School Annual Improvement Plan under the principal's leadership. In addition, STEM practices and learning are interwoven into the goals and objectives in the STEM subjects, for example, "In mathematics, the professional development goals for many in the domain have now changed to include more real problem solving". In addition, teachers reported that, "across the school there is a stronger recognition of the need to take the students out of their normal classroom experiences to connect with the community".

In addition, the success of the first iteration contributed to the school developing a common language around STEM within the school: "Presenting to the whole school staff about the first project has resulted in a greater use of a common STEM language across the three subject areas by the teachers, so that the students see the connectedness".

Assessment requirements were refined:

> "In years 8 and 9 science classes, we have been re-designing the practical report requirements to move away from students just reporting upon what they did, to describing how they designed the investigations, and more on their development of processes."
>
> "In technology, they will incorporate more on the development of the design and prototyping processes, as well as formalise assessments to incorporate these areas."

The second iteration will also allow the students to make connections to industry in an "elaboration". A local high-technology industrial company representative will visit the school to talk with students about the operation of the company and the importance of the engineering process to their business, and a number of students, representing each challenge team, will visit the business to interview key employees and record facets of the company. These students then will report back to their teams using a multimedia presentation.

What did we learn from G College?

- **Newness:** Students recognised that the STEM unit was significant and different from the normal school programme, because their class teachers from the three disciplines were teaching elements of the programme and were flagging the importance. These connections deepened student learning.
- **Making it work:** Timetable was a blocker, but this project illustrates that a teaching team can work to deliver a coordinated unit involving teachers from multiple subject areas, provided that the content is engaging and connected to the students, and sufficient planning is carried out.
- **Community links:** An "engagement" with an engineering and technology teaching space at Deakin University early in the programme reinforced the "specialness" of what they were doing and built awareness of where STEM can take them.
- **Celebrating learning:** A performance and celebration day had a positive effect on the students and acted as an incentive for students to commit to the learning process.

8.7 Discussion

The innovations reported in this chapter show that the current STEM agenda proliferating in education discussions has promise for improving education. Difficulties arise when the issues of "membership" and "definition" constrain what can be possible.

STEM is historically associated with the individual STEM disciplines but in education is being increasingly associated with integrated activities (Honey, Pearson, & Schweingruber, 2014; Stohlmann, Moore, & Roehrig, 2012; Williams, 2011). However, relegating STEM to only integration potentially excludes the work in the individual subjects – can we really say that the learning that students glean from their year 7 science or year 11 biology classes are not part of the "STEM agenda" or that the individual subjects do not play an important role in influencing students' career choices, nor their ability to participate as a STEM informed citizen? Clearly, the drop in student uptake of STEM-related subjects at the senior secondary level, and the reported decline in attitudes of students towards science in particular (Kennedy, Lyons, & Quinn, 2014), indicates that business as usual is not sparking students' interests (Tytler, 2007). The current flurry of political leadership to translate the business case for STEM into schools (Education Council, 2015; Chief Scientist, 2016) may be clearly linked to a utilitarian ideology of schooling, but with this comes funding for schools based on a recognised need for change. Schools agree and are jumping aboard the STEM train! And why wouldn't they? The teachers in the SS-STEM Program are given time to plan and support to reconstruct and reconfigure their science, maths and/or technology programmes and are gaining recognition for their work. They are excited about doing something new, and they are seeing evidence of student engagement with these novel learning experiences.

Is it reasonable that there are different versions of STEM? Pivotal to developing a clear STEM vision is recognising the diverse interpretations of STEM and then making a deliberate decision about what STEM needs to be for the context and the perceived needs. Teachers were facilitated in framing STEM by considering the "membership of STEM" and the "practices of STEM". In relation to the former, STEM can be an organisational construct that determines disciplinary membership to "the STEM team". Which subjects, therefore teachers, will be privileged? Indeed, STEM is being massaged into many forms. For example, Australian universities are beginning to refer to STEMM (extra M for medicine); primary schools often move from STEM to STEAM (A for arts) (Gardiner, 2015). Internationally, the STEM agenda is being shaped to attend to national concerns; for example, Korea's integration of the language arts, social studies and other subjects as STEAM is a response to reported low attitudes towards and interest in science (Hong & Hwang, 2013). Is this a problem? We would argue that STEM is only valuable for education if it meets a need. In primary schools it makes sense to include an often marginalised subject such as arts, as long as the learning outcomes respect the disciplinary modes of inquiry and ways of knowing. At universities, medicine is part of the science elite; therefore, membership within STEMM is justifiable.

These variations of STEM illustrate the flexibility of the construct and how it is being mobilised to meet the needs of different groups. This means that there is no common understanding of STEM, and this reinforces the need to be explicit about membership. Allowing this decision to be made at the school level, as in the SS-STEM Program, allows the teachers to work in time and spaces that are possible within their schools. For example, the intended move to STEAM at G College was considered a natural progression once the arts department saw how they could contribute to student learning outcomes in the vehicle design challenge. Looking more broadly, the meaning of STEM needs to be clearly articulated at all levels of government and education as to what is being promoted through the STEM agenda and then what the possibilities are for schools.

Another way to think of STEM is as a way of learning a set of practices that underpin the STEM disciplines, in line with Vasquez's (2015) advocacy for STEM as an approach to learning. According to this logic, the STEM practices, as collated by the SS-STEM Program above, provide the language that defines and differentiates STEM-related learning from other learning. It is possible to teach science and mathematics in a way that does not foster the development of these practices – would this learning be considered STEM? We argue that the strength of the SS-STEM Program is the common language. Teachers were clearly aligning their programmes to the STEM practices, and the language has become a mechanism by which they were empowered to communicate their initiatives to other teachers in their discipline groups, year levels, whole school-teaching staff as well as teachers beyond their school. For example, teachers from both of the case study schools gave invited presentations to teachers and principals participating in the Victorian Government-funded STEM Catalyst program. It was evident from the observation of teachers in this programme that, in order for teachers to develop a comprehensive STEM vision, there was a need for them to be clear about how their programme can be regarded as STEM.

But are these efforts likely to be sustainable in the long term? At the school level, this question rests on a number of factors. The first factor is the degree to which the new approaches are embedded within the school. G College teachers, for example, have secured a place for STEM in their Annual Improvement Plan, and the objectives and aims for a number of subjects in the school reflect the STEM practices that they see as important for improving student participation and engagement with STEM. Ultimately the test of the programme will be increased enrolments in the senior secondary STEM-related subjects, which remains to be seen.

Secondly, sustaining change at the school level requires more than simply providing funding to schools to develop new activities that integrate STEM subjects. History shows "trends" in education, such as integration in the 1980s (LaPorte & Sanders, 1995) and the Science-Technology-Society (STS) focus of the 1960s and 1970s (Yager, 1996), can be fleeting and have little sustained impact on reconfiguring curriculum from the privileged siloed position. However, education is a dynamic profession that responds to the systemic pressures that generate new theories and repackaging of old theories. In the case of STEM, a largely political agenda is being imposed by a national innovation agenda driven by economic shifts from a fossil fuel-based manufacturing economy to a knowledge economy. Such is the drive behind the funding for the SS-STEM Program in the Geelong region. Whether the STEM drive persists is perhaps dependent on the longevity of this political agenda. The challenge, therefore, is to deconstruct and reconstruct STEM in a way that is useful for schools.

Thirdly, we contend that sustainability needs a complex approach to thinking about STEM renewal, not ideologically driven but open to the needs of schools, teachers and students, and in a way that is cognizant of the enablers and blockers in making change. A strength of the SS-STEM Program is that the teachers are part of a community of practitioners coming to use a common language around STEM and where a desire for change is the common denominator. The model of teachers reporting on their projects means that teachers can gain the benefits of insights, ideas and experiences (both positives and negatives) of each other's initiatives and therefore participate in knowledge generation processes. The change exhibited by teachers between the first and second reporting days illustrates how they are also making pedagogical decisions as a result of their developing understanding of what the STEM practices are and how they can be translated into curriculum and learning experiences for their students.

The ultimate message from this research is that locally grown programmes can meet local issues. The STEM vision framework facilitates such development as it recognises that schools must make decisions about how they define STEM, which teachers are involved and how teachers might collaborate to deliver STEM. The suggestion that STEM teaching requires a teacher to make links between multiple subjects (such as B College) is but one model. Teachers working in interdisciplinary teams (such as G College) hold different opportunities but also pose additional challenges, such as timetable constraints. Case studies of different ways of working are valuable for teachers to see the possibilities for making STEM happen. The range of STEM initiatives emerging within the schools shows the power of the STEM vision framework to give autonomy and creative license to teachers. It also illustrates how school constraints drive the direction of such initiatives, such as by focusing mainly on invigorating the maths programme (B College) or by creating meaningful links between mathematics, science and technology (G College).

8.8 Conclusion

The STEM vision framework can facilitate sustained change. The framework can be used in multiple ways by different levels of education and policy. The STEM vision can be used as a vision for an approach to STEM education guiding a professional development project, as in the case of the SS-STEM Program. The framework provides a comprehensive approach to supporting teachers to embed STEM education into their schools and moves away from professional development that simply introduces new activities to teachers. It encourages professional development providers to consider how they frame STEM, what practices and pedagogies they will promote, how they expect teachers to work together to plan curriculum, how teachers will be supported to learn and support the learning of others and how they will infuse their programmes with community and industry links. This comprehensive approach, we believe, will lay the foundation for supporting sustained change in schools.

Schools can use this framework to make decisions about their response to the STEM agenda and how this can be actioned at multiple levels within the school. In addition, an individual teacher can make decisions for their own practice, including their own professional development plan.

The Education Council (2015) describes five actions in the National STEM School Education Strategy, all of which are exemplified by the SS-STEM Program. In keeping with these actions (see p.7), the ultimate goal of the SS-STEM Program is to "increase student STEM ability, engagement, participation and aspiration", although these outcomes will be reported elsewhere. The programme uses a multi-stage professional development programme to "increase teacher capacity and STEM teaching quality" (p. 7). A focus on regional teacher development by focusing on ten schools from the Geelong region through supported professional development, facilitation of a STEM teacher network in the region and capacity building of teachers as change agents within their schools helps to "support STEM education opportunities within school systems" (p. 7). Our programme is developed out of partnerships between local schools and a number of faculties within Deakin University and provides a structure for the boundary work needed to promote links between schools and local industry and companies and organisations; the programme therefore "facilitates effective partnerships" (p. 7). In keeping with the call for an evidence base, this project is underpinned by a programme of research and subject to key performance indicators. In particular, we support teachers in generating and using evidence to promote STEM and advocate for change in their schools. The fact that the STEM vision is in keeping with the actions of the National Strategy (Education Council, 2015) confirms that a multifaceted approach such as this promoted through the STEM vision framework is needed to effect real and sustained change.

Acknowledgements We acknowledge Skilling the Bay and the Victorian State Government for funding this project and also the participating teachers who are challenging themselves and their peers to innovate in response to the fast-sweeping STEM tide.

Appendices

Appendix 1: STEM Vision Template

Framing STEM
1. How is STEM currently framed for your school?
2. How will you frame STEM for your school?
STEM teaching and learning
3. What STEM practices and pedagogies are currently being used or emphasised?
4. What STEM practices and pedagogies will need to be developed?
5. Which components will you focus on in your plan for improving practice?
Teacher learning and leading change
6. What are your STEM learning needs, and what change is needed in your school?
7. How will your learning needs be met, and how will you balance personal exploration and leadership of change in your school? Consider:
(A). What strategy is needed that integrates school processes and support processes for individual teachers?
(B). What resources, knowledge and people are needed to support change?
(C). What evidence will you collect to evaluate the effectiveness of the new curriculum, the use of the STEM practices and your own learning?
(D). How will you build capacity of other teachers in your school? Consider the role of leading teachers and school leadership in supporting and enabling change.
Curriculum development
8. What teacher collaboration is currently used? (integrated or subject specific)
9. What teacher collaboration is needed (integrated or subject specific)? For what purpose?
10. What challenges are involved in developing curriculum in this way?
Community-industry links
11. How are community-industry links currently being used?
12. How might community-industry links be used more effectively?

Appendix 2: Components of Effective STEM Teaching and Learning

1 *The learning environment promotes a culture of value and respect:*

   1.1 The learning environment is characterised by a sense of common purpose and collaborative inquiry.
   1.2 Perseverance and effort are valued and lead to a sense of accomplishment.

2 *Students are encouraged and supported to be independent and self-motivated learners.*
3 *Students are challenged to extend their understandings.*
4 *Students are encouraged to see themselves as mathematical/ scientific thinkers who can use tools creatively:*

   4.1 Students are explicitly supported to engage with the processes of investigation and problem-solving.
   4.2 Students engage in mathematical/scientific reasoning and argumentation.
   4.3 Students are supported to develop an understanding of creative problem-solving and design processes.
   4.4 Students are challenged and supported to develop their own representations as a means of explaining and justifying their understanding.

5 *A range of assessment modes are used to monitor and support individual students' developing understandings:*

   5.1 Individual students' learning needs are monitored and addressed.
   5.2 Learners receive feedback to support further learning.

6 *Learning technologies are used to enhance student learning.*
7 *Content is designed to link with students' lives and tap into/elicit their interests.*

References

Akkerman, S. F., & Bakker, A. (2011). Boundary crossing and boundary objects. *Review of Educational Research, 81*(2), 132–169.

Australian Government. (2015a). *Australian government, industry employment projections 2015 report; ABS perspectives on education and training: Australian qualifications in STEM, 2010-11, Cat. 4250.0.55.005*. Retrieved from.

Australian Government. (2015b). *National innovation and science agenda: Welcome to the ideas boom*. Retrieved from Canberra.

Australian Industry Group. (2013). *Lifting our science, technology, engineering and maths (STEM) skills*. Sydney, Australia: AIG.

Bybee, R. W. (2013). *Case for STEM education: Challenges and opportunities*. Arlington, VA, USA: National Science Teachers Association. Retrieved from http://www.ebrary.com

Bybee, R. W. (2015). *The BSCS 5E instructional model: Creating teachable moments*. Arlington, Virginia: NSTA press.

Clarke, D. J. (2015). Comparative research in mathematics education: Boundary crossing and boundary creation. In K. Beswick, T. Muir, & J. Wells (Eds.), *Proceedings of the 39th conference of the international group for the psychology of mathematics education* (Vol. 2, pp. 169–176). Hobart, Australia: PME.

Dugger, W. E., Jr. (2010). *Evolution of STEM in the United States*. 6th biennial conference on technology education research at gold coast, Australia on December 8–10, 2010.
Economist, The (2012, April 21). The third industrial revolution. *The Economist*.
Education Council. (2015). *National STEM school education strategy: A comprehensive plan for science, technology, engineering and mathematics education in Australia*. Canberra, Australia: Education Council.
Ellingson, L. L. (2009). *Engaging crystallization in qualitative research: An introduction*. Thousand Oaks, CA: Sage.
Gardiner, B. (2015). Picking up STEAM: How the arts can drive STEM leadership. CIO Australia. Downloaded http://www.cio.com.au/article/585493/picking-up-steam-how-arts-can-drive-stem-leadership
Hobbs, L., Campbell, C., Chittleborough, G., Herbert, S., Jones, M., Redman, C., Kenny, J, Gilbert, A. (2015). STEPS interpretive framework. Available http://www.stepsproject.org.au
Honey, M., Pearson, G., & Schweingruber, H. (2014). *STEM integration in K-12 education: Status, prospects, and an agenda for research*. Washington, DC: National Academies Press.
Hong, S. Y., & Hwang, Y. H. (2013). Development of 3D simulation-based M&S Education Platform for smart learning. *Issues in Information Systems, 14*(2), 225–323.
Kennedy, J., Lyons, T., & Quinn, F. (2014). The continuing decline of science and mathematics enrolments in Australian high schools. *Teaching Science, 60*(2), 34–46.
LaPorte, J., & Sanders, M. (1995). Technology, science, mathematics integration. In E. Martin (Ed.), *Foundations of technology education: Yearbook #44 of the council on technology teacher education*. Glencoe/McGraw-Hill: Peoria, IL.
Morin, O., Simoneaux, L., & Tytler, R. (2015). *Engaging with socially acute questions: Development and validation of an interactional reasoning framework*. Paper presented to the meeting of European Science Educarion Research Assoication, September 31 August to 4 September 2015, Helsinki.
Office of the Chief Scientist. (2014). *Science, technology, engineering and mathematics: Australia's future*. Canberra, Australia: Australian Government.
Office of the Chief Scientist. (2016). *Australia's STEM workforce: Science, technology, engineering and mathematics*. Canberra, Australia: Australian Government.
Professional Australia. (2015). Women in STEM in Australia. Professionals Australia, downloaded http://www.professionalsaustralia.org.au
PWC. (2015). *A smart move: Future-proofing Australia's workforce by growing skills in science, technology, engineering and maths (STEM)*. Sydney, Australia: PricewaterhouseCoopers.
Scientix. (2014). *The community for science education in Europe*. Brussels, Belgium: European Schoolnet. Retrieved from http://www.scientix.eu
Stohlmann, M., Moore, T. J., & Roehrig, G. H. (2012). Considerations for Teaching Integrated STEM Education. *Journal of Pre-College Engineering Education Research, 2*(1), Article 4. https://doi.org/10.5703/1288284314653
Thompson, S., De Bortoli, L., & Buckley, S. (2013). *PISA 2012: How Australia measures up*. Camberwell, England: Australian for Educational Research.
Tytler, R. (2007). Reimagining science education: Engaging students in science for Australia's future. *Australian Educational Review*, 51.
Tytler, R, Osborne, J. F., Williams, G., Tytler, K., & Cripps Clark, J. (2008). *Opening up pathways: Engagement in STEM across the primary-secondary school transition*. Retrieved from Canberra:
Tytler, R., & Waldrip, B. (2001). *Effective teaching and learning in science: Expanding the territory*. Position Statement 4; Working Paper of the Science in Schools Research Project.
Vasquez, J. (2015). Beyond the acronym. *Educational Leadership*, 11–15.
Williams, J. (2011). STEM education: Proceed with caution. Digital and technology education. *An International Journal, 16*(1), 26–35.
Yager, R. E. (1996). *Science/technology/society as reform in science education*. Albany, NY: SUNY Press.

Chapter 9
The Importance of Diagrams, Graphics and Other Visual Representations in STEM Teaching

Peter Gates

Abstract In this chapter I look at the way we think of communication and suggest that there is an over-reliance upon linguistic and textual modes at the expense of visual and spatial modes of communication. I argue that schools fail to grasp the significance of the visual nature of communication and the implications for learning within STEM subjects. After making an argument for the importance of visuospatial forms, I provide an extensive review of the cognitive and psychological literature covering various key aspects of visualisation and how it relates to teaching STEM subjects in early secondary education. It is likely that much of this will be novel to STEM teachers yet provides us with new possibilities for opening up classroom pedagogy.

Keywords Visualisation • Spatial • Diagrams • Graphicacy • Mental representation

9.1 Introduction

> Our research has produced convincing evidence that presenting a verbal explanation of how a system works does not insure that students will understand the explanation. In our search for ways to help students understand scientific explanations, we have come to rely increasingly on what has been called multimedia learning, through presenting explanations visually as well as *verbally*. Multimedia learning occurs when students receive information presented in more than one mode, such as in pictures and words. In recent years, the once near monopoly of verbally based modes of instruction has given way to the hypothesis that meaningful learning occurs when learners construct and coordinate multiple representations of the same material, including visual and verbal representations. (Mayer, 1997, p. 1)

One of the very first things a newborn learns is to recognise faces. They have no language and know little about how the world operates, but the visual cortex kicks

P. Gates (✉)
University of Nottingham, Nottingham, UK
e-mail: Peter.Gates@nottingham.ac.uk

R. Jorgensen, K. Larkin (eds.), *STEM Education in the Junior Secondary*,
DOI 10.1007/978-981-10-5448-8_9

in with a vengeance. Within 2 years as the vision continues to develop towards 20/20 vision, we can not only recognise faces but recognise ourselves in photographs, find pictures in books, recognise and match simple objects and find our way around familiar places. A child can build towers, drink from a cup and look towards a sound. Within a very short time, an infant becomes an accomplished engineer, scientist and mathematician – all before developing language. It must be a travesty then that a mere 5 years later, when the infant gets to primary school, they find themselves bombarded with text and talk. However, when a child encounters science and technology, they are provided with frameworks for understanding and manipulating their environment. Encountering mathematics provides frameworks for making sense of the world. So the importance of learning through physical engagement immediately encapsulates a visual element through which the learner monitors, evaluates and hypothesises the world.

We live in a world that is increasingly influenced by graphical and visual images and by greater use of visual means of communication. Surely no one can fail to be impressed with Edward Tufte's work on visual display and explanations (Tufte, 1990, 1997, 2001). In a series of stunningly innovative displays, he offers Byrne's presentation of Euclid's element through shape, colour and orientation (1990, p. 84–87); Snow's identification of the cause of a cholera outbreak by mapping cases (1997, p. 27–37); and Marey's description of Napoleon's devastating losses in the Russian campaign with a spatial time series graphic (2001, p. 40–41).

However, the one prominent feature of school mathematics is a dependence on *language* and *textual* communication, largely to the exclusion of other modes of communication – most significantly, the visual. Whilst this may be partly true and to a lesser extent in science, it is not the case in technology or engineering. This derives from technology and engineering dealing directly with artefacts and "real" objects and structures, rather than the much more abstract concepts that are dealt with by mathematics – and which do only exist in the head. There are strategies that mathematics and science can appropriate from technology and engineering, in particular broadening the modes of presentation to include the manipulation and representation of physical objects and processes – too often missing from mathematics lessons.

However, these modes of presentation are processed quite differently in the brain with significant ramifications for classroom practice. It is recognised that *spoken language* and *text* are both characterised as one dimensional, sequential and sentential (Crapo, 2002) and that they are processed in the auditory centres of the brain before temporary storage in working memory (Card, Moran, & Newell, 1983). This contrasts with *visual images* which are multidimensional requiring processing in the visual cortex before temporary storage in working memory. Consequently, if our STEM teaching has rested upon the sequential and auditory channels, we need to rethink our approach if we are to adopt more multidimensional approaches to teaching and learning.

This is vital as, whilst the connection between visualisation and mathematical and scientific skills is complex and contested, much evidence points to positive associations between some aspect of visualisation and spatial skills and some elements of mathematical and scientific competencies (Cheng & Mix, 2014) and that visualisation and imagery are central to understanding and reasoning (Arcavi, 2003; Whiteley, 2004). One problem in STEM currently is the kids just don't learn the stuff:

> Why is it that a student can read or listen to every word of a scientific passage, including a cause-and-effect explanation, and yet not be able to use that information to solve problems? Our research has produced convincing evidence that presenting a verbal explanation of how a system works does not insure that students will understand the explanation. (Mayer, 1997, p. 1)

9.2 The Importance of Visuospatial Skills

How we understand the visual mode of communication is complex and involves a range of diverse elements and cognitive processes, yet evidence points to possible connections between future cognitive development and the development of early spatial skills (Kersch, Casey, & Young, 2008). However, this forces us to consider the distinction between how we think of "visualisation" (or its derivatives) and how that is distinct from the "spatial". There are a number of ways in which we might understand the "spatial" in STEM subjects. For example, *spatial relationships* might refer to the perception or proprioception of spatial objects, whereas *spatial reasoning* would refer to a process of integration of elements into a logical connection. *Spatial visualisation* might refer to the capacity to manipulate and transform mental images, to see images within others and to construct and dissect objects.

Despite this, our understanding of the nature and role of images in educational contexts is still limited (Postigo & Pozo, 2004). Part of this is because of the disproportionate prioritisation of much research into other modes such as language, textual or non-iconic forms. What we do know is that much of the engagement with visual material in schools is "superficial" (Postigo & Pozo, 2004, p. 624). It might be surprising that there is still a case for arguing that this is important since as Weidenmann (1987, p. 157) said 30 years ago: "the empirical evidence is so convincing". He goes on:

> Probably no other instructional device leads to more consistently beneficial results than does adding pictures to a text… There can be no doubt that pictures combined with texts can produce strong facilitative effects on learning and retention. (Weidenmann, 1987, p. 158)

As I suggested above, I would question the extent to which studies of the use of diagrams, illustrations, visuals, etc. as pedagogical strategies within mathematics and science education has influenced the culture of classrooms. The fragmentation of S-T-E-M education (rather than STEM) with little pedagogical or curricular crossover results in lost opportunities for greater synergy between forms of presentation and representation.

Often diagrams and visuals in textbooks (most notably in mathematics) tend to be little more than "wallpaper" offering merely some irrelevant visual distractor. Little thought goes into the creative epistemological design of text-visual components and how these contribute to the construction of mental models. There also appears to be a lack of attention given to the construction and manipulation of these mental models especially those drawing on visual-spatial processes. Dawe puts as an imperative for teachers to "consciously link visual images, verbal propositions and memories of activities, involving the manipulation of physical objects" (Dawe, 1993).

Specific visual skills useful in such activities as science and engineering might include folding, cutting and rotating (Nordina, Amina, Subaria, & Hamida, 2013) – and such skills are easily incorporated as direct strategies into STEM lessons. Other visual skills useful in STEM learning might include explicit constructing a mental image or mental model of a scientific or mathematical artefact or process, developing and using a mental representation, constructing representative diagrams and describing (representing) images and models and mental rotation – and it goes on – where the focus is on the *representation itself*, rather than the concept.

Whilst visualisation and mental imagery are cognitive processes that are evident from birth, there would appear to be different levels of individual facility; yet there is little evidence of explicit instruction at school level. One question is whether these skills are actually open to enhancement through classroom activity – but there is evidence of gains after explicit training in visualisation (Lord, 1985, 1990). The need to be proficient in visualisation is important in many fields such as engineering (Olkun, 2003; Strong & Smith, 2002), medicine and construction; in fact it may be difficult to find a field of employment where it is not important. The lack of direct instruction in visual facility in school mathematics is therefore worrying.

Lowrie and Diezmann (2005) developed the graphical literacy in mathematics test and subsequently studied elementary school children in Australia finding their performance was not particularly strong. They argued for much more explicit teaching of reading, producing, understanding and decoding information graphics. This poses problems not only for pedagogical resources but also for current test items where the forms of graphics used may detract from the underlying mathematics being tested producing inaccurate results (Lowrie & Diezmann, 2009).

There are further claims of gender and class differences in spatial skills (Linn & Peterson, 1985) and age effects (Bishop, 1978). Linn and Peterson argue there is evidence of males using a holistic approach, with females taking an analytical approach to visuospatial methods. Bishop argues that there is an interesting developmental process moving from topological, through 2D to increased sophistication in 3D. The widespread use of computer gaming by young people may also be naturally enhancing their visual and spatial ability. However as with all developmental processes, environment and social factors will play a significant part. This raises a question of what pedagogical approaches and tasks best foster a growth in visual and spatial skills and what place information technology holds in that process. More particularly, we might consider how the playing of computer games can enhance understanding of STEM through the visual activity inherent in engagement (Beck & Wade, 2006; Gee, 2007).

9.3 Achievement and the Visual

The prospect of teaching fractions, yet again, to a class of low-achieving adolescents strikes abject frustration in mathematics teachers throughout the world, and the same will be true for the other STEM subjects. Yet the reality is many young people fail to understand even basic mathematical and scientific concepts. How we get to the position after 9–10 years of formal compulsory schooling that we are still trying to convince many children that 1/4 = 2/8 is nothing short of an international scandal. Of course it is not just fractions that we fail to teach; the list goes on and on covering much of the mathematics and science curriculum. Worryingly, this is after decades of curriculum reviews, policy changes and millions spent on research.

Indeed, learning fractions provides one of the areas within mathematics curriculums around the world that persist in posing difficulties for young children (Carpernter, Coburn, Reys, & Wilson, 1976; Ellerton & Clements, 1994; Neime, 1996), although in some countries (Singapore and South Korea) international comparisons – albeit notoriously unreliable – suggest children do much better (Mullis, Martin, Fo, & Arora, 2012). One argument for the difficulty is the dependence on particular models over others (Zhang, Clements, & Ellerton, 2015), notably the area model (Cramer, Post, & delMas, 2002) and the widespread incorporation of the dubious pizza as a model of fractions. In both these cases, the visual appears very much as an afterthought rather than a carefully designed intervention aimed at supporting robust mental representations.

By looking at *visualisation* (and visual *reasoning*) as distinct from "spatial", we might see the first as a feature of cognition and the second as a feature of the physical world. Addressing the visual does not mean doing more geometry. Rather it means looking at how mental models of concepts are held and manipulated within an increasingly complex grasp of STEM subjects.

Within mathematics and science, this distinction would connect most obviously into particular concept areas (such as geometry, mechanics, measurement, graphical representation) but also into problem-solving as well as into perception and organisation of logical reasoning. Much research has been undertaken within the area of geometrical and spatial understanding and awareness. However, measurement (in 1, 2 and 3 dimensions) also provides a context for spatial relationships to be built and developed. The area of graphical representation also necessitates a use of space to represent and manipulate relationships as well as to draw on multiple representations. Spatial visualisation would be drawn on when presented with some object and needing to see it from a different perspective (rotation, scaling, inversion, etc.). In this way spatial cognition is contrasted with a more text-/linguistic-based mode of cognition and information processing, using analytical logico-deductive reasoning (Baddeley, 1998).

To *reason visually* – to use spatial representation to demonstrate a logical process – is maybe something not yet at the forefront of teachers' conceptions of STEM learning. Hence the use of diagrams and visuals may currently play a very insignificant role in classroom practice. Diagrams, graphics and other visuospatial representations have been an interest within mathematics and science education for some time but are becoming of more relevance now given the increasingly visual nature of our communications mediated through technology.

It is possible that facility with using graphics (contrasted with "graphs") might contribute to successful learning (Schnotz, Picard, & Hron, 1993). Visual information does have a number of advantages for the process of learning – notably in illustrating abstract concepts and organising complex information (Schnotz, 1993). Yet it remains questionable whether a concern for visual, mental or multiple representations has yet influenced school pedagogy and curriculum sufficiently for teachers to adopt a more multimodal approach. This may be due to the lack of an overarching coherent theoretical framework within which such work can be situated. However, it is also likely to be influenced by the culture of STEM teaching and specifically the preponderance of propositional lexical-textual forms of argumentation. Furthermore, the culture of teaching might make it more problematic incorporating a clear framework for visual information which draws on different theoretical ideas and frameworks. Dreyfus (1991) argues there are in fact two specific issues that need understanding – (a) the difficulty of visualisation and (b) the status of visualising within teaching identified by the low status accorded to visualisation by both pupils and teachers (Dreyfus, 1991, p. 34; Eisenberg & Dreyfus, 1991) which I discuss later.

9.4 Representation

The importance of mental representations for a learner within STEM is widely acknowledged as ways of constructing mental models of entities or processes. The process of instruction is surely to support that construction, but this needs to be done with some understanding of "cognitive architecture":

> Visualisation extends working memory by using the massively parallel architecture of the visual system to make an external representation function as an effective part of working memory. (Crapo, Waisel, Wallace, & Willemain, 2000, p. 220; citing Larkin & Simon, 1987)

Visual representation can thus reduce the cognitive load during engagement (Clark, Nguyen, & Sweller, 2006). Specifically, when used in a supportive way with text (as "spatial text adjuncts"), visuals, and diagrams, etc. can help to:

- *Represent* the text, providing additional nonverbal memory prompts
- *Organise* and provide structure and form to text
- *Interpret* otherwise complex text
- *Transform* text into pictorial images that can be stored more efficiently (Robinson, 2002, p. 1)

This fourfold typology can be incorporated into teaching quite easily though does require us to rethink our materials and tasks. A further typology of representation was proposed by Lesh, Post, and Behr (1987) which included five elements: static pictures, manipulative models, written symbols, real-life situations and spoken language. As with any typology, this will have its limitations but offers a structure that might be useful for classroom application. Whilst it offers a typology of external representations rather than a description of internal cognitive function, it can provide teachers with a form of classification of forms of presentation they can use to analyse the balance between different modes they use in their classrooms. This would likely demonstrate that spoken language and written symbols dominate classroom discourse. However, it can also help us to ask whether static pictures match student understanding in operationalising mathematical and scientific concepts. The interplay between diagrams, students' visual models and their representation of concepts is a very complex relationship (Anderson-Pence, Moyer-Packenham, Westenskow, Shumway, & Jordan, 2014). However, graphics and text are argued as *different* modes of representation and thus play *different* roles in fostering understanding (Schnotz et al., 1993):

> The essential point here is not only that two codes are better than one, but rather the combination of two qualitatively different principles of representation which complement one another and make possible a high efficiency of human cognition. (Schnotz, 1993, p. 248)

Studies of graphical and textual use from a psychological perspective have already provided us with evidence that each are processed differently in the brain (Schnotz et al., 1993). Whereas text rests on symbols, drawing on propositional logics, diagrams draw on more spatial forms or arrangement, and in this way they can be seen to fit into Paivio's dual coding theory (Paivio, 1976, 1978, 1986) and Baddeley's two-phase short-term memory model (Baddeley, 2003; Repovš & Baddeley, 2006). In working on text, a symbolic propositional representation is constructed on the basis of the *semantic* structure, which then constructs relationships between elements that goes on to construct a mental model. With graphical and diagrammatic representations, these are processed first as a *visual configuration* which then constructs an analogue mental model:

> In other words, text and graphics are complementary sources of information insofar as they contribute in different ways to the construction of a mental model. A text triggers the formation of a symbolic propositional representation which then serves as a basis for the construction of an analogue mental model. Conversely, a graphic can be considered as an external model which enables a more direct construction of a mental model via an analogue visual representation. (Schnotz et al., 1993, p. 183)

In this way graphics can be understood as closer to the structural form of an intrinsic mental model (Schnotz, 1993) representing some concept or process. Text on the other hand does not carry such structural features.

For too long, classroom practices on the use of graphics and visuals have rested upon outdated ideas drawn on an absence of knowledge of cognitive modelling. Where representations are concerned, it is almost "the more the merrier – sort of", as Ainsworth explains:

> Research on learning with representations has shown that when learners can interact with an appropriate representation their performance is enhanced. Recently, attention has been

> focused on learning with more than one representation, seemingly predicated on the notion 'that two representations are better than one'. Yet, as research on learning with multiple external representations (MERs) has matured, it is increasingly recognised that the issue is not whether MERs are effective but rather concerns the circumstances that influence the effectiveness of MERs. … Schnotz (2002; Schnotz & Bannert, 2003) focuses not on pictures and text per se, but on depictive (iconic) and descriptive (symbolic) representations. In this approach, mapping happens at the level of mental model construction and what results is not an integrated representation but complementary representations that can communicate with one another. (Ainsworth, 2006, p. 183-4)

Hence, as I argue elsewhere in this chapter, text and visual (or depictive and descriptive) representations are not incorporated into some complex mental model but are held and enacted *separately*.

9.5 Typology of Diagrams

Larkin and Simon (1987) suggest three main reasons why diagrams "can be superior to verbal descriptions":

- Diagrams can group together information that is used together thus avoiding textual searching.
- Diagrams use location to group information avoiding a need to match symbolic labels.
- Diagrams automatically support as large number of perceptual inferences (Larkin & Simon, 1987, p. 98).

Largely then for Larkin and Simon (1987), diagrams have a distinct advantage when considering the necessary computational requirements, but users need to know the way to use them. One area of confusion is identifying a typology of visual/diagrammatic forms which is both a theoretical question (i.e. the form, structure and semantics of visual representations) but also an empirical one (i.e. what forms are recognised and used in classroom and teaching materials). Hittleman (1985, pp. 32–33) has studied the role of illustrations and explored pedagogical implications for teachers. He offered a typology of six types of illustrations in science texts, which again might be useful for STEM teachers to incorporate in teaching:

- *Photographs*
  Accurate depictions of a scene or object
- *Realistic drawings*
  Generally related to the object
- *Representational drawings*
  Less accurate than realistic such that some elements or features are highlighted
- *Diagrams*
  Representational but drawing on symbolic representations identifying relationships that may focus on particular elements of a whole
- *Charts, graphs and figures*

Information organised spatially, representing relationships in various ways

- *Maps*
  Representation of some physical reality with some topology or spatial structure

It seems reasonable for such a typology to apply to all STEM texts and resources, though no formal analysis has taken place. Importantly "when they learn to read illustrations, [children] need to understand the various signals illustrators use to convey meaning" (Hittleman, 1985, p. 33). The sixfold typology above provides several different *signal systems*, and inducting children into these rules and conventions is likely to put them in a better position to use and interpret various modes of communication.

How do visuals convey meaning and how can we help children interpret visuals? Different types of information are stored and processed differently in the brain – some as image-like structures (Kosslyn, 1980); as a result certain styles of material are rendered easier to learn by representing in graphic form presenting concepts available for simultaneous processing. Visuals can convey meaning, but not arbitrarily so; rather through conventions and sometimes specific systems of logic (Winn, 1987) and the exploitation of use of space, arrangement, structures, linkages and graphic forms convey meaning differently and learners do need to be introduced to the conventions. A nice example of the use of visual systems is in *Geometry in Figures* (Akopyan, 2011). This book consists of some 130 pages packed with diagrams with hardly any words in the whole book. It is the job of the reader to derive the geometrical proof and statements without any text. The following is a nice example (Akopyan, 2011, p. 7). The grammar has to be derived but is relatively simple.

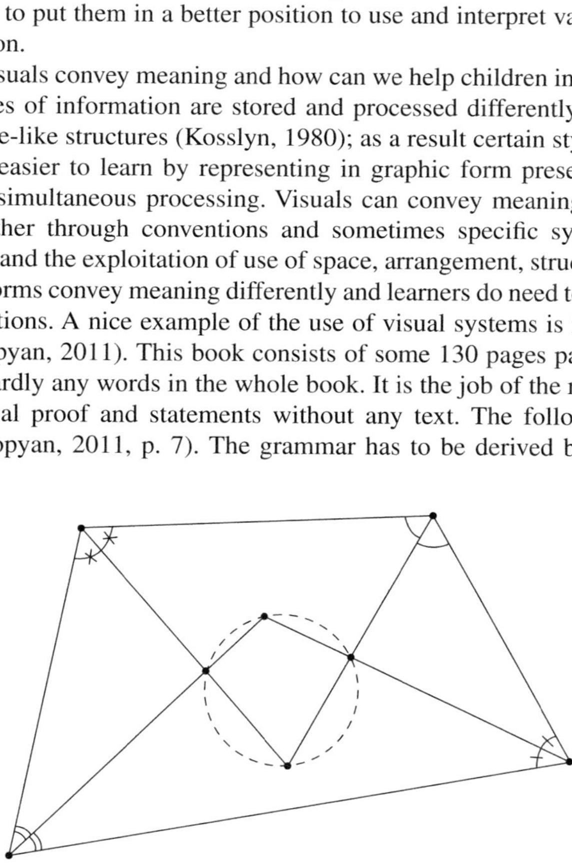

You might at this stage try to put into words the property (or theorem) that this diagram demonstrates – and how this is extracted from the visual grammar (I provide the solution at the end of the chapter).

Here is another example from an IKEA manual on how to put together a chair. The visual grammar here is very simple, depending only on the use of a  and a **X**.

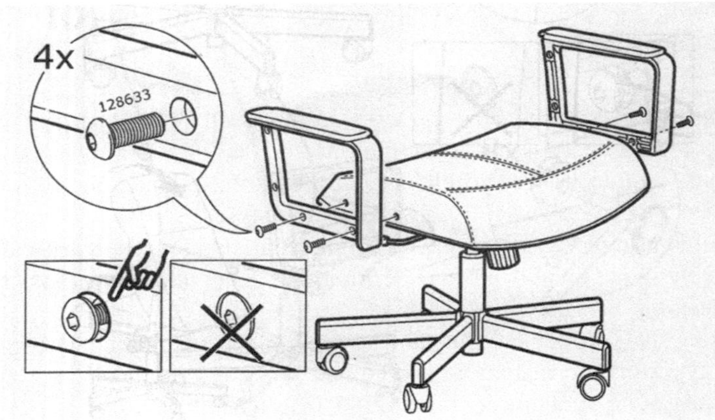

9.6 Text and Graphics

Diagrams do not "speak for themselves, but are read" (Roth, 2002, p. 20). Consequently, as with all texts, they are interpreted within a semiotic framework of meanings. Brna, Cox, and Good (2001) have argued that using and reasoning with diagrams depends on the specific task, the semantic properties of the diagram and the prior knowledge of the learner (p. 116). So a graphic might not just be an advantage but might also pose difficulties because a learner needs not only to have some grasp of the underlying concept but also be aware of the semantic visual components of diagrams which need decoding. These may be the deeper source of problems rather than misconceptions or lack of understanding of mathematical and scientific concepts (Roth, 2002).

Schnotz's study suggested that contrary to previous claims that graphics were only a secondary subordinate representation, successful learners used graphics and text in mutually constructive (adjunct) ways and that when supported and elaborated by text, graphical and visual forms contribute to more robust and effective learning. Stone and Glock (1981) also examined a very specific context of university undergraduates using simple perspective line drawings and directions for assembling a model. However, what their work suggests is that there is certain information that is best processed visually. The work of Stone and Glock (1981) illustrates a situation where studies look into very specific psychological contexts, but which provide only very limited support for the design of secondary classroom approaches. They do suggest complementarities come partly through the use of text and visual *together* to resolve ambiguities in each modality.

Mayer and Gallini (1990), in a study of illustrations in learning science, conclude that illustrations or diagrams are effective when both the text and illustrations are "appropriate" for the task. In their case, this was where text was explanative (rather than descriptive or narrative) and diagrams represented both the structure and the dynamic of the instruction. In this way we can see the structure/object and process/dynamic elements that run through mathematics and science. Their work raises the possibility that illustrations should explicitly help the learner to construct workable mental models (Mayer & Gallini, 1990). Interestingly their work raises another possibility – that explanative illustrations which embody the processes behind a mental model are more effective in engendering understanding in problem-solving, but not in verbal recall – specifically with "low prior knowledge students". Overarching this work is that the visuals are rather more than simple pictures.

In reviewing the literature on illustrations, Carney and Levin (2002) argued that "carefully constructed text illustrations generally enhance learners' performance" (p. 5) and offered ten commandments (from Levin, Anglin, & Carney, 1987) for teachers using illustrations (pp. 20–22):

1. Pictures shalt be judiciously applied to text, to remember it wholly.
2. Pictures shalt honour the text.
3. Pictures shalt not bear false fitness to the text.
4. Pictures shalt not be used in the presence of "heavenly" bodies of prose.
5. Pictures shalt not be used with text craving for images.
6. Pictures shalt not be prepared in vain.
7. Pictures shalt be faithfully created from generation to generation.
8. Pictures shalt not be adulterated.
9. Pictures shalt be appreciated for the art they art.
10. Pictures shalt be made to perform their appropriate functions.

Eitel and Scheiter (2015) present a review of 42 studies into the sequencing of text and illustration, concluding that the complexity of the content alone should determine the sequence of presentation. Scaife and Rogers (1996) discuss diagrammatic representations arguing that research "supports for the important role of diagrams as external memories, enabling a picture of the whole problem to be maintained simultaneously whilst allowing the solver to work through the interconnected parts" (pp. 193–194). Though it is apparent that cognitive science still has not provided a clear account of the neurological processes underway.

The importance of diagrams has been illustrated by various studies in cognitive psychology (e.g. through the work of Hegarty, 1992) and in mathematics education (see Clements, 1983; Lean & Clements, 1981; Presmeg 1986). These suggest a significant connection between diagram use and facility with problem-solving but only where the diagrammatic form is *representational* or *schematic* – where the visual in some way represents or models the mathematical or scientific concepts, rather than merely *pictorial* (Garderen, Scheuermann, & Poch, 2014; Hegarty & Kozhevnikov, 1999; Stylianou, 2013).

Diagrams may be presented to learners in textual material and in tasks, yet in our very visual world, images exist all around us, resulting in young people being constantly subjected to a huge array of still and moving images, icons, photographs and representations. A very common form of these within mathematics education includes pictorial images intended to provide interest or make the material more attractive. However, the studies by van Garderen et al. (2014) and Hegarty and Kozhevnikov (1999) suggest this might have some significant yet unintended disadvantages particularly when learners are encouraged to think of diagrams as pictures rather than engaging with visual representations as an epistemological and pedagogical device. This would seem to be the space where most difficulties emerge around facility with visual representations, partly because the expectation is for a diagram to be non-representational.

The benefits of using diagrams are supported by a number of elements deriving from the cognitive engagement with multiple representations and the specific affordances offered by a visual form – specifically, identifying the structure of a problem and the interconnections between elements. Diagrams also provide a means of communication between learners and between learner and teacher. Whilst this might suggest a somewhat static representation, another benefit is through illustrating the mechanism behind a problem or the problem-solving process (see Stylianou, 2002, 2010, 2013; Stylianou & Silver, 2004). Such a use as this might be termed *justificatory* (Stylianou, 2013).

However, Scaife and Rogers (1996) working within a cognitive science framework identified "a fragmented and poorly understood account of how graphical representations work, exposing a number of assumptions and fallacies" they argue for "research into graphical representations that is based on an analysis of interactivity and, thus, considers the relationship between different external and internal representations" (p. 210).

One hypothesis is the "perceptual chunking hypothesis" whereby skilled technicians are able to grasp whole chunks of circuit or representational diagrams as one entity, in the same way skilful chess players can "see" the whole board (Egan and Schwartz, 1979, p. 149).

Diagrams can thus be used as a record, as a means of communication, as a tool for doing mathematics and science or working on problems and as a device for conceptual development through mental representational forms. However, a root purpose for presenting a visual form is to offer representation in different forms such learners can discern "the common elements in many different embodiments of the same mathematics" (Dienes, 1960, p. 8), so through this the learner can "become aware of the essential sameness of the structure" (p. 42).

9.7 Graphical Understanding

Apart from geometrical understanding, one other clear area of the STEM curriculum that weaves visual processing into conceptual development is the use of graphical representations. Often the process of decoding of visuals or graphics is overlooked, yet there is evidence of the complexity posed by presenting visual graphics to learners. Graphical representations are not only functional representations within mathematics but also exist in the world outside the classroom. They are used for a variety of reasons (see the work of Tufte for example) and subject to a variety of rule sets and conventions. Graphs have particular visual properties, but learners do not come without previous readings. Visual representations are now so widespread and profound that we are no longer aware of them (Roth, 2002). Roth offers a semiotic approach to understanding graphical literacy rather than from the stance of them as mere representational or cognitive forms based on decoding separate elements. He argued for problematising the need to structure the visual field and to repeatedly shift back and forth between sign and referent (p. 4). However, in use, competent graph decoders "look at graphs and, without hesitation, see in each wiggle corresponding state in the world". One comes into "symbolic contact" with the phenomenon (Ochs, Gonzales, & Jacoby, 1996):

> Thus, when readers are very familiar with a sign system and the things it refers to, signs themselves become transparent. Readers no longer think of words, or parts of a line curve, but go directly to the things they know them to be about. This transparency is so pronounced that readers forget the distinction between sign and referent; they confuse the map with the territory (Bateson 1980, Foucault 1983). A graph simply provides the material ground that organizes competent reading; but the graph also requires competent reading to be understood and a familiarity with the situations or type of situations to which the graph refers. It is in that disappearance of the sign, the leap beyond the material basis of the text, that reading achieves its social character (Livingston 1995). (Roth, 2002, p. 6)

However, when learners are inexperienced, we can expect to see the equivalent of spelling out words, looking literally rather than looking for the meaning of the whole. Misreading a distance-time graph as a trajectory, for example:

> Texts, therefore, do not speak for themselves, for they depend on a reader's familiarity with the content domain and cultural conventions regulating the signs that make up the text. (Roth, 2002, p. 15)

In working on graphs, there are several stages one goes through (Carpenter & Shah, 1998, p. 76; Shah & Hoeffner, 2002, p. 66), namely, looking at:

(1) The characteristics of the visual display;
(2) The viewer's knowledge of graphical schemas and conventions;
(3) The content of the graph and the viewer's prior knowledge and expectations about that content.

This final stage refers not only to decoding elements of the graph legends and label but being able to imagine and live in the context (Carpenter & Shah, 1998):

> Processing a geographical map or a graph involves decoding this information by learning the codes underlying it. But in addition to this syntactic component, a further requirement is knowledge about the represented content (for example, geographical, in the case of maps) that is involved in drawing inferences, which means higher-level interpretation. In other words, interpreting a map or a graph involves describing (saying what we see, observing its distribution, or following its profile), but also explaining the reason for the configuration or profile, and the degree of elaboration will depend on the subject's knowledge. (Postigo & Pozo, 2004, p. 627)

Much material in graphicacy ignores the physiological processes in encoding graphs which are described by Shah and Hoeffner (2002) – what the eyes look for and how the brain processes what the eyes see. Specifically, they compared focussed pattern spotting with an integrative model of interpretation, finding the integrative model best supporting their data:

> Elementary-aged students are perhaps the most influenced by a graph's content. One common error is that viewers interpret abstract representations of data as an iconic representation of a real event (Bell and Janvier, 1981; Janvier, 1981; Leinhardt et al., 1990; Preece, 1990). For example, students might misinterpret a graph representing the speed of a racecar to mean the position of the racecar on a track (Janvier, 1981). This error is particularly common in contexts for which there is an obvious iconic interpretation, usually when the graph is meant to represent change (such as growth, speed) and the concrete interpretation is the value on some dimension (such as height instead of growth, location instead of speed). Although young graph readers (until around fifth grade) make this error frequently, minimal graphing instruction helps viewers overcome this error (Leinhardt et al., 1990). (Shah & Hoeffner, 2002, p. 61)

Shah and Hoeffner (2002) offer several implications for teaching graphical literacy: translating between representations, explicitly focussing on the links between visual features and meaning and making graph reading metacognitive.

9.8 Limitations of Diagrams

However, do not let us run away with the idea that visuals and diagrams are the new Jerusalem. They do have their limitations, as a pointed out by Satoy (2004) and Tversky (2010):

> [Gauss] was aware that pictures in mathematics were regarded with some suspicion during this period. The dominance of the French mathematical tradition during Gauss's youth meant that the preferred pathway to mathematical world was the language of formulas and equations. [...] For several hundred years, mathematicians had believed that pictures had the power to mislead. After all the language of mathematics had been introduced to tame the physical world. (Satoy, 2004, pp. 69–70)
>
> Mismatches between the natural interpretations of lines as paths or connections and the intended interpretations in diagrams turn out to underlie difficulties understanding and producing certain information systems designs. ... the visual trumps the conceptual and misleads. (Tversky, 2010, pp. 21–22)
>
> Another important role for visualizations of thought is to clarify and develop thought. This kind of visualization is called a sketch because it is usually more tentative and vague than a diagram. Sketches in early phases of design even of physical objects, like products and buildings, are frequently just glyphs, lines and blobs, with no specific shapes, sizes, or distances (e.g. Goel, 1995; Schon, 1983). (Tversky, 2010, p. 25)

There is also work from a philosophical, epistemic standpoint examining the visual within mathematics and science. One standpoint is to argue that visual representations and visualising are not the space within which mathematics takes place. An alternative viewpoint has been argued by Giaquinto for some time (Giaquinto, 1992, 1993, 1994, 2007) that visualisation has a part to play in the construction, argumentation and comprehension of mathematics but also that diagrams too have a significant role. In some cases, the visual and the diagrammatic present some overlapping superfluity, as the following example from a GCSE paper shows:

(b) In the diagram, A and B are points on the circumference of a circle, centre O. PA and PB are tangents to the circle.

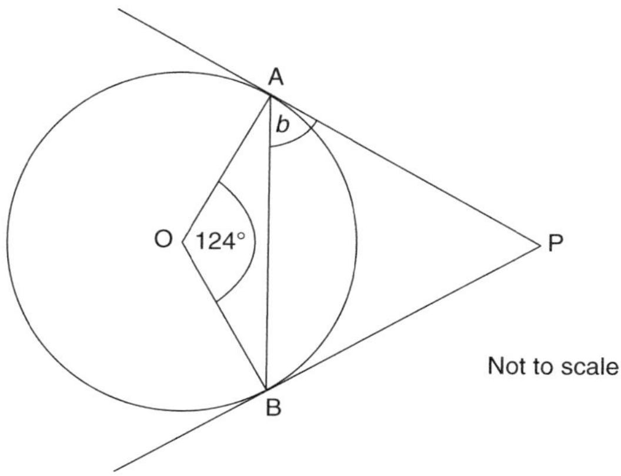

Calculate angle *b*.

Do we need the text *and* the diagram, which both give the same information? Arguably yes because the diagram makes the location of both angles 124° and b° easier to describe. The role played by the common "not to scale" is important in a pedagogical as well as epistemological point of view. A diagram has the appearance of fixing certain properties. Take the following for example:

"Let ABC be an isosceles triangle"

You will no doubt have an image already, and it may be something like this:

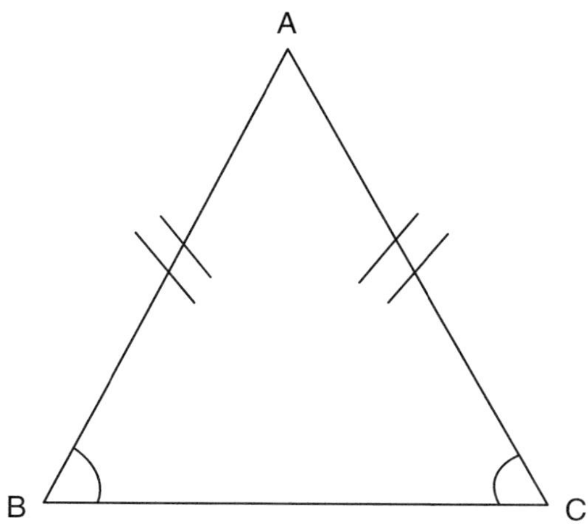

Indeed, if you Google "isosceles triangle", over 90% of the images retrieved have the base angles equal. However, you might have thought of something similar when asked:

"Let ABC be an equilateral triangle"

How do we know that an image (akin to a "fixed photograph") actually represents what and only what we want to represent, in a generic way? Some argue that philosophically, you can't:

> The danger of diagrams is that they may too easily tempt one to make unwarranted generalizations, as one's thinking may too easily depend in an unnoticed way on a feature represented in the diagram that is not common to all members of the class one is thinking about. (Giaquinto, 2007, p. 77)

This is not a trivial pedagogic problem; there will be many learners who have developed misconceptions due to seeing the particular in the general *as* the general. Whilst there are such limitations in diagrammatic representation, in presenting the general in the particular, we have also to be open to the possibility that language in the form of text (words and symbols) has its own limitations.

9.9 Teaching to Be Visual in the Classroom

Robinson points out "the information that is presented to students in classrooms appears in the form of words rather than pictures" (Robinson, 2002, p. 1). Pictures are used by teachers and in textbooks largely as illustration or decoration which is not surprising given the absence of a widely accepted framework for the use of visuals in learning.

One very particular form of visual material discussed in cognitive psychology is a *spatial text adjunct (STA)*, a visual specifically used to support and associate with some text. Robinson suggests the potential of STAs seems to be overlooked (Robinson, 2002), but they can be used as "static and animated illustrations, geographic and knowledge maps, and graphs" (p. 3); in other words they have potential as diagrammatic, videographic, spatial, topographic and graphic.

One feature of the failure of learners to achieve is the observation that school achievement is not equitably spread throughout society; children from less affluent homes do disproportionately worse than those bought up in relative affluence. Such children are at risk of sustaining a weak conceptual grasp of mathematical and scientific concepts and in numerical procedures, which hold them back from developing a more sophisticated understanding of STEM. This in turn closes off pathways to many careers and professions, but, worse, develops into anxiety and rejection of mathematics in particular, contributing instead to an identity of "I just can't do maths" (Gates, 2001).

Whilst much research has attempted to articulate this relationship, much research has simply ignored it, either through denial or in the belief that by providing good research all will benefit. Indeed, an early finding from the ICAAMS study (*http://iccams-maths.org/*) is that over 30 years:

> Attainment has not changed very much [..]. The general trend is for results to be somewhat lower than in the 1970s, although there are some exceptions to this. (Hodgen, Brown, Küchemann, & Coe, 2010, p. 8)

So, instead of trying to do the same old thing better, maybe it is time to think anew. There is some doubt that the improvements in levels of achievement in mathematics and science in the UK trumpeted by successive government have in reality been that real (Dickinson, Eade, Gough, & Hough, 2010) and undoubtedly the same holds true in other countries.

The issue of prior achievement features in the visual literature to suggest there are specific benefits in using visual forms of representation when students have experienced difficulty in their prior learning or have weak verbal skills – two features which correlate highly with a learner's SES. Indeed, Arcavi suggests it might be problem with a lack of visualisation skills which can offer an explanation for many students' particular difficulties with fractions (Arcavi, 2003). This is further supported by a recent study by Moyer-Packenham, Ulmer, and Anderson (2012) who reported an investigation the use of static and dynamic images with low-achieving students. Although this was an action research study by just one team, low-achieving students did appear to make gains in fraction leaning when provided with either virtual (computer) manipulatives or pictorial models.

A study by Mayer (1997) suggested that learners with low prior knowledge (or "low domain knowledge") might be particularly supported by visual models:

> Students who possess high levels of prior knowledge will be more likely than low prior knowledge learners to create their own mental images as the verbal explanation is presented and thus to build connections between verbal and visual representations. In contrast, students who lack prior knowledge will be less likely than high prior knowledge learners to independently create useful mental images solely from the verbal materials. Thus, low prior knowledge learners are more likely than high prior knowledge learners to benefit from the contiguous presentation of verbal and visual explanations. (Mayer, 1997, p. 15)

This is further supported by Schnotz and Bannert (2003) who studied the effect of mixing text and graphics on university students' understanding of a text. They argue that whilst "adding pictures to a text is not always beneficial … pictures facilitate learning only if individuals have low prior knowledge and if the subject matter is visualized in a task-appropriate way" (pp. 153–154):

> From the perspective of practice, the findings of our study emphasize that in the design of instructional material including texts and pictures the form of visualization used in the pictures should be considered very carefully. The question is not only which information is to be conveyed. One must also ask whether the form of visualization used in the picture supports the construction of a task-appropriate mental model. Good graphic design is not only important for individuals with low prior knowledge who need pictorial support in constructing mental models. Well-designed pictures are also important for individuals with high prior knowledge because these individuals can be hindered in their mental model construction through inappropriate forms of visualization. (Schnotz & Bannert, 2003, p. 154)

Mayer takes this further looking at those learners identified as "poor readers", who may be so because of an imbalance in text vs. visual processing, and would this benefit from a visual approach:

> Previous research on children's processing of narrative texts has shown that the poor readers profit generally more from text illustrations with regard to comprehension and learning than good readers (Cooney & Swanson, 1987; Levie & Lentz, 1982; Mastropieri & Scruggs, 1989; Rusted & Coltheart, 1979). This suggests that poor readers are able to construct a mental model from a text with pictures, whereas they would fail on the basis of a text alone. Similar results have been found for adult learners' processing of expository texts. Learners with low prior knowledge benefit from pictures in a text, whereas learners with higher prior knowledge seem to be able to construct a mental model of the described content also only from the text. (Mayer, 1997)

Furthermore, O'Donnell, Dansereau and Hall (2002, pp. 78–79) argue that knowledge maps (concept maps) are particularly useful for learners with low or weak verbal skills, though this was probably involved more than just the use of a diagrammatic representation. Their suggestions include greater integration of map and text, collaborative construction of maps, explicit work on the isomorphism between map and text and providing exemplar maps.

A study on low-attaining learners concludes "reasoning with a diagram is a difficult process that students may need more time and experience to develop"(Garderen et al., 2014, p. 147).

Hittleman (1985) argues for instruction to include a process of translation between various illustrative representation and text. There are, he argues, specific reasons why certain learners might specifically require attention in grasping the coding of illustrations:

> Children often may experience problems in reading content area illustrations because of their need to keep switching from reading the text to examining the illustrations. This punctuated or staccato reading pattern breaks up their continuous flow and processing of information. Children who experience reading difficulty in general, and those who are handicapped by breakdowns because of limited ability in conceptualizing may be confused for two reasons: 1) they lack an understanding of the nature of the information in the illustrations; 2) they are hindered by an inability to translate information from one form and organizational pattern to another. Therefore, teachers should examine and orally discuss illustrations with children before they are asked to read the accompanying text. (Hittleman, 1985, p. 34)

Hittleman problematises the "picture is worth a thousand words" dictum by pointing out that one needs to know the words and how to translate between the different languages:

> Reading is an interaction between children and authors. Too often, visual displays are considered easier to read than prose because they do not entail any words. As demonstrated, however, children's reading of illustrations requires skills that must be taught and learned. Illustrations are only representations of life, and children must learn how the illustrations picture actual objects and events. Children, especially children with difficulties manifested by disabilities in cognition and learning, need direct instruction so they can understand how a three-dimensional world is represented in a two- dimensional presentation. They need to learn how to translate those representations into spoken and written messages. A picture is worth a thousand words only if the observer (reader) already knows what those words are and has skills to relate them to the picture. (Hittleman, 1985, p. 36)

Much evidence suggests less attention is paid to visual forms in teaching (Larkin & Simon, 1987), and this results in learners misusing, or not drawing on, visual models. More specifically within the mathematics education literature, there is some evidence that how students visualise mathematical concepts "plays a pivotal role in how well students apply their …understanding in novel situations" (Anderson-Pence et al., 2014, p.???). Diagrams, in the form of static images which embody fixed images of concepts, can be crucial in supporting or hindering students' interpretations of a problem or piece of mathematics. However, diagrams are not merely accurate and unambiguous representations of mathematical objects or concepts. They require interpretation to manipulate, generate or employ in doing mathematics. For not only do diagrams embody features of mathematical concepts, but they also embody aspects of pupil misconceptions (Anderson-Pence et al., 2014, p. 14):

> The use of visualization requires a specific training, specific to visualize each register. Geometrical figures or Cartesian graphs are not directly available as iconic representations can be. And their learning cannot be reduced to training to construct them. This is due to the simple reason that construction makes attention to focus successively on some units and properties, whereas visualization consists in grasping directly the whole configuration of relations and in discriminating what is relevant in it. Most frequently, students go no further than to a local apprehension and do not see the relevant global organization but an iconic representation. (Duval, 1999, p. 14)

> Learning mathematics implies the construction of this cognitive architecture that includes several registers of representation and their coordination. Thus geometrical figures used to solve problems involves some ability in operative apprehension and awareness of how deductive reasoning works. Students do not come into such apprehension and awareness by themselves. Moreover, some coordination is required between operative apprehension, discursive apprehension and deductive reasoning. In other words, geometrical activity requires continual shifts between visualization and discourse. In order to achieve such coordination another kind of visualization is required. (Duval, 1999, p. 22)

Visual representation can draw on cognitive skills that are underused elsewhere in schools. Forty years ago, in the USA, Olsen (1977) argued that schools are biased towards verbal and textual forms. Consequently, school pedagogies may privilege certain learners – those confident and at ease with literal forms (Winn, 1987). In many but not all cases, "graphics have done more to improve the performance of low-ability students than those of high ability" (Winn, 1987, p. 169), particularly in science (Holliday et al., 1977) and mathematics – where it is claimed that visuals reduced "the reading-related working memory overload in poor readers" (Moyer, Sowder, Threadgill-Sowder, & Moyer, 1984). Though there is a claim that low-ability learners have particular difficulty with materials that are informationally rich and with redundancy (Allen, 1975).

Unfortunately, teachers don't seem to have developed the same level of appreciation. In a study of assessment practices of 11 secondary mathematics teachers, Morgan (2004) argues that whilst the teachers acknowledged the importance and value of visual representations in mathematics, they often gave them a low value within pupils' work in contrast to more abstract but non-visual (re)presentations. Indeed, she went further to argue that at times teachers would assess a piece of work more negatively if a diagram was inserted as an indication of "a concrete, practical approach rather than a more prestigious abstract approach" (Morgan, 1999).

The reluctance of learners and teachers to use diagrams, as reported by Morgan and others, is likely to be influenced by the considerable difficulty posed by working with visual representations. The mathematics itself first needs to be *decoded* (Garderen et al., 2014, p. 136) from the original textual form and then *recoded* into an appropriate visual form. This is by no means trivial and requires a sense of what information is of relevance amongst information provided (*identification* and *selection*), what *connections* exist between elements of the problem and how mathematics can represent the *structure* of a problem.

Garderen et al. have described an underlying set of skills in the context of diagram proficiency (Garderen et al., 2014, p. 137):

- To know what a diagram is and can illustrate;
- To use a visual representation to depict key component of a problem;
- To know how to represent the processes and relationships within a problem in visual form and flexibly adapt to different problem formulations;
- To be disposed to use a visual form.

This final skill is more than just an encouragement to "draw a diagram" but is a way of encouraging the learner to "see things" in a different way. It surely can be no surprise that learners who are presented in lessons, with textual information day in day out, become reluctant to use diagrams, especially if (as Morgan reports) they will get criticised for doing so. Diezmann et al. (Diezmann, 1999, 2000; Diezmann & English, 2001) have written of problems in diagram use such as:

- Not using a diagram at all;
- Using a diagram that was virtually unviable and unusable;
- Using a diagram that is imprecise, missing constraints of the problem;
- Producing inaccurate diagrams due to not noticing salient features and maybe not even being aware of salience.

A study of diagram use to solve word problems (Garderen et al., 2014) also reported differential forms of engagement between higher-ability learners and those with "learning disabilities". Garderen et al. went further to argue that in their study, students did not see the point of diagrams and claimed they did not even know what the term "diagram" meant, as well as not being aware of the diversity of visual representational forms. Furthermore, they did not see diagrams as part of doing mathematics. This is likely to derive from the forms used in school and elsewhere, limiting learners to restrictive visual representations, e.g. being told a fraction is a pizza slice, or some shaded-in shape within a triangle. This all resulted in learners not choosing to use a diagrammatical form. Garderen et al.'s argument is that weaker pupils engaged differently with diagrammatical forms through not knowing what a diagram was, was for, how it was created or was used. This might be an example of a broader problem in the process of representation in learning. If a student cannot see the structural relationships between the representation and concept being targeted, then any representation will remain rather meaningless.

However, further evidence indicates that there is a lack of explicit instruction in dealing with graphics – that unsuccessful learners would benefit from support and guidance in mapping between graphic and text information and the resulting mental models (Schnotz et al., 1993). A US study of 13 randomised control trials on learning difficulties in mathematics (see Gersten et al., 2009, p. 30 for a full bibliography) reported empirical support for using visual representations with learners who were achieving poorly in mathematics even if this was cited in some studies as providing only "moderate evidence" (Gersten et al., 2009, p. 30). They placed visuals explicitly within a framework consistent with Bruner's enactive, iconic, symbolic representation situated specifically between physical manipulatives and abstract symbolic representations. In this way, diagrams and visual representations should be used specifically to support learners' reasoning through transitions between physical models and symbolic representations. It is further argued that student understanding of these transitions can be strengthened through the use of visual representations of mathematical concepts (Hecht, Vagi, & Torgesen, 2007):

> A major problem for students who struggle with mathematics is weak understanding of the relationships between the abstract symbols of mathematics and the various visual representations. (Gersten et al., 2009, p. 30)

They go on to argue that materials specifically for pupils with difficulties "provide very few examples of the use of visual representations" (p. 36). We can see the same reluctance to place visual reasoning in reteach studies examining instructional strategies – for example, Darch, Carnine, and Gersten (1984) who offer "explicit instruction" with no attempt to consider any visual forms between word problems and solution.

A conclusion for mathematics educators is to foster an approach with teachers to recognise and respect the visual and diagrammatical form as a pedagogical tool to represent and work on mathematics. Low attainers seem to have greater difficulty seeing the salience in a problem than can be represented in multiple ways – particularly the visual – or even to have a disposition so to do. They conclude "reasoning with a diagram is a difficult process that students may need more time and experience to develop" (Garderen et al., 2014, p. 147). In addition, we do not have an understanding of the way in which diagrammatic competence develops over time, maybe because we have little idea of what we mean by diagrammatic competence and have rarely used it as a legitimate pedagogical device within mathematics. For it is only once we recognise the difficultly and "lack of transparency …Can we begin to identify and adopt strategies to support students" (Rubenstein & Thompson, 2013, p. 550).

Schnotz argues that a mental representation is never a representation *per se* but is dependent on the context within which the model was created and the purposes for which it is used. As a result, the use of external representations does need to "take into account the interplay between the representation and the task demands" (Schnotz, 2002, p. 104). Descriptive (made up of symbols with an arbitrary and conventional connection to the content) and depictive representations (iconic signs associated with the content through structural features) thus have different functions and purposes within mathematics. For example, "the angles in a triangle add up to 180°, compared to a diagram in which the three angles just happen to add to 180°". However depictive representations can carry much more information and reduce cognitive load. Given this, the use of diagrams and visual representations within mathematics teaching rests upon complex cognitive operations. Yet how they feature in pedagogy is largely unknown, with little work in to how visual modalities are used by teachers, to what end. Furthermore, there is little theoretically sound examples of where visual modalities can be used to improve the quality of mathematics teaching.

Schnotz describes the ways in which learners engage with descriptive and depictive representations:

- *Textual representation* – First one is presented with linguistic information, in the form of words and symbols, from which specific syntax needs to be understood. Then semantic content needs to be interpreted and mapped onto the referential content. To make sense of this, the learner relates to their own domain-specific world knowledge. Thereafter there comes a stage of communicating the content and an appreciation of the forms of communication.
- *Picture representation* – First one encounters a visual stimulus that one encodes the perceptual surface structure, drawing on common structural features between the picture and the reference. Next, one encodes the information carried within the stimulus by drawing on one's awareness of pictorial communication.

However, the interplay and specific ways in which the learners enact these operations is still a matter of conjecture within cognitive neuroscience (Schnotz 2002). Anderson-Pence et al. (2014) argue that merely presenting learners with "visual static models" does not guarantee that they will be able to use and incorporate the model into their own understanding. Notably these diagrams will not necessarily expose misconceptions or allow learners to focus on their understanding sufficiently enough to use the diagram to work on mathematics. This needs to be incorporated into instructional resources that support learners to use, develop and adapt diagrams, which in turn improves their visualisation skills (p. 14).

What people attend to makes a big difference to how they interpret and what they notice (Whiteley, 2004). Whiteley discusses the processes used in proving, within geometry, and highlights several strategies that experts use which may not be accessible to the novice:

> Of course the expert has learned to do the animations and sequences in the mind's eye, with shifting attention, with mental movements and comparisons of pieces, with the shifts from parts to whole and back again…Too often we do not teach the skills or even explicitly model the skills in a way that the apprentices can observe and imitate. (Whiteley, 2004, p. 290)

We can, as teachers, influence what learners attend to, this after all is the very essence of teaching.

9.10 Overview

In this chapter I have underlined the importance of engaging with visual models as part of a process of learning where text and visuals are both key elements of the way our brain interprets and manipulates information. One root cause of much underachievement in STEM in early secondary school can be traced back to a heavy reliance upon textual and lexicographical representation and communication at the expense of more visuospatial representations. By recognising the diversity of use we can make of visuals and diagrams, we can begin to look for a range of ways of incorporating both in classroom materials and tasks such that we place more emphasis upon the cognitive reconstructing of STEM ideas using cooperation between different forms. Such cooperation might be facilitated by greater opportunities for crossover CPD between teachers of each component of STEM, such that divergent forms of representation along with differing incorporation of visual forms might help all STEM teachers broaden their awareness of the importance of connecting concept-artefact representation in a more eclectic multimodal way.

Using a multimodal approach has the potential of supporting children from less affluent backgrounds who may well have a more limited linguistic experience by the time they transfer to secondary school. However, using diagrams and other visual representation is not instinctive but requires and responds to explicit instruction. Hence teachers can support learners in STEM by focussing more on representations as carriers of concepts by fostering in learners the skills in constructing and using visuospatial representations.

We still have a lot to learn about how our brain works with text and graphics and whether there is a "best" way of using them in fostering learning. What we do know however is that it is advantageous taking a different look at our pedagogy and exploring ways of offering a visual channel of communication. This would mean incorporating diagrams and other graphics into our pedagogy not merely as visual *representations* to illustrate but also as visual *forms* for the learner to activate and interpret and where the learner creates the visual as a way of strengthening their conceptual architecture. However, it means going further than this. It also means using *spatial representations* more widely as a mode of communication and cognition. This might include concept mapping, worked examples, writing frames, etc. – though it will not be restricted to these.

As a conclusion I want to suggest we recognise and respect the visual and diagrammatical form as a major pedagogical tool to represent STEM concepts and to work on STEM processes. We still do not have an understanding of the way in which diagrammatic competence develops over time, maybe because we have little idea of what we mean by diagrammatic competence in learning and have rarely used it as a legitimate pedagogical device within the classroom. We all have a lot to learn – even me, who has written a chapter praising and encouraging visualisation – with only four diagrams. In the words of one of my teachers in the 1960s, "Peter could do much better if he applied himself more".

The geometrical property in Sect. 9.5 is – the four angle bisectors of any quadrilateral form a cyclic quadrilateral. IKEA are suggesting you should not tighten up the screw, or you won't get the back in!

References

Ainsworth, S. (2006). DeFT: A conceptual framework for considering learning with multiple representations. *Learning and Instruction, 16*, 183–198. doi:10.1016/j.learninstruc.2006.03.001.

Akopyan, A. (2011). *Geometry in figures*. Available from: http://www.mccme.ru/~akopyan/papers/EnGeoFigures.pdf

Allen, W. (1975). Intellectual abilities and instructional media design. *AV Communication Review, 23*, 139–170.

Anderson-Pence, K., Moyer-Packenham, P., Westenskow, A., Shumway, J., & Jordan, K. (2014). Relationships between visual static models and students' written solutions to fraction tasks. *International Journal for Mathematics Teaching and Learning, 15*, 1–18.

Arcavi, A. (2003). The role of visual representations in the learning of mathematics. *Educational Studies in Mathematics, 52*, 215–241.

Baddeley, A. (1998). Imagery and working memory. In M. Dennis, J. Engelkamp, & J. Richardson (Eds.), *Cognitive and Neuropsychological approaches to mental imagery, NATO ASI Seriues: Series D. Behavioural and social science, no. 42* (pp. 169–180). Dordrecht, The Netherlands: Martiunus Nijhoff Publishing.

Baddeley, A. (2003). Working memory and language: An overview. *Journal of Communication Disorders, 3*, 189–208.

Beck, J., & Wade, M. (2006). *The kids are alright. How the gamer generation is Changing the workplace*. Boston: Harvard Businbess School Press.

Bishop, J. (1978). Developing students' spatial skills. *Science Teacher, 45*, 20–23.

Brna, P., Cox, R., & Good, J. (2001). Learning to think and communicate with diagram: 14 questions to consider. *Artificial Intelligence Review, 15*, 115–134.

Card, S., Moran, T., & Newell, A. (1983). *The psychology of human-computer interaction*. Hillsdale, NJ: Lawrence Erlbaum Associates.

Carney, R., & Levin, J. (2002). Pictorial illustrations still improve students' learning from text. *Educational Psychology Review, 14*(1), 5–26.

Carpenter, P., & Shah, P. (1998). A model of the perceptual and conceptual processes in graph comprehension. *Journal of Experimental Psychology: Applied, 4*(2), 75–100.

Carpernter, T., Coburn, M., Reys, R., & Wilson, J. (1976). Notes from national assessment: Addition and subtraction with fractions. *Arithmetic Teacher, 23*, 137–141.

Cheng, Y.-L., & Mix, K. (2014). Spatial training improves children's mathematics ability. *Journal of Cognition and Development, 15*(1), 2–11. doi:10.1080/15248372.2012.725186.

Clark, R., Nguyen, F., & Sweller, J. (2006). *Efficiency in learning. Evidence-based guidelines to manage cognitive load*. San Francisco, CA: Pfeiffer.

Clements, K. (1983). The question of how spatial ability is defined, and its relevance to mathematics education. *Zentralblatt fur Didaktik der Mathematik, 1*(1), 8–20.

Cooney, J., & Swanson, H. (1987). Memory and learning disabilities: An overview. In H. Swanson (Ed.), *Memory and learning disabilities: Advances in learning and behavioral disabilities* (pp. 1–40). Greenwich, CT: JAI.

Cramer, K., Post, T., & delMas, R. (2002). Initial fraction learning by fourth-and fifth-grade students: A comparison of the effects of using commercial curricula with the effects of using the rational number project curriculum. *Journal for Research in Mathematics Education, 33*(2), 111–144.

Crapo, A. (2002). *A cognitive-theoretical approach to the visual representation of modelling context*. Doctoral dissertation, Rensselaer Polytechnic Institute.

Crapo, A., Waisel, L., Wallace, W., & Willemain, T. (2000). Visualization and the process of modelling: A cognitive-theoretic view. In *Proceedings of the sixth ACM SIGKDD international conference on Knowledge discovery and data mining*. Boston.

Darch, C., Carnine, D., & Gersten, R. (1984). Explicit instruction in mathematics problem solving. *Journal of Educational Research, 77*(6), 351–359.

Dawe, L. (1993). Visual imagery and communication in the mathematics classroom. In M. Stevens, A. Waywood, D. Clarke, & J. Izard (Eds.), *Communicating mathematics: Perspectives from classroom practice and research* (pp. 60–76). Hawthorne, VIC: Australian Council for Educational Research.

Dickinson, P., Eade, F., Gough, S., & Hough, S. (2010). Using realistic mathematics education with low to middle attaining pupils in secondary schools. In M. Joubert & P. Andrews (Eds.), *Proceedings of the 10th British congress for mathematics education* (pp. 73–80).

Dienes, Z. (1960). *Building up mathematics*. London: Hutchinson Educational.

Diezmann, C. (1999). Assessing diagram quality. Making a difference to representation. *Proceedings of the 22nd annual conference of the Mathematics Education Research Group of Australasia* (pp. 185–191). Adelaide: MERGA.

Diezmann, C. (2000). The difficulties students experience in generating diagrams for novel problems. In T. Nakahara & M. Koyama (Eds.), *Proceedings of the 25th annual conference of the international groups for the psychology of Mathematics education* (pp. 241–248). Hiroshima: PME.

Diezmann, C., & English, L. (2001). Promoting the use of diagrams as tools for thinking. In A. Cuzco & F. R. Curcio (Eds.), *The roles of representation in school Mathematics: 2001 yearbook* (pp. 77–89). Reston, VA: National Council of Teachers of Mathematics.

Dreyfus, T. (1991). On the status of visual reasoning in mathematics and mathematics education. In F. Furinghetti (Ed.), *Proceedings of the 15th conference of the international group for the psychology of mathematics education* (Vol. 1, pp. 32–48). Assisi.

Duval, R. (1999). Representation, vision and visualization: Cognitive functions in mathematical thinking. Basic issues for learning. In *Proceedings of the annual meeting of the North American chapter of the international group for the psychology of mathematics education* (pp. 3–25).

Egan, D., & Schwartz, H. (1979). Chunking in recall of symbolic drawings. *Memory & Cognition, 7*, 149–158.

Eisenberg, T., & Dreyfus, T. (1991). On the reluctance to visualise in mathematics. In W. Zimmerman & S. Cunningham (Eds.), *Visualisation in teaching and learning mathematics*. Washington, DC: Mathematical Association of America.

Eitel, A., & Scheiter, K. (2015). Picture or text first? Explaining sequence effects when learning with pictures and text. *Educational Psychology Review, 27*, 153–180. doi:10.1007/s10648-014-9264-4.

Ellerton, N., & Clements, M. (1994). Fractions: A weeping sore in mathematics education. *SET: Research Information for Teachers, 2*, 93–96.

Garderen, D. V., Scheuermann, A., & Poch, A. (2014). Challenges students identified with a learning disability and as high-achieving experience when using diagrams as a visualization tool to solve mathematics word problems. *ZDM Mathematics Education, 46*, 135–149. doi:10.1007/s11858-013-0519-1.

Gates, P. (2001). What is an/at issue in mathematics education? In P. Gates (Ed.), *Issues in mathematics teaching*. London: RoutledgeFalmer.

Gee, J. P. (2007). *What video games have to teach us about learning and literacy*. Basingstoke, UK: Palgrave.

Gersten, R., Beckmann, S., Clarke, B., Foegen, A., Marsh, L., Star, J., et al. (2009). *Assisting students struggling with mathematics: Response to Intervention (RtI) for elementary and middle schools (NCEE 2009–4060)*. Washington, DC: National Center for Education Evaluation and Regional Assistance, Institute of Education Sciences, U.S. Department of Education.

Giaquinto, M. (1992). Visualisation as a means of geometrical discovery. *Mind and Language, 7*, 381–401.

Giaquinto, M. (1993). Visualising in arithmetic. *Philosophy and Phenomenological Research, LIII*, 385–396.

Giaquinto, M. (1994). Epistemology of visual thinking in elementary real analysis. *British Journal for the Philosophy Science, 45*, 789–813.

Giaquinto, M. (2007). *Visual thinking in mathematics. An epistemological study*. Oxford, UK: Oxford University Press.
Hecht, S., Vagi, K., & Torgesen, J. (2007). Fraction skills and proportional reasoning. In D. Berch & M. Mazzocco (Eds.), *Why is math so hard for some children? The nature and origins of mathematical learning difficulties and disabilities* (pp. 121–132). Baltimore: Paul H. Brookes Publishing Co.
Hegarty, M. (1992). Mental animation: Inferring motion from static displays. *Journal of Experimental Psychology: Learning, Memory, and Cognition, 18*(5), 1084–1102.
Hegarty, M., & Kozhevnikov, M. (1999). Types of visual-spatial representations and mathematical problem solving. *Journal of Educational Psychology, 91*(4), 684–689.
Hittleman, D. (1985). A picture is worth a thousand words…if you know the words. *Childhood Education, 62*(1), 32–36.
Hodgen, J., Brown, M., Küchemann, D., & Coe, R. (2010). *Mathematical attainment of English secondary school students: A 30-year comparison*. Paper presented at the British Educational Research Association conference. http://www.kcl.ac.uk/sspp/departments/education/research/cppr/Research/pastproj/TISME/Events/Papers/iCCAMS-paper-for-BERA-symposium-2010.pdf
Holliday, W., Brunner, L., & Donais, E. (1977). Differential cognitive and affective responses to flow diagrams in science. *Journal of Research in Science Teaching, 14*, 129–138.
Kersch, J., Casey, B., & Young, J. M. (2008). Research on spatial skills and block building in girls and boys. The relationship to later mathematics learning. In O. Saracho & B. Spodek (Eds.), *Contemporary perspectives on mathematics im early childhood Educatioin* (pp. 233–251). Charlotte, NC: Information Age Publishing.
Kosslyn, S. (1980). *Image and mind*. Cambridge, MA: Harvard University Press.
Larkin, J., & Simon, H. (1987). Why a diagram is (sometimes) worth ten thousand words. *Cognitive Science, 11*(1), 65–99.
Lean, G., & Clements, K. (1981). Spatial ability, visual imagery, and mathematical performance. *Educational Studies in Mathematics, 12*(3), 267–299.
Lesh, R., Post, T., & Behr, M. (1987). Representation and translations amongst representations in mathematical learning. In C. Janvier (Ed.), *Problems of representation in the teaching and learning of mathematics* (pp. 33–40). Hillsdale, MI: Lawrence Earlbaum Associates.
Levie, H., & Lentz, R. (1982). Effects of text illustrations: A review of research. *Educational Communication and Technology Journal, 30*, 195–232.
Levin, J. R., Anglin, G. J., & Carney, R. N. (1987). On empirically validating functions of pictures in prose. In D. M. Willows & H. A. Houghton (Eds.), *The psychology of illustration: I. Basic research* (pp. 51–85). New York: Springer.
Linn, M., & Peterson, A. (1985). Emergence and characterisation of gender differences in spatial ability: A meta-analysis. *Child Development, 56*, 1479–1498.
Lord, T. (1985). Enhancing the visual-spatial aptitude of students. *Journal of Research in Science Teaching, 22*, 395–405.
Lord, T. (1990). Enhancing learning in the life sciences through spatial perception. *Innovative Higher Education, 15*(1), 5–16.
Lowrie, T., & Diezmann, C. (2005). Fourth-grade students' performance on graphical languages in mathematics. In H. Chick & J. Vincent (Eds.), *Proceedings of the 29th annual conference of the International Group for the Psychology of mathematics education* (Vol. 3, pp. 265–272). Melbourne, VIC: PME.
Lowrie, T., & Diezmann, C. (2009). National Numeracy Tests: A graphic tells a thousand words. *Australian Journal of Education, 53*(2), 141–158.
Mastropieri, M., & Scruggs, T. (1989). Constructing more meaningful relationships: Mnemonic instruction for special populations. *Educational Psychology Review, 1*, 83–111.
Mayer, R. (1997). Multimedia learning: Are we asking the right questions? *Educational Psychologist, 32*(1), 1–19.
Mayer, R., & Gallini, J. (1990). When is an illustration worth ten thousand words? *Journal of Educational Psychology, 82*(4), 715–726.

Morgan, C. (1999). What is the role of diagrams in communication of mathematical activity? In P. Gates (Ed.), *Proceedings of thre BSRLM day conference, London Institute of Education* (vol. 14, pp. 80–92).

Morgan, C. (2004). What is the role of diagrams in communication of mathematical activity? In B. Allen & S. Johnston-Wilder (Eds.), *Mathematics education: Exploring the culture of learning* (pp. 134–145). London: RoutledgeFalmer.

Moyer, J., Sowder, L., Threadgill-Sowder, J., & Moyer, M. (1984). Story problem formats: Drawn versus verbal versus telegraphic. *Journal for Research in Mathematics Education, 15*(5), 342–351.

Moyer-Packenham, P., Ulmer, L., & Anderson, K. (2012). Examining pictorial models and virtual manipulatives for third grade fraction instruction. *Journal of Interactive Online Learning, 17*(2), 179–212.

Mullis, I., Martin, M., Fo, P., & Arora, A. (2012). *TIMSS 2011 international results in mathematics*. Boston: TIMSS and PIRLS International Study Centre.

Neime, D. (1996). A fraction is not a piece of pie: Assessing exceptional performance and deep understanding in elementary school mathematics. *Gifted Child Quarterly, 40*(2), 70–80.

Nordina, M. S. B., Amina, N. B. M., Subaria, K. B., & Hamida, M. Z. B. A. (2013). Visualization skills and learning style patterns among engineering students at Universiti Teknologi Malaysia. *Procedia – Social and Behavioral Sciences, 93*, 1769–1775. doi:10.1016/j.sbspro.2013.10.114.

O'Donnell, A., Dansereau, D., & Hall, R. (2002, March). Knowledge maps as Scaffolds for cognitive processing. *Educational Psychology Review, 14*(1), 71–87.

Ochs, E., Gonzales, P., & Jacoby, S. (1996). 'When I come down I'm in the domain state': Grammar and graphic representation in the interpretive activity of physicists. In E. Ochs, E. Scheglo, & S. Thompson (Eds.), *Interaction and Grammar* (pp. 328–369). Cambridge, MA: Cambridge University Press.

Olkun, S. (2003). Making connections: Improving spatial abilities with engineering drawing activities. *International Journal of Mathematics Teaching and Learning,* http://www.cimt.plymouth.ac.uk/journal/sinanolkun.pdf

Olsen, D. (1977). The languages of instruction: On the literate bias of schooling. In R. Anderson, R. Spiro, & M. Montague (Eds.), *Schooling and the Acquisition of Knowledge* (pp. 65–69). Hillsdale, MI: Erlbaum.

Paivio, A. (1976). *Imagery and verbal processes*. New York: Holt Reinhardt and Winston.

Paivio, A. (1978). A dual coding approach to perception and cognition. In H. Pick & E. Saltzman (Eds.), *Modes of perceiving and processing information* (pp. 39–52). Hillsdale, MI: Erlbaum.

Paivio, A. (1986). *Mental representations. A dual coding approach*. New York: Oxford University Press.

Postigo, Y., & Pozo, J. I. (2004). On the road to Graphicacy: The learning of graphical representation systems. *Educational Psychology, 24*(5), 623–644. doi:10.1080/0144341042000262944.

Presmeg, N. (1986). Visualisation in high school mathematics. *For the Learning of Mathematics, 6*(3), 42–46.

Repovš, G., & Baddeley, A. (2006). The multi-component model of working memory: Explorations in experimental cognitive psychology. *Neuroscience, 139*(1), 5–21.

Robinson, D. (2002, March). Spatial text adjuncts and learning. *Educational Psychology Review, 14*(1), 1–3.

Roth, W.-M. (2002). Reading graphs: Contributions to an integrative concept of literacy. *Journal of Curriculum Studies, 34*(1), 1–24. doi:10.1080/00220270110068885.

Rubenstein, R., & Thompson, A. (2013). Reading visual representations. *Mathematics Teaching in the Middle Years, 17*(9), 544–550.

Rusted, J., & Coltheart, M. (1979). Facilitation of children's prose recall by the presence of pictures. *Memory and Cognition, 7*, 354–359.

Satoy, M. D. (2004). *The music of the primes. Why an unsolved problem in mathematics matters*. London: Harper Perennial.

Scaife, M., & Rogers, Y. (1996). External cognition: How do graphical representations work? *International Journal of Human – Computer Studies, 45*, 185–213.

Schnotz, W. (1993). Some remarks on the commentary on the relation of dual coding and mental models in graphics comprehension. *Learning and Instruction, 3*, 247–249.

Schnotz, W. (2002). Commentary e Towards an integrated view of learning from text and visual displays. *Educational Psychology Review, 14*(1), 101–120.

Schnotz, W., & Bannert, M. (2003). Construction and interference in learning from multiple representations. *Learning and Instruction, 13*(2), 141–156.

Schnotz, W., Picard, E., & Hron, A. (1993). How do successful and unsuccessful learners use texts and graphics? *Lcaming and Instruction, 3*, 181–199.

Shah, P., & Hoeffner, J. (2002). Review of graph comprehension research: Implications for instruction. *Educational Psychology Review, 14*(1), 47–69.

Stone, D., & Glock, M. (1981). How do young adults read directions with and without pictures? *Journal of Educational Psychology, 73*(3), 419–426.

Strong, S., & Smith, R. (2002). Spatial visualisation: Fundamentals and trends in engineering graphics. *Journal of Industrial Technology, 18*, 1–6.

Stylianou, D. (2002). On the interaction of visualisation and analysis – The negotiation of a visual representation in problem solving. *Journal of Mathematical Behavior, 21*(3), 303–317.

Stylianou, D. (2010). Teachers' conceptions of representation in middle school mathematics. *Journal of Mathematics Teacher Education, 13*(4), 325–434. doi:10.1007/s10857-010943-y.

Stylianou, D. (2013). An examination of connections in mathematical processes in students' problem solving: Connections between representing and justifying. *Journal of Education and Learning, 2*(2), 23–35. doi:10.5539/jel.v2n2p23.

Stylianou, D., & Silver, E. (2004). The role of visual representations in advanced mathematical problem solving: An examination of expert-novice similarities and differences. *Journal of Mathematical Thinking and Learning, 6*(4), 353–387.

Tufte, E. (1990). *Envisioning information*. Cheshire, CT: Graphics Press.

Tufte, E. (1997). *Visual explanations*. Cheshire, CT: Graphics Press.

Tufte, E. (2001). *The visual display of quantitative information*. Cheshire, CT: Graphics Press.

Tversky, B. (2010). Visualizing thought. *Topics in Cognitive Science, 2*, 1–37. doi:10.1111/j.1756-8765.2010.01113.x.

Weidenmann, B. (1987). When good pictures fail: An information-processing approach to the effect of illustrations. In H. Mandl & J. Levin (Eds.), *Knowledge acquisition from text and pictures* (pp. 157–170). Amsterdam: Elsevier.

Whiteley, W. (2004). Teaching to see like a mathematician. In G. Malcolm (Ed.), *Multidisciplinary approaches to visual representations and interpretations, Studies in Multidisciplinarity* (Vol. 2, pp. 279–292). Amsterdam: Elsevier.

Winn, W. (1987). Charts, graphs and diagrams in educational materials. In D. Willows & H. Houghton (Eds.), *The psychology of illustration* (pp. 152–193). New York: Springer.

Zhang, X., Clements, M. A. K., & Ellerton, N. (2015). Conceptual mis(understandings) of fractions: From area models to multiple embodiments. *Mathematics Education Research Journal, 27*, 233–261. doi:10.1007/s13394-014-0133-8.

Peter Gates has been a lecturer and researcher at the University of Nottingham in England for 25 years. He has worked in math education for 40 years, as a teacher, teacher educator and researcher. He has a commitment to social justice and equity, and this feeds into everything he does in his professional and personal life. In recent years he has been particularly interested in working with low SES and disadvantaged young people looking to support their learning through a move away for a highly literacy-based curriculum, looking at offering an entry to an otherwise secret garden.

Peter has two daughters and two dogs, only two of which run to the door every time he comes home from work!

Chapter 10
Digital Technologies and Junior Secondary: Learning *with* and *About* Digital Technologies

Glenn Finger

Abstract This chapter focuses on digital technologies within STEM, referred to in this book as an integrated study of Science, Technology, Engineering and Mathematics. A distinction is made between learning *with* and learning *about* digital technologies. As well as acknowledging the importance of digital technologies assisting with teaching and learning in Junior Secondary schools in STEM subjects, it makes the case for digital technologies to be seen as a study in its own right. This is evidenced through, for example, identification of some schooling systems making coding and robotics mandatory areas of study in primary and secondary schools. Key trends, challenges and developments in technology are discussed. An implication for teachers is for them to develop technological knowledge (TK) so that they have the Technological Pedagogical Content Knowledge (TPACK) capabilities to effectively teach *with* and *about* digital technologies. Given that future teachers need these capabilities, there are implications for initial teacher education programmes preparing the next generation of Junior Secondary teachers. The chapter concludes with a summary of the *2016 Queensland Digital Technologies Summit: Initial Teacher Education*, including the co-constructed philosophy and strategies for action for future teachers developed at that Summit.

10.1 Introduction

Throughout this book, STEM is referred to as an integrated study of Science, Technology, Engineering and Mathematics within a coherent paradigm based on real-world applications. This chapter, focusing largely on technology within that STEM paradigm, examines learning *with* and *about* digital technologies in Junior Secondary schooling.

This chapter commences with an examination of learning *with* and *about* digital technologies in Junior Secondary school contexts. Importantly, while other chapters in this book appropriately focus on Science, Engineering and Mathematics and how

G. Finger (✉)
Griffith University, Gold Coast, QLD, Australia
e-mail: g.finger@griffith.edu.au

R. Jorgensen, K. Larkin (eds.), *STEM Education in the Junior Secondary*,
DOI 10.1007/978-981-10-5448-8_10

digital technologies might contribute to those subject areas, this chapter examines how digital technologies might be a subject for study in its own right within the Junior Secondary school curriculum. While contexts and curriculum will differ through various international schooling systems, this chapter draws upon, and makes specific reference to, the *Australian Curriculum: Technologies Learning area* (ACARA, 2016a) and the two subject areas within that curriculum, namely, *Design and Technologies* (ACARA, 2016b) and *Digital Technologies* (ACARA, 2016c). It focuses primarily on the *Digital Technologies* (ACARA, 2016c) subject area while also understanding that, in the *Australian Curriculum*, the *ICT Capability* (ACARA, 2016e) is one of seven general capabilities expected to be developed across all learning areas, including Science, Technology, Engineering and Mathematics. As these curriculum areas require teachers to have technological knowledge (TK), the Technological Pedagogical Content Knowledge (TPACK) conceptualisation (Mishra & Koehler, 2006) is discussed briefly.

Subsequently, the chapter summarises findings from the report *From Print to Pixel: The role of videos, games, animations and simulations within K-12 education* (Project Tomorrow, 2016a), and the Speak Up findings which identified *10 Things Everyone Should Know about K-12 Students' Digital Learning* (Project Tomorrow, 2016b) are discussed. These give a sense of the lived experience of students with digital technologies. The voices of students are then complemented by a discussion of key trends, significant challenges and important developments in digital technologies through reference to the *New Media Consortium (NMC)/Consortium for School Networking (CoSN) Horizon Report 2016 K-12 Edition* (Adams Becker, Freeman, Giesinger Hall, Cummins, & Yuhnke, 2016). Given these trends and the voice of students as indicators of digital technology use, the concluding section of the report examines what this means for current and future teacher capabilities. Specific reference is made to the co-constructed philosophy and strategies for action for future teachers, which was developed at the *2016 Queensland Digital Technologies Summit: Initial Teacher Education* (Finger et al., 2016).

As this chapter is read, I encourage readers to engage more broadly with technologies and teaching within a STEM agenda in Junior Secondary schooling and to heed Selwyn's (2015) advice that 'there are no simple answers or predetermined narratives waiting to unfold' (p. 191) and to be 'prepared to ask the difficult questions of how digital technologies are actually finding a place in educational settings and educational contexts' (p. 191). In thinking about the issues that I raise in this chapter, an underpinning premise is that a set of critical questions need to be asked (Selwyn, 2015):

- What is actually new here?
- What are the unintended consequences or second-order effects?
- Who is pushing these ideas in education? What are their reasons for doing so? What wider agendas are attached to these conversations?
- What is being said about education that might be useful? What is being said about education that teachers might wish to challenge and talk back to?

The chapter draws largely on the policy literature around the implementation of STEM, with a specific focus on digital technologies. This focus provides a context within which to locate and better understand how Junior Secondary schools, students and teachers are being shaped by the policy domain. In this chapter, I draw on two main policy domains to illustrate this. The first is the national curriculum for the technologies learning area of this curriculum, which is a relatively new initiative in the Australian policy domain and has sought to develop a common curriculum across Australian schools. Some specific reference is made to the Queensland context, an educational system within Australia, which, like other state educational systems, has assumed responsibility for the implementation of that national curriculum. Developments in Australia are elaborated upon to illustrate as an example of one country's approach within broader international developments. The second is the initial teacher education (ITE) policy domain, which is concerned with the preparation of future teachers with the capabilities to teach *with* and *about* digital technologies. Again, specific reference is made to ITE in Australia as an example to be considered within an understanding of a diverse range of responses that might be evident in other contexts internationally.

10.2 Complexifying Digital Technologies and Junior Secondary School Students

By adopting a position which complexifies digital technologies, this chapter goes beyond what might be regarded as advocacy for using digital technologies in often simplistic, superficial ways whereby digital technologies or information and communication technologies (ICT) – both terms are used interchangeably in this chapter – are viewed simply as values-neutral tools for use with existing curriculum, pedagogy and assessment. Importantly, once digital technologies are introduced, they impact upon curriculum, pedagogy and assessment.

Many of the superficial discussions about technologies reflect a technological determinist view of the future (Finger, Russell, Jamieson-Proctor, & Russell, 2007). Rather, adopting a technochoice perspective draws upon an understanding that educational rationale should inform the selection and use of digital technologies for learning and teaching in Junior Secondary school contexts. Furthermore, an ICT discourse is only one of a range of important, possible discourses about the role of technology. For example, Milojevic (2005) has noted that an ICT discourse is one of the five relatively recent discourses that have permeated the field of education, with others being globalisation, feminist, indigenous and spiritual discourses.

As well as understanding that the use of technologies in teaching is complex, it is also critically important to understand that there is neither a single 'Junior Secondary' school context nor a stereotypical 'Junior Secondary' student. Thus, for example, assumptions about equity of access, participation and outcomes for Junior Secondary students, in relation to digital technologies, need to be challenged. For example, Auld and Djabibba (2015) warn that, in an Australian context, 'Given the complex history of Aboriginal and Torres Strait Islander communities you can see that it is difficult to identify one approach towards technology that works for all contexts in Australia' (p. 58). That complexity can be extrapolated to other communities within Australia and indeed worldwide. Similarly, issues and challenges relating to digital technologies and equity are evident in relation with gender and rurality (Anderson, 2015). Anderson (2015) notes that these equity themes impact on the educational use of digital technologies. He highlights, for example, the under-representation of female students in digital technologies subjects in secondary schools and in higher education.

At the heart of any discussion on digital technologies and Junior Secondary school, students must have a discourse that understands education is a human endeavour and technologies have been developed, modified and appropriated by humans. Therefore, digital technologies are not value-neutral tools and have become part of a more complex digital ecosystem in which students connect, communicate, collaborate, share ideas and undertake formal and informal learning. Digital technologies now play a role in impacting upon and shaping the identity formation and life experiences of those Junior Secondary students.

10.3 Learning *with* and *About* Digital Technologies

Two important distinctions about digital technologies are made here. That is, there are distinct differences between learning *with* and learning *about* technologies within STEM in Junior Secondary schooling.

Firstly, learning *with* digital technologies highlights that digital technologies can be used for learning and teaching purposes. This can be applied not only in STEM-related subjects but throughout all learning areas of the curriculum. For example, there is significant literature relating to both online enhancement of largely face-to-face and fully online teaching and learning. In most instances, such studies have dealt with questions relating to how we might enhance or transform teaching and learning through using digital technologies. In particular, in addition to more traditional synchronous (face-to-face) teaching, digital technologies have enabled asynchronous (learning anywhere, anytime, anyplace) and what Dalgarno (2014) suggests are quasisynchronous or polysynchronous learning experiences. Dalgarno defines polysynchronous learning as 'the integration of learner-learner, learner-content and learner-teacher interaction through a blending of multiple channels of face to face, asynchronous online and synchronous online communication' (p. 676).

To illustrate with an example of research into online learning, Patrick and Powell (2009) undertook a meta-analysis and identified effectiveness studies (e.g. Barbour & Reeves, 2009; Smith, Clark, & Blomeyer, 2005) in that meta-analysis and concluded that:

> Online learning has the potential to transform teaching and learning by redesigning traditional classroom instructional approaches, personalizing instruction and enhancing the quality of learning experiences. The preliminary research shows promise for online learning as an effective alternative for improving student performance across diverse groups of students. (Patrick & Powell, 2009, p. 9)

In a more recent review, Shattuck (2015) noted that the K-12 literature throughout the period from 2008 to mid-2013 had focused on studies which compared face-to-face with online learning, whereas research throughout 2013–2015 had focused more 'on teacher training, professional development, and leadership for more effective online learning' (Shattuck, 2015, p. 1). According to Shattuck, for Junior Secondary students who are learning *with* digital technologies, interaction and learner engagement are important. But these need to be deliberately encouraged and supported through 'assuring a logical and explained connection among the learning objectives, learning materials, activities, learning necessary technologies, and assessments, to optimal effect' (Shattuck, 2015, p. 12).

In parallel with these examples of research about learning *with* digital technologies, there is the complementary challenge of teaching *with* digital technologies. Again, this discussion can apply specifically to teaching STEM subjects in Junior Secondary but can relate to all curriculum areas. A growing research interest in teaching with technologies is evident through the research on Technological Pedagogical Content Knowledge (TPACK). In brief, Shulman (1986) made a significant contribution to teaching as a profession by proposing that teachers required pedagogical content knowledge (PCK) which is drawn upon to inform decisions about how teachers might best represent content for learning in a given context. However, while teachers might have used technologies previously – such as blackboards, overhead projectors and whiteboards – the more dynamic technologies available to teachers in recent times have resulted in teaching with digital technologies being seen as a 'wicked problem' (Mishra & Koehler, 2006). In this context, it needs to be remembered that Shulman's seminal work was conducted before the Internet became available in schools and well before Google, Twitter, iPads, mobile phones with their current capabilities, social media and the like. To this end, the impact of digital technologies on learning and teaching needs to be seriously considered as they now permeate most, if not all, aspects of the learning contexts in schools.

Due to the significant technological changes noted above, Mishra and Koehler (2006, 2007), more than a decade ago, suggested that ICT had changed so significantly that Shulman's concept of PCK needed to be built upon to include technological knowledge (TK). The intersection of technological knowledge (TK) with content knowledge (CK) and pedagogical knowledge (PK) is known as TPACK (Thompson & Mishra, 2007). A search of the LearnTechLib (see https://www.learntechlib.org/) using 'TPACK' as a search word identified 1219 research publications since 2006. Given the amount of research, it is a reasonable expectation that teachers, who are teaching *with* digital technologies in Junior Secondary teaching contexts, will understand the ways in which technology, pedagogy and content interrelate.

Secondly, learning and teaching *about* digital technologies find expression through defined content or subject areas which Junior Secondary students might study. These subjects offered in Junior Secondary are likely to vary markedly in different countries and schooling systems. While those differences are likely, as digital technologies emerged, early curriculum has tended to focus on students learning how computers worked (Zagami, 2015). This shifted when 'some educators (for example, Papert, 1980) saw in this technology the potential for new ways of thinking and understanding the world through the study of programming' (Zagami, 2015, p. 170).

My own experience as a leader in schools in the early 1990s is consistent with this, as *The Queensland Sunrise Centre* – a 4-year project – was established at Coombabah State School, in Queensland, Australia, where I was the Deputy Principal at that time. I conducted my doctoral study (Finger, 1996), with full ethics approval, by completing a case study of that initiative. The *Queensland Sunrise Centre* was a 'lighthouse project' established in one site to inform other schools about laptop programmes and about the value which programming might add to student learning. The initiative involved students in Years 6, 7 and 8 each having a laptop computer for 3 years, and they used LogoWriter, which was the first programming language written especially for children, to programme and solve problems. This project was heavily influenced by Papert's (1980) thinking. Interestingly, in the second edition of his book *Mindstorms: Children, Computers, And Powerful Ideas* (Papert, 1993), Papert had observed that many students 'were now able to use programming as an expressive medium to study other topics rather than as a skill to be learned for the sake of learning it' (Papert, 1993, p. xvii). This was found to be the case in the study that I undertook and in the research undertaken by Ryan (1991) in the first year of the project. Students became very confident in using computers and became competent at programming, not merely in a technical or procedural sense, but they regarded computers as integral to their lives. The students demonstrated that some long-held beliefs underestimated what students might learn at these ages. For example, Ryan (1991) and Finger (1996) observed students undertaking complex programming, demonstrating knowledge of computational variables and displaying innovative and creative forms of written and graphical expression.

However, despite isolated examples such as *The Queensland Sunrise Centre* and other sites drawing upon Papert's thinking, there was a recognisable shift since the 1990s from understanding how programming, information, network and communication systems worked to learning how to use software applications such as desktop publishing, word processing, spreadsheets and databases. Zagami (2015) suggests that this shift from students as creators of programmes and content to end users of computer programmes contributed to a decline in computer science and information technology studies.

In more recent times, there has been a resurgence in computer science as a study, largely reflected in governments and education systems recognising the importance of the STEM agenda and a renewed call for STEM skills needed for future jobs. Government policies on computer science have called for a revitalisation of digital technologies in schools, some of which is based on learnings from other nations. For example, the Queensland Government, in *#coding counts A discussion paper on coding and robotics in schools* (Queensland Government, 2015), noted that the UK has introduced mandatory computer programming for all students aged 5–16 years, that Finland will integrate coding across subjects for students aged 7–15 years and that 'The Queensland Government is committed to making sure that every student will learn the new digital literacy of coding and have the opportunity to apply these skills through robotics' (Queensland Government, 2015, p. 9).

What these policies are highlighting is that there is a greater demand for a wide range of computer skills and technologies than had been in the past and that these skills are instrumental in the preparation of students for the worlds they will enter when they leave schools. A further example, outside the Australian context, is the *K-12 Computer Science Framework* (K-12 Computer Science Framework Steering Committee, 2016) in the USA which reinforces the contemporaneous skills that are needed for students when they exit schools, particularly in terms of the inclusion of the diversity within the society. The document is seeking computer skills for all students rather than an elite few and:

> ...comes at a time when our nation's education systems are adapting to a 21st century vision of students who are not just computer users but also computational-literate creators who are proficient in the concepts and practices of computer science. ...The framework provides a unifying vision to guide computer science from a subject for the fortunate few to an opportunity for all. (p. 4)

The US framework includes core concepts to be studied – computing systems, networks and the Internet, data and analysis, algorithms and programming and impacts of computing – and core practices to be developed, fostering an inclusive computing culture, collaborating around computing, recognising and defining computational problems, developing and using abstractions, creating computational artefacts, testing and refining computational artefacts and communicating about computing. The framework, representing an example outside Australia, outlines expectations, for example, for grade bands, which are Junior Secondary, and also highlights the relationships between computer science, science and engineering and mathematics practices. This aligns with other national documents, including Australia, as I discuss in the ensuing sections in greater detail.

In an Australian context, *The Australian Curriculum: Technologies* (ACARA, 2016a) is a relatively recently approved curriculum for study by students from Prep to Year 10. It consists of two subject areas, namely, *Design and Technologies* (ACARA, 2016b) and *Digital Technologies* (ACARA, 2016c, 2016d). Clear expectations for students in Junior Secondary are stated in year-level bands, including for Junior Secondary, through Content Descriptions and Achievement Standards for Years 7–8 and for Years 9–10.

To illustrate, the following are extracts from the Achievement Standards (ACARA, 2016c, 2016d) for Junior Secondary school students in the *Digital Technologies* subject:

> Year 7 and 8 Achievement Standard

By the end of Year 8…

…Students plan and manage digital projects to create interactive information. They define and decompose problems in terms of functional requirements and constraints. Students design user experiences and algorithms incorporating branching and iterations, and test, modify and implement digital solutions. They evaluate information systems and their solutions in terms of meeting needs, innovation and sustainability. They analyse and evaluate data from a range of sources to model and create solutions…

- Year 9 and 10 Achievement Standard

By the end of Year 10…

…Students plan and manage digital projects using an iterative approach. They define and decompose complex problems in terms of functional and non-functional requirements. Students design and evaluate user experiences and algorithms. They design and implement modular programs, including an object-oriented program, using algorithms and data structures involving modular functions that reflect the relationships of real-world data and data entities. They take account of privacy and security requirements when selecting and validating data. Students test and predict results and implement digital solutions. They evaluate information systems and their solutions in terms of risk, sustainability and potential for innovation and enterprise…. (ACARA, 2016c, 2016d, p. 1)

In learning *about* digital technologies, the *Digital Technologies* (ACARA, 2016c, 2016d) subject area develops students' systems thinking, computational thinking and design thinking. Junior Secondary school students are expected to undertake project-based learning in designing digital solutions. This involves new knowledge and skills for many students, particularly in relation to designing coding solutions, robotics solutions and information solutions. Project-based learning and these ways of thinking have implications for pedagogical approaches and for the professional identity of teachers. Zagami (2015), for example, appropriately warns that *The Review of the Australian Curriculum* (Wiltshire & Donnelly, 2014) 'was particularly dismissive of a pedagogical reform agenda in the Australian Curriculum and strongly advocated for a return to more explicit forms of direct instruction' (Zagami, 2015, p. 177). Such a move in potentially reverting to older approaches highlights the pedagogical tensions facing teachers and students in Junior Secondary, as learning *about* digital technologies requires more than direct or explicit instruction. Adopting a predominantly direct or explicit instruction of pedagogical approach compromises Papert's warning that it is inappropriate for students to be learning programming for its own sake.

While making reference to *The Review of the Australian Curriculum*, it is worthwhile noting that the review, referring to digital technologies as a subject area, made the recommendation that 'This learning area should be introduced from Year 9' (p. 211). There was no research-based or evidence-based justification provided to support this recommendation. Fortunately, the *Digital Technologies* subject area has been designed to provide the scope and sequence from Prep to Year 10 that enables systems thinking, computational thinking and design thinking to be developed throughout the year levels leading up to Junior Secondary school. The implication for Junior Secondary is that, in future, students should transition from primary school to Junior Secondary school with the appropriate Achievement Standards developed. Currently, as this subject area has not yet been implemented in primary schools, teachers in Junior Secondary schools are finding that they have to assume that those students come with little or no knowledge about digital technologies as a subject area.

Part of the tension in the approaches that underpin the philosophical basis to the teaching approaches advocated in the Australian Curriculum could be related to the advice sought in the establishment of the curriculum framework. Particular people have particular biases in how they see the world and how subjects should be taught. Interestingly, Wiltshire and Donnelly (2014) engaged a 'subject matter specialist' in *The Review of the Australian Curriculum*. That specialist made the following problematic suggestion that flew in the face of much of the research and advocacy that has occurred in the field of ICT learning in Australia and internationally:

> Consideration should be given for renaming 'digital technologies'. It is a name that is not readily identifiable as a commonly known term in the IT industry, Australian tertiary education or education systems in Canada, Finland, Singapore or the UK. (Wiltshire & Donnelly, 2014, p. 210)

This suggestion was certainly problematic in relation to tertiary education, and the term is also in widespread use in both primary and secondary schools in many educational contexts. No supporting evidence was provided as to why that defective assertion was made. Indeed, digital futures are evident in the strategic planning and priorities of tertiary institutions in Australia. An example is Deakin University's (2015) *Live the Future Agenda 2020: 2015–2017 Triennium*, which states that 'Deakin will harness the power, opportunity and reach of the digital world in all that it does' (p. 16).

Similarly, the recommendation tends to look backwards, and, given the dynamic technological changes and innovation, digital technologies require a future orientation. For example, Niemi, Multisilta, Lipponen and Vivitsou (2014) provide evidence of changes in Finnish curriculum in *Finnish Innovations and Technologies in Schools: A Guide towards New Ecosystems of Learning*. They indicate that 'The expansive use of digital technologies in education has generated the need for fresh perspectives and approaches in the development of pedagogical methods and models' (p. x). Thus, while the 'subject matter specialist' indicated that 'digital technologies' were not a readily identifiable term in Finland, there is a new curriculum emphasis on digital literacies, including coding in schools in Finland. Again, it is wise to ask the questions suggested by Selwyn (2015), such as: What is being said about education that might be useful? What is being said about education that teachers might wish to challenge and talk back to?

A distinction needs to be made between *The Australian Curriculum: Technologies* (ACARA 2016a) which consists of the *Digital Technologies* (ACARA, 2016c, 2016d) subject area and the ICT Capability (ACARA, 2016e) which is a general capability intended to be developed across all learning areas. ICT Capability is expected to be developed, as students learn:

> …to use ICT effectively and appropriately to access, create and communicate information and ideas, solve problems and work collaboratively in all learning areas at school and in their lives beyond school. ICT capability involves students learning to make the most of the digital technologies available to them, adapting to new ways of doing things as technologies

> evolve and limiting the risks to themselves and others in a digital environment. (ACARA, 2016a)

It is anticipated that through the curriculum, Junior Secondary students should develop the ICT Capability, which is organised around five interrelated elements in a learning continuum. Those elements include applying social and ethical protocols and practices when using ICT, investigating with ICT, creating with ICT, communicating with ICT, communicating with ICT and managing and operating ICT. In essence, developing this ICT Capability within STEM subjects can include approaches that assist learning *with* and learning *about* digital technologies.

As an exercise, consider how relevant the chapters in this book deal with digital technologies within their STEM focus. Do they provide guidance for learning *with* digital technologies and/or learning *about* digital technologies? In Junior Secondary schooling, there is the danger that, without digital technologies being a subject itself, it can tend to be treated only as learning *with* digital technologies or as a general capability to be developed within each learning area. This is the long-standing tension between having a stand-alone subject as opposed to an integrated subject. Both of those are desirable, but neither is sufficient within a strong STEM agenda. Those approaches are unlikely to develop systems thinking, computational thinking and design thinking needed in a project-based learning approach whereby students create digital solutions to real-world problems.

In summary, learning *about* digital technologies in Junior Secondary needs to be understood within the curriculum developments occurring within various education systems internationally. There are clearly implications for rethinking pedagogy and managing the tensions evident in the diverse range of perspectives about how teachers should teach *about* digital technologies and how students should learn *about* digital technologies. The TPACK research suggests that teachers need technological knowledge, pedagogical knowledge and content knowledge. Directly relevant to this discussion about learning *with* and learning *about* digital technologies, Zagami (2015) indicates that teacher identity and teacher confidence are likely to be challenged.

Hattie (2012), from his meta-analysis of educational research, has established that teachers do make a difference and that 'expert' teachers make a greater difference than other teachers. An expert teacher, according to Hattie, has the following five attitudes and beliefs:

1. Expert teachers identify the most important ways to represent the subjects they teach.
2. Expert teachers create an optimal classroom climate for learning.
3. Expert teachers monitor learning and provide feedback.
4. Expert teachers believe all students can reach the success criteria.
5. Expert teachers influence a wide range of student outcomes not solely limited to test scores.

In relation to digital technologies, insights into the 'expert teacher' have been suggested by Zagami (2015) who has argued that effective change is needed to occur, through teachers needing to have the ability to:

- Support the development of student ICT general capabilities.
- Achieve the explicit curriculum goals of the digital technology subject.
- Make effective use of educational technologies to improve teaching and learning.
- Support pedagogical reform through the supportive use of ICT (Zagami, 2015, p. 176).

What Zagami's work has raised is that there needs to be support for teachers to develop this range of skills to become the expert teachers, as suggested by Hattie (2012) in digital technologies, in order to effect positive change, transitions and skills in their pedagogy so that Junior Secondary students are able to build the range of skills and dispositions in ICT that are needed for their future lives.

10.4 Junior Secondary Students and Digital Technologies: Trends, Challenges and Technologies

In this section, trends, challenges and technologies in relation to Junior Secondary school students and digital technologies are explored through examining student's voice from Project Tomorrow's *Speak Up Research Project for Digital Learning* (Project Tomorrow, 2016a) and complemented by examining the *NMC/CoSN Horizon Report: 2016 K-12 Edition* (Adams Becker et al., 2016).

As cautioned in the introduction to this chapter, it is problematic to adopt a technocentric approach to technology and teaching, and a critical approach is encouraged to be able to ask (and answer) the questions posed by Selwyn (2015). Therefore, it becomes important to consider Junior Secondary school students and how there is a need to challenge assumptions that underpin the approaches taken in schools. Such challenges include whether or not such technologies are purely beneficial for teaching and learning that all students have equitable access to technologies, and/or Junior Secondary students are all confident and competent users of digital technologies when they complete their mandated studies.

To illustrate, one of the key topics that needs to be considered is whether or not these students, who have been referred to as 'digital natives' (Prensky, 2001) and through this construction, are characterised as different from their teachers who were seen to be 'digital immigrants'. According to Prensky, this difference could be attributed to these students having been born and immersed in a world that was technology-rich. Similarly, more recent generational stereotypes, such as 'Millennials' or 'Generation Me' being narcissistic and lazy, the Facebook generation and all about instant gratification, are being contested – often by those who were born in those generations. This differentiation of the digital native is evident when Angone (2014) argued that there are considerable differences among the estimated 1.3 billion Millennials in countries as diverse as China and the USA. Angone suggests that the diverse cohorts represent complex, diverse and unique individuals, so it becomes impossible to homogenise the generation. Angone points out that:

> …Some Millennials will have an IV of technology hooked to their veins. Some still like the feel and smell of a paper book…
> …Some are immersed in social media and their iPhone, and yet at the same time feel very much alone…
> …Some Millennials embrace being called a Millennial. Others can't stand it. (Angone, 2014)

Anecdotally, from my numerous discussions with Junior Secondary teachers and initial teacher education students undertaking their professional experiences in Junior Secondary schooling, it seems that many Junior Secondary students are presented with curriculum, assessment and technologies that are challenging and unfamiliar to them. This suggests that homogenising the Junior Secondary student as a 'digital native' is highly problematic and quite simplistic as they vary considerably in terms of their relationship and fluency with digital tools. Moreover, some Junior Secondary students are disengaged with the curriculum, unless it is presented in relevant, meaningful ways, which is symptomatic of the digital native discourse. Teachers also related stories that some of their students were either naïve or unaware of key aspects of the ICT Capability (ACARA, 2016e) expected of students in Year Levels 7–8 and 9–10. To illustrate, major issues have arisen due to Junior Secondary school students not applying social and ethical protocols and practices when using ICT and not communicating appropriately with ICT. These anecdotal reports are supported by research (Cowan, 2011; Thompson, 2013; Waycott, Bennett, Kennedy, Dalgarno, & Gray, 2010), referred to by Johnson (2015), that demonstrates that the 'digital natives' who have entered secondary schools, and even universities, 'are no better or more adept at using technology than previous cohort of students' (Johnson, 2015, p. 12), thus challenging the all-assuming position of Prensky. Clearly, there are differences in this cohort of students, and it is simplistic to label them all as digital natives.

Within this contested space, it becomes important to gain deeper and more complex, rather than superficial, understandings of Junior Secondary school students and their use of digital technologies. It is not sufficient to label them all as digital natives but rather to see them as a diverse cohort, like previous generations. Collecting and analysing data from students, through methodologies such as digital technologies inventories and surveys, seeking their voice about their use of digital technologies for their personal use and for learning purposes, is desirable, if not essential. In this way, a much richer understanding of them as learners and users of ICTs is then possible rather than to assume they are all digital natives.

Undertaken in the USA, the report *From Print to Pixel: The role of videos, games, animations and simulations within K-12 education* (Project Tomorrow, 2016a, p. 2) provides interesting and useful insights into the views of teachers and students. It is a comprehensive study that was informed by *Project Tomorrow's Speak Up Research*

Project for Digital Learning (Project Tomorrow, 2016b, p. 2), drawing upon the views of students, teachers and community members from 7600 schools and 2600 districts in the USA and reports around the world. The report affirmed that there is an 'increasing use [of] videos, games, animations and simulations across all segments of the population to support both informal learning and entertainment' (Project Tomorrow, 2016b, p. 3), and 'students do not see learning as only happening from 8 to 2:30 each day' (Project Tomorrow, 2016b, p. 3).

When compared with 2012 data, teachers had increased their use of digital content, particularly online content, in the classroom, and there had been an increased use of online videos, games, curriculum and textbooks, as well as animations, virtual field trips, self-created videos and simulations. The importance of this study is that it highlights how young people are using digital media and tools in their lives and the links that this has with schooling.

These Speak Up findings presented the *10 Things Everyone Should Know about K-12 Students' Digital Learning* (Project Tomorrow, 2016c), and these have relevance for our understanding of Junior Secondary students' use of digital technologies. Those '10 things everyone should know' are summarised here:

1. **Learning via YouTube**. Students (38%) reported that online videos help them with their homework; 27% regularly watch teacher-created videos.
2. **Students are mobilists**. Personal access to mobile devices has reached significant tipping points with 72% (Grades 6–8), and 86% (Grades 9–12) of students are smartphone users.
3. **More games please**. Almost two thirds want to use digital games for learning in school; 53% believed they had better grades by using technology for learning.
4. **Students want to code – especially girls**. Fifty percent of girls in Grades 6–8 want to code.
5. **Teacher – I have a question!** Students are regularly using digital tools outside of school to communicate with their teachers about school; 48% email and 15% text.
6. **Tweet-Tweet?** Forty-seven percent of students in Grades 9–12 are now using Twitter, compared with only 11% in 2011.
7. **I'll take my learning mobile**. Seventy-six percent of students think every student should have access to a mobile device during the school day to support learning.
8. **Watching online videos**. Seventy-four percent of students in Grades 6–8 say they watch online videos for schoolwork, and they mostly do this for science.
9. **Change in social media use**. Forty-three percent of Grades 6–12 students indicated that they never use Facebook, but students are spending more time with content creation sites; 54% indicated that they use YouTube all the time.
10. **Goodbye 1:1!** Students use different tools for different tasks; e.g. laptops top the students' lists for writing a report, taking online tests and working on group projects, while smartphones are first for connecting with classmates and accessing social media.

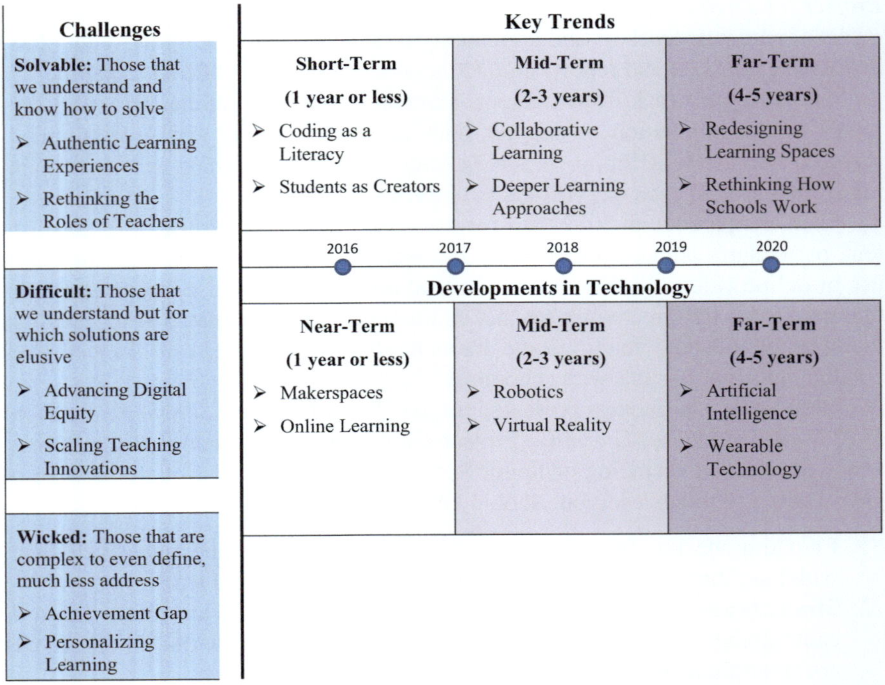

Fig. 10.1 Overview of trends, challenges and technology developments in K-12 education (Adapted from Adams Becker et al., 2016, p. 2)

The New Media Consortium's Horizon Project, which examines technologies and their potential impact on teaching, learning and creative inquiry in educational contexts, has a sustained track record of 15 years of research and publications. Of relevance is the *NMC/CoSN Horizon Report: 2016 K-12 Edition* (Adams Becker et al., 2016) which identified key short-term, midterm and long-term trends accelerating technology adoption in K-12 education, challenges impeding technology adoption and important developments in educational technology over three Time-to-Adoption horizons. These are visually displayed in Fig. 10.1 across the period from 2016 to 2020 and provide the basis for considering their implications, transferability and potential impact in Junior Secondary schools.

Both learning *with* and learning *about* digital technologies within the STEM paradigm find expression in the identified developments in technology across the three adoption horizons of near-term (1 year or less), midterm (2–3 years) and far-term (4–5 years). Specifically, as shown in Fig. 10.1, the trends *Coding as a Literacy* and *Students as Creators* are becoming increasingly evident where digital technologies are being studied in Junior Secondary schools. There are examples of schools with students studying coding as an integral component of project-based learning whereby students create digital solutions to real-world problems.

The developments in technology, shown in Fig. 10.1, while not providing an exhaustive, definitive list of all technologies in Junior Secondary school, highlighted makerspaces, online learning in the near-term horizon, robotics and virtual reality in the midterm horizon and artificial intelligence and wearable technology in the far-term horizon. Those developments are elaborated upon in that report through providing explanations and examples of these occurring in schools, including Junior Secondary school contexts.

Furthermore, as shown in Fig. 10.1, the suggested solvable challenges – those that we understand and know how to solve – were authentic learning experiences and rethinking the role of teachers. TPACK is referred to here and reinforces the importance of this conceptualisation for STEM. Adams Becker et al. (2016) argue that 'Educators can also benefit from innovative learning models such as the Technological Pedagogical and Content Knowledge (TPACK) framework which describes the types of knowledge teachers need to effectively integrate technology into curricula' (p. 25). The report highlighted the creation of the *Practitioner's Guide to TPACK*, which is a site developed by the National Technology Leadership Coalition.

Each of these trends, challenges and technologies is elaborated upon in that report and case studies are provided. Some elaboration and examples of those case studies in Junior Secondary schooling for *Coding as a Literacy* and *Students as Creators* are discussed here to illustrate their importance and the digital technologies being used for teaching and learning.

10.4.1 Coding as a Literacy

Coding as an area for curriculum study has been argued as essential for current and future employability and workforce needs, and the case is being made in many school systems for coding to be embedded into K-12 curricula. Adams Becker et al. (2016) noted that 'Schools worldwide are developing coding programs in which students collaboratively design websites, develop educational games and apps, and design solutions to challenges by modelling and prototyping new products' (p. 16).

The rise of coding signals the shift, outlined earlier in this chapter, from students studying how to use computers, software applications and programmes to coding and programming that enables students to control how those devices interact with them (Adams Becker et al., 2016; Zagami, 2015).

The *NMC/CoSN Horizon Report: 2016 K-12 Edition* (Adams Becker et al., 2016) noted that coding is becoming a part of the curriculum in parts of Europe, such as Estonia, the UK has mandated coding in primary and secondary school, and, in 2016, as outlined earlier in this chapter, Finland requires primary school students to learn coding. As shown in Table 10.1, in the *Digital Technologies* (ACARA, 2016c, 2016d) subject area of *The Australian Curriculum: Technologies* (ACARA, 2016a), students from Prep to Year 10 are expected to develop concepts, program-

Table 10.1 *Digital Technologies* subject area concepts, programming and test and debug expectations

By the end of:	Concepts	Programming	Test and debug
Year 2	N/A	N/A	N/A
Year 4	Branching (decisions) and user input	Visual programming	
Year 6	Iteration (repetition)	Visual programming	
Junior secondary expectations: Years 7–10			
Year 8	User interfaces and functions	General purpose text programming	In algorithm content descriptor
Year 10	Modularity, algorithms and data structures	Object-oriented programming	In algorithm content descriptor

ACARA (2016c, 2016d)

ming and testing and debugging capabilities with increasing sophistication throughout those year levels. For example, students in Years 7 and 8 are expected to 'design user experiences and algorithms incorporating branching and iterations and test, modify and implement digital solutions', while students in Years 9–10 are expected to 'design and implement modular programs, including an object-oriented program, using algorithms and data structures involving modular functions that reflect the relationships of real-world data and data entities' (ACARA, 2016c, 2016d).

10.4.2 Students as Creators

While there has been a lot of rhetoric about students creating knowledge and content, this is very possible in the digital environment. The *NMC/CoSN Horizon Report: 2016 K-12 Edition* (Adams Becker et al., 2016) observed that 'A shift is taking place in schools all over the world as learners are exploring subject matter through the act of creation rather than the consumption of content' (p. 18). Digital technologies are supporting this shift. In particular, the growing accessibility of mobile technologies and social media apps, such as Instagram and Snapchat, enables students to share photographs and videos and investigate, create and produce stories. This also enables game development, making and programming in ways in which students are learning to be entrepreneurial and inventors. Digital technologies include free and increasingly affordable platforms, such as Socrative, Kahoot, Nearpod and Google Forms which can capture and save learning progress evidence. The most recent Horizon Report (Adams Becker et al., 2016) provided an example at South Miami Middle Community School where students tracked climate change by downloading satellite image data from the NASA NEO website. Similarly that report highlighted how Minecraft enables students to 'create visual representations and simulations of concepts they are studying while learning problem-solving skills' (Adams Becker et al., 2016, p. 19).

Teachers in STEM subjects in Junior Secondary schools might consider how learning experiences are being designed, or might be designed, to enable students to be creators, as well as consumers of content. This design challenge requires considerations of the curriculum, learning objectives and outcomes and the needs of the learners, and these should inform the selection of the digital technologies. In relation to learning about digital technologies, it is possible to have 'unplugged' learning experiences, but these will be insufficient by themselves for Junior Secondary students to create content, for example, using sound, animation, video and digital images.

As Selwyn (2015) suggests, we should ask questions here, such as – What is actually new here? What are the unintended consequences or second-order effects? It's worthwhile noting that Brown and Gormley (2016) have provided critical reflections, for example, in relation to the Horizon Reports in Higher Education in Ireland in asking – What did we learn from the exercise? They cautioned that the report relied upon select voices and there was limited dialogue in its methodology; it was pedagogically barren but had more than 3000+ downloads and reinforced the need for infrastructure. The *NMC/CoSN Horizon Report: 2016 K-12 Edition* (Adams Becker et al., 2016), while employing a similar methodology, is far from pedagogically barren and highlights the challenges and trends, including the 'wicked' problems, as well as the developments in technology being forecasted. Thus, a balanced view is that, while the K-12 Horizon Reports might be subjected to some criticism, they are useful in terms of dissemination of the distilled ideas for those involved in teaching with digital technologies.

As teacher capabilities, and especially those of future teachers, are important, the following section examines initial teacher education in relation to digital technologies. This is discussed in the following section, through presenting a summary of the *Queensland Digital Technologies Summit 2016: Initial Teacher Education*, conducted on 15 June 2016 in Brisbane, Queensland, Australia.

10.4.3 Digital Technologies Summit: Initial Teacher Education

The *Queensland Digital Technologies Summit 2016: Initial Teacher Education* focused on initial teacher education (ITE) in Australia but was situated within international developments and research. It was deliberately designed to be distinct from a conference or 'talk fest', by resulting in the construction of a Summit communique with strategies for action.

The resulting *Communique: Queensland Digital Technologies Summit 2016: Initial Teacher Education* (Finger et al., 2016) was achieved through interactive engagement and live polling using technology which captured participant responses and questions. Presenters included guest speakers, panel sessions which stimulated group discussions and case studies which provided stories of digital technologies in practice. Consistent with the discussion in this chapter, the Summit made the distinctions between learning and teaching *with* digital technologies and learning *about* digital technologies in relation to *The Australian Curriculum: Technologies Learning Area* the Technologies Learning Area, which includes the *Design and Technology* and *Digital Technologies* subject areas, and the ICT Capability as a general capability to be developed across all learning areas.

The School of Education and Professional Studies, Griffith University, with the support of the Queensland Council of Deans of Education and the Queensland College of Teachers (QCT), hosted the Summit which brought together key stakeholders to:

- Identify and prioritise digital technologies challenges and issues in ITE.
- Co-construct a shared digital technologies philosophy in ITE.
- Co-construct a shared digital technologies framework for ITE.
- Identify shared actions and strategies for digital technologies learning and teaching in ITE.

Participant responses, in rank order of importance, to the question – What do you consider is the highest priority digital technologies issue/challenge in initial teacher education? – were:

1. Flexible, open, creative mindset for school students and ITE students (agile/resilient/coping with change).
2. Resources/access/infrastructure for the classroom/technology.
3. Alignment between ITE in universities and school practices.
4. Practical examples and preparing initial teacher education ITE students for the realities of teaching.
5. Digital technologies finding expression in curriculum, pedagogy and assessment.
6. ITE students and school students need to be creators as well as users.

The conceptualisation displayed in Fig. 10.2 was used as a stimulus for panel discussions and case studies to develop a shared digital technologies philosophy and framework and inform the agreed strategies arising from the Summit. That conceptualisation referred to contexts of use; ethical wellbeing and citizenship; curriculum, teaching and learning; communication and collaboration; creativity and innovation; and initial teacher education (ITE).

The key strategies for digital technologies learning and teaching in initial teacher education identified by participants at the Summit are presented in Table 10.2.

The Summit achieved its objectives by identifying and prioritising digital technologies challenges and issues in initial teacher education, co-constructing a shared digital technologies philosophy in initial teacher education, co-constructing a shared digital technologies framework for initial teacher education and identifying the shared actions and strategies for digital technologies learning and teaching in initial teacher education. Participants expressed a very strong commitment to action informed by these outcomes.

In terms of digital technologies within a STEM agenda, those actions have the potential to build the capabilities needed for future teachers to impact positively

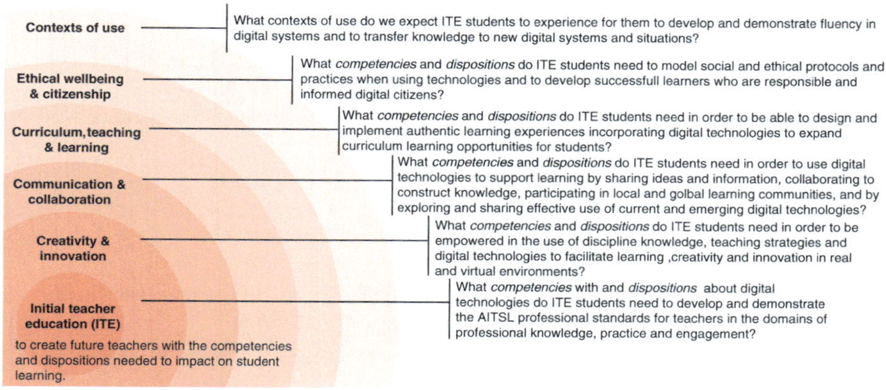

Fig. 10.2 Summit stimulus: conceptualisation for co-constructing a shared digital technologies philosophy, framework and actions in initial teacher education

where it matters most, that is, to impact positively upon their students learning *with* and learning *about* digital technologies. In terms of Junior Secondary, ITE programmes need to embrace this call for action so that future teachers in Junior Secondary have the capabilities and teacher efficacy needed.

10.5 Conclusion

This chapter focused on technology and, more specifically, on digital technologies, within the STEM paradigm viewed as an integrated study of Science, Technology, Engineering and Mathematics.

Through commencing the chapter with an examination of learning *with* and *about* digital technologies in Junior Secondary school contexts, the case was made that digital technologies are being recognised as a study in its own right within the Junior Secondary school curriculum. That is, examples such as the UK, Finland and Australia were provided, whereby learning *about* digital technologies is finding expression through many schooling systems now requiring this to be a mandatory subject for study for students from primary school to secondary school.

As students learn and develop systems thinking, computational thinking and design thinking in primary schools, as well as being introduced in secondary schools, this has obvious implications for Junior Secondary school teachers. It was suggested that learning with digital technologies and learning about digital technologies now require teachers to develop Technological Pedagogical Content Knowledge (TPACK) (Mishra & Koehler, 2006).

Key trends, significant challenges and important developments in digital technologies were discussed by drawing upon student voice about their use of digital technologies provided by the Speak Up findings (Project Tomorrow, 2016a, 2016b).

Table 10.2 Actions and strategies for digital technologies learning in ITE

Area	To ensure ITE student digital technologies learning by
Contexts of use	Employing real connections with real-life issues
	Exposing ITE students to professional networks through practical experience
	Fostering university-industry partnerships
	Requiring ITE students to create with digital technologies
	Employing ITE students in roles as mentors
Ethical wellbeing and citizenship	Applying ethical considerations in all learning areas
	Modelling the development and debate of ethical and policy positions as a problem-solving process
	Requiring ITE students to develop a professional profile/digital footprint that supports professional practice
	Requiring ITE students to understand and teach the appropriate uses of data (in particular, student data)
	Understanding the expectations of university and school policies relating to ethical issues
Curriculum, teaching and learning	Promoting problem-solving in engagement with digital technologies in real-world contexts, case studies and examples
	Developing teaching strategies that allow ITE students to critically evaluate and justify
	Focusing assessment as journey rather than end product
	Making curriculum, teaching and learning to be relevant to ITE student needs and the school curriculum
Communication and collaboration	Having ITE students develop a thorough knowledge of the curriculum
	Forming makerspaces with local schools
	Reflecting the use of digital mediums through professional
	learning networks
	Working with school-based digital champions
Creativity and innovation	Modelling and providing opportunities for ITE students to create and innovate
	Providing opportunities to explore digital technologies
	Valuing adaptability, flexibility, change, collaboration, communication and critical thinking
	Immersing ITE students in problem-based learning tasks across the curriculum in real-life and virtual contexts
	Liaising with schools to see how it works in schools
	Including creativity as an assessable aspect of all assessments
Initial teacher education	Encouraging collaboration, modelling and demonstrating the pedagogies necessary to embed digital technologies
	Providing authentic learning experiences and high-quality examples
	Ensuring depth of learning with regard to the curriculum
	Providing access to quality mentoring

Student voice was complemented by the *NMC/CoSN Horizon Report 2016 K-12 Edition* (Adams Becker et al., 2016).

Collectively, the discussion and analysis presented in this chapter require initial teacher education programmes to develop future Junior Secondary teachers with the capabilities to effectively teach *with* and *about* digital technologies. Specific reference was made to the outcomes of the *2016 Queensland Digital Technologies Summit: Initial Teacher Education* (Finger et al., 2016) and the presentation of the co-constructed philosophy and strategies for action for future teachers developed at that Summit.

These are times of significant educational challenge for Junior Secondary schools, their teachers and their students, situated in complex contexts characterised by new and emerging digital technologies. The trends, challenges and digital technologies provide opportunities for Junior Secondary students to create digital solutions to real-world problems. Learning *with* and *about* digital technologies has the potential to enhance student learning in all areas of a well-integrated STEM paradigm.

References

Adams Becker, S., Freeman, A., Giesinger Hall, C., Cummins, M., & Yuhnke, B. (2016). *NMC/CoSN Horizon report: 2016 K-12 edition*. Austin, Texas: The New Media Consortium. Retrieved November 5, 2016, from https://www.nmc.org/publication/nmc-cosn-horizon-report-2016-k-12-edition/

Anderson, N. (2015). Digital technologies and equity: Gender, digital divide and rurality. In M. Henderson & G. Romeo (Eds.), *Teaching and digital technologies: Big issues and critical questions* (pp. 46–56). Port Melbourne, Australia: Cambridge University Press.

Angone, P. (2014). *Millennials are not the problem – The Stereotypes are!* Retrieved November 5, 2016, from http://allgroanup.com/featured/enough-with-the-millennial-stereotypes/

Auld, G., & Djabibba, L. (2015). Using digital technologies with Aboriginal and Torres Strait Islander students. In M. Henderson & G. Romeo (Eds.), *Teaching and digital technologies: Big issues and critical questions* (pp. 57–70). Port Melbourne, Australia: Cambridge University Press.

Australian Curriculum, Assessment and Reporting Authority (ACARA). (2016a). *Australian curriculum: Technologies learning area*. Retrieved November 5, 2016, from http://www.australiancurriculum.edu.au/technologies/key-ideas

Australian Curriculum, Assessment and Reporting Authority (ACARA). (2016b). *Design and technologies*. Retrieved November 5, 2016, from http://www.australiancurriculum.edu.au/technologies/design-and-technologies/curriculum/f-10?layout=1

Australian Curriculum, Assessment and Reporting Authority (ACARA). (2016c). *Digital technologies*. Retrieved November 5, 2016, from http://www.australiancurriculum.edu.au/technologies/digital-technologies/curriculum/f-10?layout=1

Australian Curriculum, Assessment and Reporting Authority (ACARA). (2016d). *Information and Communication Technology (ICT) capability*. Retrieved November 5, 2016, from http://www.australiancurriculum.edu.au/generalcapabilities/information-and-communication-technology-capability/introduction/key-ideas

Australian Curriculum, Assessment and Reporting Authority (ACARA). (2016e). *Information and communication capability learning continuum.* Retrieved November 5, 2016, from http://www.acara.edu.au/_resources/General_capabilities_-_ICT_-_learning_continuum.pdf

Barbour, M. K., & Reeves, T. C. (2009). The reality of virtual schools: A review of the literature. *Computers in Education, 52*(2), 402–416.

Brown, M., & Gormley, P. (2016). *Critical reflections on the Irish horizons report: What did we learn from the experience?* Concise paper at the next generation digital learning research symposium, Dublin City University, 1st November. Retrieved November 5, 2016, from http://www.slideshare.net/mbrownz/critical-reflections-on-the-horizon-report-what-did-we-learn-from-the-exercise

Cowan, B. (2011). Why 'digital natives' aren't necessarily digital learners. *The Chronicle of Higher Education*, *58*(12), n.d.

Dalgarno, B. (2014). Polysynchronous learning: A model for student interaction and engagement. In B. Hegarty, J. McDonald & S. K. Loke (Eds.), Rhetoric and reality: Critical perspectives on educational technology. Proceedings ASCILITE, Dunedin 2014, (pp. 673–677). Retrieved from http://ascilite2014.otago.ac.nz/files/concisepapers/255-Dalgarno.pdf

Deakin University Australia. (2015). *Live the future Agenda 2020: 2015–2017 Triennium.* Retrieved November 5, 2016, from https://www.deakin.edu.au/__data/assets/pdf_file/0007/363364/2015_live_the_future_printable.pdf

Finger, G. (1996). *Evaluating the integration of learning technology in Queensland State Schools: A case study of the Queensland Sunrise Centre.* Unpublished doctor of philosophy thesis, Griffith University.

Finger, G., Pendergast, D., Zagami, J., Brown, R., Jamieson-Proctor, R., Kinnane, A., Jorgensen, M., Thompson, K., & Reynolds, J. (2016). *Communique: Queensland digital technologies summit 2016: Initial teacher education.* Queensland digital technologies summit 2016, 15th June 2016, Pulman hotel, Brisbane, Australia. Retrieved November 5, 2016, from https://www.griffith.edu.au/__data/assets/pdf_file/0011/908480/Qld-Digital-Technologies-Summit-2016--Communique.pdf

Finger, G., Russell, G., Jamieson-Proctor, R., & Russell, N. (2007). *Transforming learning with ICT: Making IT happen*. Frenchs Forest, NSW: Pearson Education Australia.

Hattie, J. (2012). *Visible learning for teachers: Maximizing impact on learning*. New York: Routledge.

Johnson, N. F. (2015). Digital natives and other myths. In M. Henderson & G. Romeo (Eds.), *Teaching and digital technologies: Big issues and critical questions* (pp. 11–21). Port Melbourne, Australia: Cambridge University Press.

K-12 Computer Science Framework Steering Committee. (2016). *K-12 computer science framework.* Retrieved November 5, 2016, from https://k12cs.org/wp-content/uploads/2016/09/K%E2%80%9312-Computer-Science-Framework.pdf

Milojevic, I. (2005). *Educational futures: Dominant and contesting visions*. London: Routledge.

Mishra, P., & Koehler, M. J. (2006). Technological pedagogical content knowledge: A framework for teacher knowledge. *Teachers College Record, 108*, 1017–1054.

Mishra, P., & Koehler, M. J. (2007). *Technological Pedagogical Content Knowledge (TPCK): Confronting the wicked problems of teaching with technology.* In Paper presented at the society for information technology and teacher education international conference 2007, San Antonio, Texas, USA.

Niemi, H., Multisilta, J., Lipponen, L., & Vivitsou, M. (Eds). (2014). *Finnish innovations and technologies in Schools: A guide towards new ecosystems of learning.* Rotterdam: Sense Publishers. Retrieved November 5, 2016, from http://www.cicero.fi/files/Cicero/site/2121-finnish-innovations-and-technologies-in-schools_ToC.pdf

Papert, S. (1980). *Mindstorms: Children, computers, and powerful ideas*. New York: Basic Books.

Papert, S. (1993). *Mindstorms: Children, computers, and powerful ideas* (2nd ed.). New York: Basic Books.

Patrick, S., & Powell, A. (2009). *A summary of research on the effectiveness of K-12 online learning. International association for K-12 online learning*. Retrieved from http:// http://www.k12.com/sites/default/files/pdf/school-docs/NACOL_ResearchEffectiveness-hr.pdf

Prensky, M. (2001). Digital natives, digital immigrants. *On the Horizon, 9*(5), 1–6.

Project Tomorrow. (2016a). *From print to pixel: The role of videos, games, animations and simulations within K-12 education. Speak up 2015 national findings*. Retrieved November 5, 2016, from http://www.tomorrow.org/speakup/pdfs/SU15AnnualReport.pdf

Project Tomorrow. (2016b). *Project tomorrow's speakup research project for digital learning*. Retrieved November 5, 2016, from http://www.tomorrow.org/speakup/

Project Tomorrow. (2016c). *Ten things everyone should know about K-12 students' digital learning*. Retrieved November 5, 2016, from http://www.tomorrow.org/speakup/pdfs/10things_students2015.pdf

Queensland Government. (2015). *#codingcounts a discussion paper on coding and robotics in Queensland schools*. Retrieved November 5, 2016, from http://advancingeducation.qld.gov.au/SiteCollectionDocuments/Coding-and-robotics-booklet.pdf

Ryan, M. (1991). *The Queensland sunrise centre: A report of the first year*. Hawthorn, Australia: The Australian Council for Educational Research Ltd.

Selwyn, N. (2015). Never believe the hype: Questioning 'digital disruption' and other big ideas. In M. Henderson & G. Romeo (Eds.), *Teaching and digital technologies: Big issues and critical questions* (pp. 182–194). Port Melbourne, Australia: Cambridge University Press.

Shattuck, K. (2015). *Review of K-12 online and blended education research literature*. Retrieved November 5, 2016, from https://www.qualitymatters.org/node/3136/download/K-12%20review%20of%20research%20summary-updated%2012-31-2015-FINAL.pdf

Shulman, L. S. (1986). Those who understand: Knowledge growth in teaching. *Educational Researcher, 15*(2), 4–14.

Smith, R., Clark, T., & Blomeyer, R. (2005). *A synthesis of new research in K-12 online learning*. Naperville, IL: Learning Point Associates.

Thompson, A. D., & Mishra, P. (2007). Breaking news: TPCK becomes TPACK! *Journal of Computing in Teacher Education, 24*(2), 38–64.

Thompson, P. (2013). The digital natives as learners: Technology use patterns and approaches to learning. *Computers & Education, 65*, 12–33.

Waycott, J., Bennett, S., Kennedy, G., Dalgarno, B., & Gray, K. (2010). Digital divides: Student and staff perceptions of information and communication technologies. *Computers & Education, 54*, 1202–1211.

Wiltshire, K., & Donnelly, K. (2014). *Review of the Australian curriculum. Students first*. Retrieved November 5, 2016, from https://docs.education.gov.au/system/files/doc/other/review_of_the_national_curriculum_final_report.pdf

Zagami, J. (2015). Digital technologies in the curriculum: National and international. In M. Henderson & G. Romeo (Eds.), *Teaching and digital technologies: Big issues and critical questions* (pp. 169–181). Port Melbourne, Australia: Cambridge University Press.

Chapter 11
STEM Education in the Brazilian Context: An Ethnomathematical Perspective

Milton Rosa and Daniel Clark Orey

Abstract Our ethnomathematical curriculum perspective helps students to demonstrate effective mathematical processes as they reason, solve problems, communicate mathematical ideas, chose appropriate representations through the development of mathematical practices, as well as recognize connections between mathematical concepts and the STEM disciplines. Our proposal is based on the *trivium curriculum for mathematics* by Ubiratan D'Ambrosio and provides communicative, analytical, and technological instruments necessary for a twenty-first-century reality. STEM education for a Junior Secondary school mathematics curriculum in Brazil proposes pedagogical action that deals with problem-solving, modeling, critical judgment, and making sense of mathematical and non-mathematical contexts, which involve distinct ways of thinking, reasoning, and knowing mathematics in practical contexts. This ethnomathematical perspective proposes a transformative pedagogy exposing its power to transform learners into critical and reflective citizens of change within the society.

Keywords Brazilian context • Ethnomathematics • Ethno-STEM curriculum • Junior secondary school • Literacy • Matheracy • Pedagogical action • STEM education • Technoracy • Trivium curriculum

11.1 Introduction

Science, technology, engineering, and mathematics (STEM) education is a universal concern. Improving the teaching and learning processes in STEM has become an economic factor for many developing countries, emerging economies, as well as the established economies such as the United States, Japan, and Germany. There seems to be a growing demand for STEM skills to meet the needs of specific sectors of the economy such as business analytics, design, smart, and high-tech industries and the

M. Rosa (✉) • D.C. Orey
Universidade Federal de Ouro Preto, Ouro Preto, Brazil
e-mail: milrosa@hotmail.com; oreydcead@gmail.com

R. Jorgensen, K. Larkin (eds.), *STEM Education in the Junior Secondary*,
DOI 10.1007/978-981-10-5448-8_11

expanding gas and oil sectors; however, the skill supply is far from meeting these demands (Marginson, Tytler, Freeman, & Roberts, 2013). Various forms of STEM have become an essential part of the solution in the educational systems development and enable many educators to transmit enthusiasm to their students to study science, mathematics, and technology.

In many modern economies with a high demand for qualified researchers and technicians, STEM is a major thematic domain. While Brazil has enjoyed strong economic growth, much of the workforce lacks the necessary skills to participate in an increasingly competitive global economy. In this context, the main educational policies in Brazil focus on increasing the levels of formal education across the entire population. In an effort to strengthen higher education and increase international cooperation in the STEM fields, the Brazilian government launched, in 2011, the exchange program named *Ciência Sem Fronteiras* (Science Without Borders), which is a training scientific research exchange opportunity. This is one of the strategies used to develop STEM that is found mainly at the higher educational level.

The central theme in Brazilian educational policy is a focus on the quality of education at all levels from preschool to university. Currently, the main challenges are related to the access for all learners, children, young people, and adults to educational resources independent of their educational level. Found primarily at the higher educational levels in Brazil, there are some emerging strategies for the ongoing development of STEM. At the same time, some policies are beginning to focus on changing engineering education by adopting a problem-based learning approach (Horta, 2013).

Reflecting on the imperatives of poverty reduction and equitable education, Brazil has developed policies that focus mainly on enhancing the quality of its educational system, industrial development, and science and technology in general. For example, the Brazilian *Educational Development Plan 2011–2020* focused on improving school education through enhanced teaching quality and teacher career pathways (Marginson et al., 2013).

As well as the offering of degrees that combine the presence of students in class and distance learning, some of the policies and initiatives used to improve quality are clearly associated with the STEM fields and with fostering the engagement of students in technical-oriented education (Horta, 2013). STEM participation in Brazil is "framed in terms of improving participation in basic education, and putting in place a qualified teaching workforce; and issues of socioeconomic equity and building human capital in previously excluded populations have greatest resonance" (Marginson et al., 2013, p. 57).

In accordance with this context, the imperatives of poverty reduction and equitable education in the developing economies of Brazil, Argentina, and South Africa have created national policies that focus on improving the quality of the "education systems and emerging industry development, rather than STEM-specific policies" (Marginson et al., 2013). The importance of a knowledge-based ethnomathematics and a trivium curriculum could be used to develop and conceptualize aspects of STEM in the Brazilian context.

This chapter is designed to ascertain ways in which minority students and their teachers can make use of local wisdom and heritage in relation to the trivium curriculum and ethnomathematics practices in regard to STEM. The richly specialized technology, mathematics, and science knowledge woven into the life ways of students often goes unrecognized as a developed, sophisticated, and functioning mathematical, scientific, or technological cultural context by educators.

Students need opportunities in mathematics classrooms that enable them to build bridges between their own personal context and to their schools and classrooms. Yet, systems and curriculum remain in places that impede the ability of teachers to identify, explore, and expand the connections between these knowledge traditions.

The literature review shows that some of the policies and initiatives to improve the quality of education in Brazil are clearly associated with STEM fields and with fostering and educating of learners in technical and engineering-oriented educational pedagogies. The establishment of a trivium curriculum, an ethnomathematics program, and the STEM program for Junior Secondary students can become a way to encourage the use of existing traditions and techniques and develop innovative technological resources to solve problems students face in their own communities.

11.2 The Brazilian Educational System

In the last 10 years, the Brazilian government has become concerned with ensuring the access to public schools as well as the overall quality of the education provided (Brasil, 2007). The general access to STEM-based disciplines that Junior Secondary Brazilian students receive in schools is insufficient to prepare them to develop sophisticated scientific, mathematical, and technological capabilities.

The current Brazilian educational system is based on the 1988 Constitution, which highlighted education as a universal right that should be promoted and protected by the federal, state, and municipal governments. The Lei de Diretrizes e Bases da Educação (LDB) is the National Education Guidelines and the Framework that requires a common national base for curriculum in primary and secondary education in Brazil. When it was enacted, these guidelines increased the length and the number of teaching days, accounted for the evaluation of courses and institutions at all educational levels, allowed for the integration of vocational education, and elaborated considerations for special and indigenous education (Horta, 2013).

In Brazil, the basic educational system is composed of early childhood education; fundamental education I and II, which is mandatory for children between the ages of 6/7 and 14/15; middle education, which is also free but is not mandatory; and higher education, which is free at public universities. In general, Junior Secondary level is an under-researched area, yet it is the significant transitional phase in the lives of young people, particularly in relation to STEM.

With the current emphasis on STEM, it is a great, yet little understood, field of schooling, especially in relation to the mathematics curriculum. According to Rosa (2010), curriculum is the strategy for educational action that needs to provide the essential tools for the development of citizenship, so the autonomy of mathematics in the curriculum and its central role as a dominating discipline must be reconsidered.

As mentioned earlier, basic education in Brazil is mandatory for all children from 6/7 to 14/15 years old. According to federal law, the state is obliged to offer a free and universal fundamental education. The mandatory nature of fundamental education means that this is a minimum educational standard for all citizens irrespective of their age or social class. This aspect of the law highlights the concern of the Brazilian government with the need for the adult population to attend school in order to improve qualifications and abilities to learn. Upon the conclusion of this education, students should be able to read, write, and perform calculus.

Other important objectives are needed to develop student capacity to understand and comprehend the natural and social environment, the political system, technology, and the arts. The basic curricula guidelines define compulsory disciplines such as mathematics and the Portuguese language as well as a set of subjects of the physical and natural world, including general science, which encompasses notions of biology, physics, and chemistry. Typical Junior Secondary students in Brazil enroll in these disciplines.

All Brazilian students are required to study mathematics through to the end of upper secondary level. The mathematics "curriculum includes advanced level mathematical sub-disciplines/knowledge areas" (Marginson et al., 2013, p. 81). In this mathematics curriculum, students need to understand the importance of the use of technology and its innovation. New competencies require students to be prepared to both communicate and use innovative technologies in order to install new systems, assimilate information, and propose and solve problems (Brasil, 1997).

Even though its characteristics and its concepts may be abstract, mathematics originated in the real world. Mathematical vitality is found in distinct applications in other areas and in numerous practical aspects of daily life such as in the contexts of industry, commerce, and technology. Hence, sophisticated mathematical tools are essential to the development of physics, chemistry, engineering, and astronomy (Brasil, 1997). The relevance of a curricular reference for Brazil is to guarantee the right of every student to use and enjoy accumulated sets of knowledge that are scientifically and historically elaborated in a way that is articulated within regional characteristics and which are essential to the effective exercise of citizenship.

The development of student projects has been practiced by many schools in Brazil. These projects provide contexts that generate the necessity and the possibility to organize mathematical content in a way that gives meaning in accordance to themes students choose to develop in classrooms (Brasil, 1997). Students are able to develop projects involving issues relevant to their communities, such as consumer education that allows them to study mathematical content related to percentage, interest, and the monetary system.

To develop mathematical meaning from problems and situations is an expression for resolving problems and describing reality in the social context and across other knowledge areas. The *Brazilian Curricular Parameters for Mathematics* (Brasil, 1997) states that students must demonstrate interest in, investigate, explore, and interpret mathematical concepts and procedures in different daily life contexts and across other knowledge areas.

11.3 Defining STEM Education

The STEM acronym usually represents the subjects of science, technology, engineering, and mathematics. However, as a concept, STEM is not limited to those subjects because it often includes other domains and forms of literacy, including language arts, social studies, and the arts (Bybee, 2010). The basis of STEM involves integration of these subjects by breaking down compartmented disciplines that students experience in schools with the objective of making connections to the context of their daily lives (NRC, 2014).

Over the last decade, STEM has become an international topic of discussion in educational institutions. The changing global economy and workforce drive this discussion in relation to the global shortage of STEM-prepared workers and educators. STEM perspectives introduce connections to vital knowledge fields in relation to global competitiveness, innovation, economic growth, and productivity (Carnevale, Smith, & Melton, 2011; Myers & Pavel, 2011).

Even workers in non-STEM jobs need to possess some basic STEM competencies to remain competitive and survive in an increasingly technological society (BHEF, 2011). It is important to emphasize STEM connections at an early school level in order to encourage student curiosity and career exploration as they continue through their education.

The investigation of STEM is growing in importance in today's school systems around the world. It is a process of teaching and learning that has a project-based focus on solving real-world problems. The main objective of this approach is to foster creative, critical, and reflective thinking in all students in order to emphasize innovation, collaboration, inquiry, and the development of analytical skills by addressing how learners perceive and experience the world around them (NRC, 2014).

Many STEM teaching and learning opportunities rest on inquiry, technology, and project-based learning activities and lessons that relate to the real world around the learner. A diverse and interdisciplinary curriculum is necessary to best prepare students for success in a global society through the development of their citizenship, in order to transform society. In the STEM classrooms, learning must reach beyond the walls of the school in order to include connections to local mathematics contexts developed by the members of distinct cultural groups.

Currently, there is a necessity to emphasize how many STEM experiences lack the connections to racial, ethnic, cultural, and gender diversity required to address the technological, mathematical, and scientific needs of contemporary society. Much of the STEM literature does not account for the unique political, social, and economic realities and backgrounds of the students. It is vital that STEM subjects are connected to historical, social, cultural, political, environmental, and economic contexts in order to allow learners to see the holistic and lifelong nature of learning with experiential indicators that occur outside of the classrooms. The more familiar and humanized STEM becomes to the students, the more likely they are to picture themselves in these fields.

STEM must include the application of curricular activities across all grade levels in both formal and informal classroom settings (Gonzalez & Kuenzi, 2012). In this context, Bybee (2013) argues that STEM helps students to develop:

(a) Necessary knowledge, attitudes, and skills to identify questions taken from real-life situations, explain the natural and designed world, and draw evidence-based conclusions about these problems
(b) Understanding of the characteristic features of science, technology, engineering, and mathematics as forms of human knowledge, inquiry, and design
(c) Awareness of how these disciplines shape material, intellectual, social, technological, and cultural environments
(d) Willingness to engage in STEM-related issues in order to help students to become critical, constructive, and reflective citizens

This context shows that it is necessary to elaborate STEM-related experiences that excite and engage the interest of students in the Junior Secondary years. School-based factors that influence their success in schools include parental involvement and support, availability of bilingual support, culturally relevant pedagogy, early exposure to STEM fields, interest in STEM careers, self-efficacy in STEM subjects, and STEM-related field opportunities and support programs (Museus, Palmer, Davis, & Maramba, 2011).

This definition may offer new insights into how to make STEM more interesting to students in order to fully engage them in the development of curricular activities in classrooms. It is necessary to articulate diversity in STEM in order to enhance complex thought development by allowing students to solve mathematical and scientific problems in new ways (Chubin & Malcom, 2008). "Effective instruction actively engages students in science, mathematics, and engineering practices throughout their school" (NRC, 2011, p. 18) by providing "students with opportunities to learn science and engineering by addressing problems that have real-world applications" (Chiu, Price, & Ovrahim, 2015, p. 7).

Currently, there is a need to reform existing STEM activities (Roehrig, Moore, Wang, & Park, 2012) in order to improve mathematics education. Knowledge based on mathematical modeling has the potential to contribute to the development of these activities (English, Hudson, & Dawes, 2013). Similarly, a more promising path to the development of STEM can be found in knowledge fields such as ethnomathematics, trivium curriculum, and ethnomodeling[1] that contribute to the evolution of this emerging study area.

[1] Ethnomodeling is the process of elaboration of problems and questions that grow of the practical contexts that form an image or sense of an idealized version of the *mathema*. The focus of this perspective constitutes a critical analysis of the generation and production of knowledge (creativity) in order to critically discuss the social mechanisms of institutionalization of knowledge (academics) and its diffusion through generations (education) (Rosa & Orey, 2013).

According to Eglash, Bennett, O'Donnell, Jennings, and Cintorino (2006), ethnocomputing[2] also translates from the STEM concepts and practices embedded in local cultural practices and vernacular activities to their contemporary equivalents. It is necessary then to connect STEM to innovative programs and research fields rather than simply strengthening traditional mathematics and science curriculum and instruction.

11.4 Ethnomathematics as a Pedagogical Action for STEM Education

The Brazilian mathematician Ubiratan D'Ambrosio defined ethnomathematics as the intersection between culture, historical traditions, sociocultural roots, and mathematics. An ethnomathematics program encourages the investigation and adaptation of these concepts within and outside of classrooms (2006).

The goal of this program is to acknowledge different cultural systems and distinct frameworks that have been developed throughout history and to help teachers discover new pathways that foster student engagement through developing and supporting a high quality teaching of mathematical competencies. A strong proponent of finding relevance in real-world applications builds experiential and service learning into each field study. An important influence is that culturally STEM practices are recovered and integrated in curricular activities that help students to enrich the global pedagogical base through an ethnomathematical approach (Furuto, 2014).

A STEM-based program for a Junior Secondary school mathematics curriculum in Brazil needs to propose a "pedagogical action that deals with problem solving, modelling, critical judgment, and making sense of mathematical and non-mathematical contexts, which involves distinct ways of thinking, reasoning, and knowing mathematics and its uses in practical contexts" (Rosa & Orey, 2015, p. 597).

This ethnomathematical perspective proposes a transformative pedagogy for Junior Secondary mathematics curriculum because it exposes its power to transform learners into critical and reflective citizens of change within the society. This perspective coupled with the trivium curriculum provides communicative, analytic, and technological instruments that help students to socialize the quantitative and qualitative ways of dealing with their surrounding reality.

The importance of ethnomathematics in relation to STEM is the development of its practical and pedagogical implications. In an effort to erase arrogance, inequity,

[2] Ethnocomputing can be used as a way of creating content-aware democratic spaces in which local knowledge systems and culturally relevant pedagogy become a sort of cultural capital as way of countering prevalent internalized pathologies of the myths of genetic and cultural determinisms that give rise to STEM avoidance. Cultural educational tools represent a constructionist pedagogy that has been uniquely designed to fill potentially harmful spaces with culturally significant computational capital (Babitt, 2014).

and bigotry in society, it is necessary to develop forms of curriculum innovation in relation to teaching, teacher education, and in policy-making. In this context, ethnomathematics is a research program that focuses on epistemology, philosophy, history, science, and mathematics with obvious implications for education (D'Ambrosio, 1999). This perspective encourages the investigation and adaptation of these concepts within formal and informal environments.

The goal is to acknowledge how diverse cultural systems and frameworks have served many cultures well and to help educators to become empowered to discover and develop pathways and connections that foster student engagement through conceptualizing and supporting multiple approaches to learning. A strong component is finding relevance in real-world applications through the physical, environmental, spiritual, and cultural capacities. For example, members of many indigenous cultures in Brazil make canoes in response to their need to travel and to transport food on water. The study of the construction of canoes embraces the STEM approach because it incorporates elements of science, technology, mathematics, and engineering.

An important goal of STEM is to support students in their Junior Secondary years to develop necessary abilities for their success in the Brazilian society. These skills relate to collaboration, communication, problem-solving, and inquiry. Through STEM, students learn how to work with project-based learning methods[3] that aim to build content understandings and its application (Lantz, 2009) while they develop their creative, critical, and reflective thinking.

It is necessary that in a Junior Secondary mathematics curriculum, the STEM subjects are taught as connected and integrated subject areas, so students become aware of diverse often-disconnected content in order to develop a systematic comprehensive view of the world around them. It is important to teach these subjects in a broader sociocultural context in order to facilitate STEM disciplines which break down barriers between the school and the outside world with the objective to reduce the gap between school and real-life learning environments. Thus, it is necessary to develop a dynamic and synergic relation between two or more of the contributing knowledge fields of STEM in the mathematics curriculum.

Ethnomathematics absolutely relates to STEM because it incorporates a holistic approach embedded in different subjects. It is also concerned with the understanding and comprehension of the context, nature, and creation of mathematical knowledge in response to the scientific and technological needs of distinct cultural group members; with local perceptions, ideas, notions, procedures, practices; and with the underlying cosmologies developed by the members of distinct cultural groups. It is an articulation of a particular culture, describing often-unique systems of local

[3] Project learning methods are instructional methodologies that include challenging questions or phenomena involving students' problem-solving, decision-making, and investigative abilities. There is also a critical reflection component that involves teachers as facilitators of this process. These projects integrate *knowing* and *doing* in order to support the notion that students construct their own learning by driving questions that encourage them to discover central concepts and principles of mathematical and scientific concepts through hands-on learning (Markham, 2011).

knowledges and technologies characteristic of local cultural groups (Rosa & Orey, 2015).

In this regard, the "emergence of ethnomathematics, as a research field, is the result of the recognition that every cultural group develops, as a result of its rationality, its own ways and styles of explaining, understanding and coping with their environment" (D'Ambrosio, 2006, p. 79). The *Brazilian Curricular Parameters for Mathematics* (Brasil, 1997) shows that an ethnomathematics program is an alternative proposal for pedagogical action. An ethnomathematics program seeks to understand thinking processes and ways to explain, comprehend, and act in the reality according to the cultural context of the students. Therefore, an ethnomathematics-based program uses reality to come to its pedagogical action in a natural manner with a cognitive focus and strong cultural foundation.

The construction and utilization of mathematical knowledge are not only developed by mathematicians, scientists, or engineers but in diverse ways by members of distinct cultural groups who have developed different or unique ways to count, locate, measure, draw, represent, play, and explain as a function of their own necessities and interests. It is necessary to value this mathematical knowledge (including both intuitive and cultural forms) in order to translate it to academic forms of mathematics.

It is necessary to establish the connection and interdependence between diverse forms of knowledge that include communication, language, religion, arts, science, mathematics, technology, and engineering, all of which are elements of a theory of the *cycle of knowledge* in the development of an ethnomathematics curriculum. In D'Ambrosio's (2006) point of view, this approach recognizes the relevance of the cultural dynamics of the encounters, which brings new light into the understanding of how mathematical and scientific ideas are generated and how they have evolved throughout the history.

When mathematics is considered only as a knowledge field produced exclusively by members of *Western cultural groups or societies*, it ignores the contributions of diverse members of humanity. An ethnomathematics perspective is important as it allows us to make explicit the historical-social dynamics and the production of mathematical knowledge (Brasil, 1997) developed by all peoples throughout history.

11.5 An Ethno-STEM Curriculum

Classrooms and learning environments cannot be isolated from the communities in which they are embedded. They are part of a school community with defined cultural practices. Thus, classrooms are environments that may facilitate the development of pedagogical practices, which are applied by using an ethnomathematical approach. When students come to school, they bring with them the values, norms, and procedures that they have acquired in their own sociocultural context, which often includes distinct forms of knowledge such as the mathematical, the scientific, and the technological. However, mathematical concepts and practices included in the school curriculum are often presented in a way that is not related to the students' cultural backgrounds.

It has been hypothesized that low attainment in mathematics could be due to the lack of *cultural consonance* in the curriculum (Rosa, 2010), which is the degree to which individuals, in their own beliefs and/behaviors, approximate widely shared cultural models. This concept and its associated measurement model connect cultures to social practices at the individual level in order to provide a measure of how well they match their own cultural models (Dressler, 2011).

Moreover, the inclusion of cultural aspects of mathematics in the curriculum has long-term benefits for all students. This approach contributes to recognizing mathematics as part of students' daily life which enhances their ability to make meaningful connections and deepen their understanding. Pedagogical work toward an ethnomathematics perspective allows for a broader analysis of the school context in which pedagogical practices come to transcend the classroom environment as these practices embrace the sociocultural context of students (Rosa, 2010). Thus, the pedagogical elements necessary to develop an ethno-STEM curriculum can be found in the school community (Damazio, 2004), where there is a recognition that ethnomathematics is a research program that guides pedagogical practices in STEM.

There is a need to examine the embeddedness of mathematics in culture by drawing from a body of literature that takes on the cultural nature of knowledge production in a mathematics curriculum. Mathematics as part of the school curriculum reinforces and values the cultural knowledge of students rather than ignoring it or negating it. This curriculum integrates the cultural mathematical background of the members of the community and facilitates the acquisition of formal mathematical knowledge (Rosa & Orey, 2007).

The trend toward ethnomathematical approaches in curriculum reflects a comprehensive development for mathematics education. Ethnomathematical approaches are intended to make school mathematics more relevant and meaningful to students and to promote the overall quality of education. It is necessary to plead for a more culturally sensitive view of mathematics to be incorporated into the school curriculum (Rosa & Orey, 2015).

For example, it is necessary to propose the elaboration of a mathematics curriculum based on student knowledge and which allows teachers to have more freedom and creativity to choose academic mathematical topics to be covered in the lessons. Through dialogue with the students, teachers are able to apply mathematical themes that help them to elaborate the mathematics curriculum. This context allows teachers to engage students in the critical analysis of the dominant culture as well as of their own culture (Rosa & Orey, 2007).

The investigation of conceptions and mathematical practices developed by the members of distinct cultural group by the students themselves is necessary to the incorporation of these traditions into the mathematics curriculum (Rosa, 2010). This involves a connection between academic mathematics and ethnomathematical knowledge and contributes to the process of social change. Curriculum based on an ethnomathematical perspective infuses the cultural backgrounds of students in the learning environment in a holistic manner and provides them opportunities to relate their new learning experiences to knowledge they have previously acquired (Rosa, 2010).

This approach supports the view that "mathematics (…) is conceived as a cultural product which has developed as a result of various activities" (Bishop, 1988, p. 182). The objective of this approach is to make mathematics more relevant to students and for them to see how every culture has developed mathematical responses to phenomenon they experience in a day-to-day fashion. A classroom using this type of curriculum would be full of examples that draw on their own experiences that they see as common to their sociocultural experience (Rosa & Orey, 2007).

Another possibility is the integration of mathematical ideas, procedures, and practices originating in the local cultural context with those of the formal academic mathematics. This ethnomathematical curriculum takes students' culture and uses it explicitly to integrate these outside experiences into the conventional mathematics curriculum. In such a classroom environment, students build on what they know as well as on the experiences they have acquired from their own sociocultural environments (Rosa, 2010). These experiences are part of the understanding of how mathematical ideas are developed and how they are built into systems, formulated, and applied in various ways within culture.

Therefore, links are made to familiar practices and concepts by realizing and understanding the need for mathematical characteristics such as accuracy and formal reasoning in both academic mathematics and in real-life situations. The understanding of conventional mathematical knowledge then feeds back and contributes to a broader understanding of culturally based principles (Lipka, 2002).

It is assumed that a curriculum of this nature motivates students to recognize mathematics as part of their everyday life and enhances their ability to make meaningful mathematical connections by deepening their understanding of all forms of mathematics (Adam, 2002). For example, Duarte (2004) investigated the uniqueness of mathematical knowledge produced by workers in the home construction industry through a study of mathematical ideas and practices that they develop in construction sites. The results of this study reflected on the mathematical knowledge possessed by the members of this working class in order to value their knowledge and determine the pedagogical and curricular implications that are inferred in the process of production of this kind of knowledge.

The overall objective for developing this curriculum model for classrooms is to assist students to become aware of how different people mathematize and think mathematically in their own cultures, to use this awareness to learn about formal mathematics, and to increase the ability to mathematize in any context (Duarte, 2004). This curriculum leads to the development of a sequence of instructional activities enabling students to become aware of potential practices in mathematics in their own culture so that they are able to understand the nature, development, and origins of academic mathematics. Students are able to value and appreciate their own previous mathematical knowledge, which allows them to understand and experience cultural activities from a mathematical point of view, thereby allowing them to make the link between school mathematics and the real world (Rosa & Orey, 2015).

Students improve their understanding of mathematics as they become aware of how it was developed and is used in their own context. With increased awareness, students see it as a human activity rather than just a set of abstract symbols, numbers, and figures. Diverse cultural (mathematical) practices can be related to conventional mathematical systems and vice versa by looking at symbolizing, generalizing, abstracting, and making logical connections. These are easily facilitated by seeing mathematics used in various contexts which enables the learning of mathematics through practical examples and investigations (Rosa & Orey, 2007).

One possible bridge is to understand how both teachers and students realize the connections between academic mathematics and the real world. This includes the examples teachers use in their instruction and the characteristics of the informal and academic mathematics they choose to explore in classroom activities. This is relevant because it introduces an understanding about the nature of mathematics so that when students come to understand it, they acquire the tools to better comprehend the relevance of the application of knowledge in the various aspects of their everyday lives (Rosa & Orey, 2007).

The establishment of cultural connections is a fundamental aspect in the development of strategies in teaching and learning mathematics; it allows students to perceive mathematics as a significant part of their own cultural identity (Rosa & Orey, 2007). This curriculum focuses on mathematics as a process rather than as a collection of facts, and it is based on the idea that mathematics is a human creation that emerges as people attempt to understand their own world.

Therefore, mathematics is considered a process and as human activity, rather than just as a set of academic content (D'Ambrosio, 2007). This approach implies that this curriculum is not just about the application of relevant contexts in the teaching and learning process, but it is also about generating knowledge from the development of mathematical ideas, notions, and procedures. This cultural approach to the mathematics curriculum attempts to make academic mathematics better understood, appreciated, and meaningful to learners.

Teachers must analyze the role of students' *ethnoknowledge* in the mathematics classroom. Ethnoknowledge is acquired by students in the pedagogical action process of learning mathematics in a culturally relevant educational system. In this process, the discussion between teachers and students about the efficiency and relevance of mathematics in different contexts should permeate instructional activities. The ethnoknowledge that students develop must be compared to their academic mathematical knowledge. In this regard, the role of teachers is to help students to develop a critical view of the world by using mathematics (Borba, 1990).

According to Rosa (2010), teachers also need to develop a different approach to mathematics instruction that empowers students to understand mathematical power more critically by considering the influence of culture on the development of mathematical knowledge, and they must work with their students to uncover the distorted and hidden history of the mathematical knowledge. This methodology is essential in developing the curricular practice of ethnomathematics and the development of the trivium curriculum for mathematics.

11.6 A Trivium Curriculum for STEM

The proposal for a *trivium curriculum for mathematics*, in which *literacy*, *matheracy*, and *technoracy* are its basic components (D'Ambrosio, 1999), supports specialized preparation programs that enable students to meet academic needs and improve their STEM knowledge, competencies, abilities, and skills. This curriculum helps students to demonstrate effective mathematics processes as they reason, solve problems, communicate mathematical ideas, chose appropriate representations through the development of mathematical practices (Rosa & Orey, 2015), as well as recognize connections between mathematical concepts and other STEM disciplines. This proposal for a new concept for mathematics curriculum provides the communicative, analytical, and technological instruments.

It is important to state that this form of literacy is different from its usual meaning; matheracy is not the same as mathematics; and technoracy is different from technology. Literacy essentially means empowerment using communicative instruments that justify reading and writing; matheracy means empowerment using analytic instruments that justify mathematics; and technoracy means empowerment through the use of technology or technological instruments. In these three words, the suffix *racy* means the essence or root and its implications to empowerment. In this case, literacy is equal to the essence of communicative instruments; this is what justifies reading and writing. Matheracy is equal to the essence of analytic instruments; this justifies mathematics. Technoracy is equal to the essence of technology or technological instruments.

The trivium curriculum is related to STEM since it takes into account the student context and culture and its contribution to their ways of learning. STEM projects are designed to integrate science, technology, and mathematics in the classroom with the aim to teach students to think critically. It also focuses on an engineering or design approach toward real-world problems while building on their mathematics and science base (Jolly, 2014). Accordingly, an argument is that even though teachers may develop STEM projects with their students, they may not be sufficient for providing space for the culture of the leaners to emerge, so there is a need for the development of the trivium curriculum for mathematics.

This *trivium* is a proposal for a curriculum based on developing a broad perception of the world and modern society and providing the instruments to engage with this complexity. This curriculum opposes excessive emphasis on the quantitative data, which may be detrimental to the equally important emphasis on the qualitative data. This proposal is an answer to the criticism of this lack of equilibrium in the mathematics curriculum. The trivium curriculum for mathematics proposed by D'Ambrosio (1999) is an important innovative ethnomathematics approach that needs increased investigation in order to address pedagogical purposes (Rosa & Orey, 2015).

D'Ambrosio (2007) argues that mathematics curriculum is the strategy for the educational action that should offer three instruments in order to provide what is essential for the development of citizenship in a world moving swiftly toward a planetary culture developed by the implementation of transdisciplinarian and transcultural educational approaches, by restoring the dignity of its members. These are the communicative instruments, the analytic/symbolic instruments, and the technological instruments, which constitute the modern *trivium* composed by *literacy*, *matheracy*, and *technoracy* components.

11.6.1 Literacy

Literacy is the use of communicative instruments. It was essential for the development of mercantilism and modern science as it became the imprint of the contemporary world. It is the critical capability of processing information such as the use of written and spoken language, signs, gestures, and numbers. In this regard, critical reading means to read with the goal of finding deep understanding in order to comprehend a diversity of informational and communicational materials. It is the act of analyzing and evaluating the reading materials as students make their way through the texts or as they reflect back upon their reading.

Reading has a new meaning today since students are distracted by a wide variety of media, including games, movies, and TV programs. To read and understand instructions for video games or school tasks and interpret data available in newspapers, magazines, and books give meaning to the information available in these media contexts. They also need to be able to comprehend statements of employee benefits, payment schedules, tax tables, mileage charts, and sports league standings that are depicted in graphics. These instruments provide unlimited resources for mathematical and scientific information.

Numbers, figures, and signs are communicative instruments that enrich the capability of the discourse and conversation as well as sway our opinions. These instruments are embedded in the tasks and activities that members of distinct cultural groups develop in their own environments. These members are capable of using calculators, computers, and mobile devices. Therefore, it is necessary to use traditional technologies such as pencils, pens, papers, chalk, and blackboards. This approach reconciles differences in diverse communicative forms.

In the last three decades, the social, political, and cultural concept of literacy have changed because it also includes the power of numeracy or quantitative literacy. It involves the interpretation of graphs, tables, charts, diagrams, figures, and other ways of achieving understanding of the condensed language of codes (D'Ambrosio, 1999). Graphs are an integral part of both the workplace and daily life since they provide necessary information for the completion of job-related and academic tasks.

Literacy is composed of numbers, figures, graphs, and signs that are related to the *communicative instruments*, which enrich the capability of discourse, conversation, and description. Critical familiarity with these instruments, embedded in diversified cultural environments, is part of dealing with literacy. This communicative trend cannot be reversed in the school system. Thus, the power of literacy does not only relate to the ability to read and write, but rather it relates to the capacity of the student to learn the skills for shaping the course of their own life.

With the insight that genuine literacy involves *reading the word and the world*, this helps to open the door to a broader understanding of the term, one that moves from a strict decoding and reproducing of language into issues of economics, health, and sustainable development (Freire, 1987). Whether it is the words of a language, the symbols in a mathematical system, or the images posted to the Internet, literacy transforms lives.

Because "mathematics is so often conveyed in symbols, oral and written communication about mathematical ideas is not always recognized as an important part of mathematics education. Students do not necessarily talk about mathematics naturally; teachers need to help them to do so" (NCTM, 1996, p. 60). Knowing how to use the unique symbols that make up the shorthand of mathematical statements, such as numerals, operation signs, and variables that stand in for numbers, has always been part of what mathematics teachers are expected to teach. In this direction, students need to:

> (…) learn to use language to focus and work through problems, to communicate ideas coherently and clearly, to organize ideas and structure arguments, to extend their thinking and knowledge to encompass other perspectives and experiences, to understand their own problem-solving and thinking processes as well as those of others, and to develop flexibility in representing and interpreting ideas. At the same time, they begin to see mathematics, not as an isolated school subject, but as a life subject — an integral part of the greater world, with connections to concepts and knowledge encountered across the curriculum. (Martinez & Martinez, 2001, p. 47)

The trivium curriculum helps students to become literate in opposition to the traditional forms of mathematics that deprive students of becoming critical, reflective, and conscious citizens. The *National Governors Association* (2011) reported that STEM literacy is the ability to adapt to and to accept changes driven by new technology, to work with others across borders, to anticipate the multilevel impact of their actions, to communicate complex ideas to a variety of audiences, and to find measured creative solutions to problems faced by society.

Mathematics is often referred to as language that allows the discussion of abstract concepts of numbers and space. The power of this language is to enable students to construct mathematical poems or metaphors called *ethnomodels*, which help students to think critically about physical phenomena and explore in depth their underlying ideas. Ethnomodels are described as cultural artifacts that are pedagogical tools used to facilitate the understanding and comprehension of systems taken from the reality of the members of distinct cultural groups. They are small units of information that compose the entire representation of these systems (Rosa & Orey, 2013).

The term literacy is usually interpreted as the ability to read and write (Kintgen, Kroll, & Rose, 1988). However, extensions of this term to visual and media literacy, computer literacy, cultural literacy, political literacy, and STEM literacy suggest that its semantic aspect is powerful. Literacy is usually used in a descriptive manner, which is the mastery of a body of knowledge that provides an understanding of intended meaning.

11.6.2 Matheracy

Matheracy provides the analytical instrument and symbolic analysis of *mathema* as proposed by the classical Greek mathematicians. In the sense given in epistemological Greek meanings, mathema was appropriate for philosophers to be concerned with higher objectives in order to explain the world by an approach based on reason and evidences.

The broader concept of matheracy comes closer to the way in which mathematics was present in both classical Greece and local cultures. In this context, the concern of the members of these cultural groups relates to the capabilities, abilities, and competencies of counting and measuring, but also relates to divination and aims at explaining, understanding, and comprehending reality and its phenomena, which are the elements of the common knowledge they developed over time (D'Ambrosio, 2007).

Matheracy is also present in critical capabilities of inferring results to reach conclusions based on known facts that relate to the formation of opinion from evidence. It is the proposition hypotheses and interpretation of information in order to allow students to draw conclusions from data and from the results of mechanical calculations. This set of capabilities gives meaning to reading, understanding, comprehending, and interpreting data. It is related to the deeper reflection about humankind and society.

Matheracy is a symbolic analysis that forms the central idea behind the origins of mathematics. It is not only the result of the appropriation of skills, as it has always been a competency acquired over time and used to analyze data and information available daily. It is also a first step toward an intellectual, critical, and reflective posture in relation to solving problems faced by society (D'Ambrosio, 2007). An intellectual posture can be considered as the students' engagement in the critical study of problems they face daily in order to propose its solutions as well as their deep reflection about reality of society.

Currently, this intellectual posture helps to participate in politics, either to defend concrete propositions or to denounce injustices, usually by producing or by extending an ideology and by defending a system of values, which support them to adopt an attitude or take an official position on the problems that affect their own lives. The absence of this posture in the school systems, mainly, in the mathematics classrooms, allows students to perform mechanical calculations that are unrelated to the activities in their daily lives (D'Ambrosio, 2007).

Regrettably, even the so-called problem-solving activities, modeling tasks, and projects as developed in many classrooms are really a set of mechanic techniques that only allow students to manipulate numbers and to operate mathematical rules. This actually ends up impoverishing mathematical instruction by restricting it to purely manipulative techniques and procedures that are necessary for the development of utilitarian purposes (D'Ambrosio, 2007).

According to this context, matheracy is not only the result of the appropriation of skills; its notions and concepts are acquired through the development of competencies to enable learners to analyze data and deal with information available in different types of media such as newspapers, books and magazines, recorded music, film, radio, television, and the Internet.

11.6.3 Technoracy

Technoracy is the familiarity with technology, which in many cases are inaccessible to the individuals. Nevertheless, the basic ideas behind technological resources, the possibilities and dangers of the application of its devices, and the morality supporting its uses are essential issues to be raised among students at a very early age. History shows that ethics and values are intimately related to technological progress. It is important to recognize the special role of technology in the human species and its implications for the development of scientific and mathematical knowledge.

There is a need to recognize the relevance of the history of technology since it is an essential element for furthering ideas, concepts, and theories relating to science and mathematics (D'Ambrosio 2004). This recognition proposes critical reflections about the role of technology in mathematics education. Once the role of technology in the development of mathematics is recognized, reflections about the future of mathematics pose important questions about the role of technology in mathematics education. The rapid development of electronic technology has allowed for the acceleration and dissemination of the acquisition of mathematical concepts, procedures, and techniques that are necessary to solve problems faced in daily life (D'Ambrosio, 2012).

By connecting to the Internet, students can obtain wider access to sources of mathematical and scientific knowledge. They can explore economics as well as physics by making ethnomodels and simulations, and the rigor of mathematics can be extended to areas that were previously inaccessible. In this context, the National Curricular Parameters for Mathematics (Brasil, 1997) states that innovative competencies, abilities, and skills require new knowledge because the world requires that the members of distinct cultural groups use different technologies and languages (beyond oral and written communication), installing new rhythms of production, rapid assimilation of information, and proposing and solving problems collaboratively and cooperatively.

In accordance to this context, it is important to emphasize that technoracy resonates with the concept of *technacy* (Seemann, 2000), which "provides a framework for considering science and technology within a socio-environmental context" (p. 7). It transcends the competency in using technology since it "relates to a holistic view of problem solving, communication and practice that includes consideration of social technical and environmental resources and constraints" (p. 7).

There is a recognition that the members of every cultural group have the right to access technological tools and instruments in order to communicate, cope with reality, understand and explain phenomena, and provide tools for critical and reflective thinking in order to define strategies for actions. In terms of outcomes, technological resources seek "to develop skilled, holistic thinkers and doers who can select, evaluate, transform and use appropriate technologies that are responsive to local contexts and human needs" (Seemann, 2000, p. 2).

The development of technological resources and the facility of information retrieval through distinct media reveal that there is no place for the *propaedeutic* character of mathematics education (D'Ambrosio, 2006), which may be defined as the necessary knowledge for the preparation of learning, but not for its proficiency.

11.7 Connections Between Ethnomathematics, Trivium, and STEM

Cultures are anchored in mathematics, which is, for the scientific community, the dorsal fin of the modern world. This aspect leads to focus the concerns about the nature of mathematics. It is difficult to deny that mathematics provides important instruments for social analyses because Western civilizations rely on data control and management. For example, the "world of the twenty-first century is a world awash in numbers" (Steen, 2001, p. 1). Social critics may find it difficult to argue without an understanding of basic quantitative mathematics.

For learners, it is necessary to highlight new directions for the development of the mathematics teaching and learning process that result from the sociocultural contexts. Thus, it is necessary to develop new mathematical structures that consist of flexible hypotheses (Gromov, 1998). According to this assertion, these:

> (…) remarkable ideas, although very difficult, clearly indicate that the new generation of scientists, engineers, and, obviously, mathematicians will need broader attitudes towards mathematics. The challenging problems require, besides new mathematical techniques, the training of a new generation of researchers in the mathematical sciences. (D'Ambrosio, 2012, p. 205)

This context encourages the "creation of a new breed of mathematical professionals able to mediate between pure mathematics and applied science. The cross-fertilization of ideas is crucial for the health of science and mathematics" (Gromov, 1998, p. 847).

Many schools recognize the need for students to gain a broad worldview of society and the mathematics it has developed. This is where the ethnomathematics program and a trivium curriculum for mathematics play a crucial role. Given standardized requirements in mathematics content, certain mathematical competencies, abilities, and skills must be taught, and mathematical procedures, concepts, and practices must be covered, but often even these can be treated from the point of view of mathematics from around the world (global) and from the structures of local cultures.

The links between an ethnomathematics program, the trivium curriculum, and STEM certainly describe the possibility for an innovative pedagogy for Junior Secondary students. It is important to comprehend the major shift in perception and understanding that is required to help teachers to learn ethnomathematics techniques as well as to acquire the confidence and courage to release themselves from years of traditional thinking. In order to enrich its pedagogical base, the influence of ethnomathematical ideas, procedures, and practices can be integrated into the STEM curriculum.

Ethnomathematics and the trivium curriculum can be important arguments for the application of mathematics that requires STEM representation in Junior Secondary education. Full knowledge and mastery of Western mathematics is necessary for the development of humanity and even creative thought. Nontraditional/Western mathematical knowledge developed by members of distinct cultural groups helps students to understand problems faced by society. It shows them how mathematics developed and is used in alternative ways they might never study in the traditional Western curriculum. It is powerful, both as a way of deepening the understanding of mathematics and as a cultural conduit. It also allows for creativeness in mathematics to be made manifest. When this objective is achieved, students gain access to meaningful curricular opportunities that promote critical thinking skills that can be applied to their academic as well as everyday lives.

The focus on STEM brings a greater pressure on schools and teachers to develop pedagogical ways to provide programs that address the integration of curricular areas, the implementation of engineering design principles, and the use of workplace technologies (Mayes & Koballa, 2012). Both traditional mathematics and ethnomathematics provide communicative instruments that help students to socialize quantitative and qualitative information they need to acquire in order to comprehend the mathematical context of their realities (D'Ambrosio, 2006).

It is the responsibility of the school community to prepare its students to creatively explain, understand, comprehend, and solve problems and phenomena they face in society. This approach requires abstractions and conceptualizations that are the essence of analytic instruments, which help students to move into the future equipped with strategies and action. Ethnomathematics provides such instruments by enabling students to study, research, use, combine, improve, and invent technological artifacts and instruments in order to allow the members of a specific cultural group to communicate with the members of other cultures (Rosa & Orey, 2015).

In educational systems, ethnomathematics can critically and reflectively provide communicative, analytic, and technological opportunities that enhance STEM work. Critical familiarity with these opportunities is an important objective of education. Both traditional mathematics activities and ethnomathematics are intrinsic to the use of such instruments. STEM aims to build content understandings and applications of knowledge (Lantz, 2009). However, success with content knowledge is not enough for students to succeed in the development of their citizenship in order to transform society. Twenty-first-century abilities and skills that are required and essential to the development of citizenship include collaboration, communication, problem-solving, critical thinking, and working in diverse groups (Schlechty, 2011).

A focus on a diversity of cultures, languages, creativity, and innovation is necessary for the development of learners. In this regard, it is necessary to engage "students in work that results in their need to learn material that is essential to their education as citizens in a democracy and to their right to claim to be well educated human beings is the primary business of schools" (Schlechty, 2011, p. 8). In this context, it is necessary to emphasize that:

> Issues of socio-economic equity and building human capital in previously excluded populations have greatest resonance in these nations, where participation in good quality upper secondary and tertiary education (indeed, participation in the modern economy) is by no means universal. (Marginson et al., 2013, p. 57)

Technologically interested citizens demand that individuals be able to solve real-world problems through the processes of gaming, investigation, model building, data analysis, presentation of evidence-based reasoning, and communication of findings (Moon & Singer, 2012). Students should emerge from Junior Secondary science and mathematics education experiences fully prepared to transition into high school as well as to become participating citizens in their communities.

STEM provides teachers with guidelines for middle grades in order to develop the necessary content and pedagogical knowledge as well as helping develop the personal dispositions necessary to develop critical and reflective learners. The more familiar and humanized STEM subjects become to the students, the more likely they are able to picture themselves in these fields. Therefore, it is important to reconcile the differences in communicative norms, integrate identity, incorporate community, and locate STEM subjects in a historical, social, political, economic, and cultural context that best suits the needs of the community.

Historically, societies both absorb and create new mathematical, scientific, and technological innovations. It is important to understand the way material and intellectual innovation permeates the critical and reflective thinking of the citizens. The technology component allows for a deeper understanding of science, mathematics, and engineering since it allows students to apply learned content by using computers and other technological resources, which can help students to explore STEM subjects. It is important to demonstrate how the pedagogical action of the ethnomathematics program can be linked to the study of STEM and to the proposal for a modern *trivium* curriculum for mathematics, which is composed by literacy, matheracy, and technoracy.

Including an ethnomathematics program and trivium curriculum into STEM will create multiple opportunities to develop interdisciplinary activities. It is necessary that educators and teachers understand, comprehend, and respect the diverse ways of explaining, knowing, and doing of their students. This approach relates to transcultural views of the history and philosophy of science, technology, and mathematics, with particular attention to the members of distinct cultural groups and their contribution.

A trivium curriculum, when based upon the principles of ethnomathematics, facilitates a broad understanding about the importance of mathematics to pedagogical activities developed in the mathematics classrooms, and they directly link to the reality of the learners. It is necessary to highlight here that most of the traditional mathematics curricula focuses on the mastery of skills, grammar, rules, accumulation of facts, and the ability to perform algorithms that are extremely necessary for standardized examinations. Since this curriculum is experienced as mathematical content, most students leave school feeling negative and thinking that mathematics is done only at school and that it has no relevance to their lives (Rosa, 2010).

The development of a trivium curriculum for mathematics is related to the principles of STEM. Creating innovative pedagogical actions that embrace multiple levels of the curriculum in one class or subject area is extremely vital, indeed necessary, for students to embrace a path that is continually moving from a scientific perspective to a mathematical conceptualization and flows into engineering and technology careers. The ethnomathematics pedagogical action diminishes students' mathematics avoidance and can help them to experience the connections between science, mathematics, and technology with personal experience and cultural heritage (Rosa & Orey, 2015). The trivium curriculum functions as a bridge that permits students to perceive the interrelation of the study of mathematics and science with their cultural background.

11.8 The Role of STEM in the Junior Secondary Years

Pedagogical experiences should be connected to students' lives in meaningful ways. Science and mathematics camps provide a place-based context for students to develop "cognitive abilities to engage in STEM content and problem solving activities" (DeJarnette, 2012, p. 80). Proficiency in mathematics means much more than counting, measuring, sorting, comparing, and solving problems aimed at drilling. Regrettably, even conceding that problem-solving, modeling, and projects are practiced in some mathematics classrooms, the main importance is usually given to developing skills, particularly in the manipulation of numbers and operations.

Problems and situations present in daily life are often new and unexpected. Students should be prepared to tackle the new, to engage in applying their growing understanding of mathematics to their increasing understanding of the world around them. STEM is an area of study, but it is also a way of teaching and learning that is project-based, collaborative, and focused on solving real-world problems and emphasizes innovation, problem-solving, critical thinking, and creativity. A specific challenge to advance STEM actively incorporates a diversity of technologies and engineering into school programs; unfortunately the rate at which technology and engineering appears in school curricula is still quite low (Bybee, 2010).

It is necessary to integrate engineering into curricula because it is "becoming more prevalent in Junior Secondary schools and can provide great problem solving opportunities for students to learn about mathematics, science, and technology while working through the engineering design process" (Stohlmann, Moore, & Roehrig, 2012, p. 30). In the classroom, constructivist approaches, problem-based learning, and making connections to the real world often characterize effective STEM work when using inquiry-based strategies. Techniques such as active learning and learning to work in cooperative learning groups are central to achieving the most important outcomes of STEM (Smith, Douglas, & Cox, 2009).

Technologies used to enhance teaching and learning contribute to major changes in curriculum, expectations for student achievement, and the role of mathematics teachers in the education enterprise. Thus, it is necessary that students comprehend the importance of using technology and monitor and follow its ongoing renovation (BRASIL, 1997). In this regard, it is important to relate mathematics and technology. Concurrent with the development of revised standards and assessments is the increasing number of new technologies that show great promise not only for changing how Junior Secondary students come to explore and learn new concepts, skills, and reasoning methods but also how their classrooms and time for learning, both in and out of school, can be restructured (Johnson, Adams, & Cummins, 2012).

The combination of formal and informal education is a useful way of looking at STEM because its philosophy can be adapted to different and equally important pedagogical actions. First, the formal environment ensures students an unrestricted education, allowing everyone to have access and share content and ideas. Second, the informal environment plays a complementary role, using the previously established knowledge in formal education to offer students a context of greater autonomy. With the relatively recent focus on STEM, it is important that educational systems provide programs that address the integration of curricular areas in an interdisciplinary fashion.

Consequently, it is necessary to understand the scenarios as complementary: one environment that offers security and stability and the other where a more open-ended, somewhat chaotic but always dynamic creativity and innovation abounds. Although methodologically different, these environments are capable of absorbing STEM philosophy as an ideology and as a social responsibility, which provides students with a complex and integrated form of information.

It is important to make the mathematics and science curriculum more meaningful and accessible to the students, and an ethno-trivium-STEM collaboration can do this. Thus, it is necessary to verify how to explicitly find and then use the student's own knowledge, how to draw experiences from the community they live in, and how to determine their understanding of culture-based mathematics and science. In this context, students' attitudes may change in relation to mathematics when activities are task-oriented and related more closely to their lives (Rosa, 2010). Hence, science and mathematics must be identified as a natural connector between learning and local practices.

This form of STEM encourages the spheres of socioeconomic status, ethnicity, culture, and linguistic differences to come together and examine the mathematics, the science, and the engineering produced and used by diverse groups of people. It suggests an educational alternative for the students who are more likely to become marginalized during their Junior Secondary schooling. Marginalization is manifested as a lack of interest in mathematics, low performance in mathematics classes, and underrepresentation in science (mathematics)-related careers (Buxton & Lee, 2010). Therefore, it is necessary to develop innovative ways of improving mathematics learning since the majority of students do not have access to this type of resource. This form of STEM activity can do this.

11.9 Final Considerations

In an effort to address issues of equity and quality education, STEM subjects need to be explored in order to design and implement lessons and activities grounded in the ethnic, historical, and cultural diversities of Brazil. It is necessary to accomplish these goals through conferences, continuous orientation, professional development, in-services, and workshops. This is the philosophy behind a movement aimed at influencing Brazil's educational system and economy from the bottom up, by teaching the teachers.

Recognizing the importance of nurturing strong STEM abilities and the skills needed to compete in today's global economy, both developed and developing countries are exploring ways to teach math and science in more engaging ways. It is in the Brazilian government's interest to promote the full use of human resources in mathematics, science, engineering, and technology to ensure the full development and use of the talents, abilities, and skills from distinct ethnic, racial, and economic backgrounds of all learners (Horta, 2013). Thus, it is necessary to guide public school teachers in Brazil to teach Junior Secondary students methods that spark their interest in learning STEM-related subjects in order to apply skills for practical projects by using mathematics concepts learned both inside and outside of the classroom.

Pedagogical experiences toward the development of ethnomathematics and the trivium curriculum have been proposed as ways of improving mathematics learning. STEM experiences are those in which students learn mathematics in a way that reflects how scientists perform mathematics rather than what is reflected in conventional activities. They call for high quality learning while helping students develop a critical consciousness, reflective thinking, and cultural competence (Rosa, 2010).

By applying the STEM approach, ethnomathematics and the trivium curriculum support students in learning to think about mathematical and scientific knowledge in a different way. This curriculum is an interdisciplinary approach that provides relevant learning experiences for students that go beyond the mere transference of knowledge. This learning process engages students and equips them with critical and reflective thinking, problem-solving, and creative and collaborative abilities.

This curriculum also helps students understand and apply mathematical and scientific content, which are the foundations for success in college and careers. STEM education attempts to transform the typical teacher-centered classroom by encouraging a curriculum that is driven by problem-solving, discovery, and exploratory learning and requires students to actively engage a situation in order to find its solution.

In this curriculum literacy relates to the capacity students have to process information present in their daily lives. Matheracy is the capacity students have to interpret and analyze signs and codes in order to help them to propose ethnomodels to find solutions for problems they face daily. Technoracy is the capacity students have to use and combine different technological instruments to help them to solve these problems. These communicative, analytical, and technological instruments that constitute a trivium mathematics curriculum are based on the development of the complexity of society and provide the instruments to deal with problems society faces daily. Proficiency in mathematics means more than just the testing of counting, measuring, sorting, comparing, and solving drilling problems (D'Ambrosio, 1999).

In this regard, it is important to state that some forms of problem-solving, modeling, and projects are practiced in many mathematics classrooms. The main importance for this is usually given to skill development, particularly in regard to the manipulation of numbers and operations. Students should be prepared to tackle problems and situations present in daily life. The combination of the instruments in the trivium curriculum for mathematics using an ethnomathematical perspective is essential for the development of Junior Secondary students in Brazil.

References

Adam, S. (2002). Ethnomathematics in the Maldivian curriculum. In M. Monteiro (Ed.), *Proceedings of the 2nd International Congress on Ethnomathematics (ICEm2)*. Ouro Preto, MG: Lyrium Comunicação Ltda.

Babitt, W. E. (2014). *Ethnocomputing: The design and assessment of culture-based learning software for math and computing education*. Published doctorate dissertation. Troy, NY: Rensselaer Polytechnic Institute.

BHEF. (2011). *Meeting the STEM workforce challenge: Leveraging higher education's untapped potential to prepare tomorrow's STEM workforce. BHEF policy brief*. Washington, DC: Business Higher Education Forum.

Bishop, A. J. (1988). *Mathematics enculturation: A cultural perspective on mathematics education*. Dordrecht: Kluwer.

Borba, M. C. (1990). Ethnomathematics and education. *For the Learning of Mathematics, 10*(1), 39–43.

Brasil. (1997). *Parâmetros curriculares nacionais*: matemática [National curricular parameters: Mathematics]. Brasília, DF: MEC/SEF.

Brasil. (2007). *Decreto no. 6.094 de 27 de abril de 2007*: plano de metas compromisso todos pela educação [Decree no. 6094, April 27, 2007: commitment goals plan all for education]. Brasília, DF: Casa Civil.

Buxton, C., & Lee, O. (2010). *Diversity and equity in science education*. New York: Teachers College Press.

Bybee, R. W. (2010). Advancing STEM education: A 2020 vision. *Technology and Engineering Teacher, 70*, 30–35.

Bybee, R. W. (2013). T*he case for STEM education: Challenges and opportunities*. Arlington, VA: NSTA Press.

Carnevale, A. P., Smith, N., & Melton, M. (2011). *STEM: Science, technology, engineering, mathematics*. Washington, DC: Georgetown University Center on Education and the Workforce.

Chiu, A., Price, A. C., & Ovrahim, E. (2015). *Supporting elementary and middle school STEM education*. Chicago: Museum of Science and Industry.

Chubin, D. E., & Malcom, S. M. (2008, October). Making a case for diversity in STEM fields. *Inside Higher Education*, 1–4.

Damazio, A. (2004). *Especificidades conceituais da matemática da atividade extrativa do carvão* [Conceptual specifications of mathematical activities of coal extraction]. Coleção Introdução à Etnomatemática [Introduction to Ethnomathematics Collection]. Natal: UFRN.

D'Ambrosio, U. (1999). Literacy, matheracy, and technoracy: A trivium for today. *Mathematical Thinking and Learning, 1*(2), 131–153.

D'Ambrosio, U. (2004). *Educação Matemática da teoria à prática* [Mathematics education from theory to practice]. São Paulo, Brazil: Papirus.

D'Ambrosio, U. (2006). The program ethnomathematics and the challenges of globalization. *Circumscribere, 1*, 74–82.

D'Ambrosio, U. (2007). Peace, social justice and ethnomathematics. *The Montana Mathematics Enthusiast, Monograph, 1*, 25–34.

D'Ambrosio, U. (2012). A broad concept of social justice. In A. A. Wager & D. W. Stinson (Eds.), *Teaching mathematics for social justice: Conversation with educators* (pp. 201–213). Reston, VA: NCTM.

DeJarnette, N. K. (2012). America's children: Providing early exposure to STEM (Science, Technology, Engineering and Math) initiatives. *Education, 133*(1), 77–84.

Dressler, W. W. (2011). Cultural consonance: Linking culture, the individual and health. *Preventive Medicine, 55*(5), 390–393.

Duarte, C. G. (2004). Implicações Curriculares a partir de um olhar sobre o mundo da construção civil [Curricular implications concerning the world of civil construction]. In G. Knijnik, F. Wanderer, & C. J. Oliveira (Eds.), *Etnomatemática*: Currículo e Formação de Professores [Ethnomathematics: Curriculum and teacher education] (pp. 195-215). Santa Cruz do Sul, RS,: EDUNISC.

Eglash, R., Bennett, A., O'Donnell, C., Jennings, S., & Cintorino, M. (2006). Culturally situated designed tools: Ethnocomputing from field site to classroom. *American Anthropologist, 108*(2), 347–362.

English, L. D., Hudson, P. B., & Dawes, L. A. (2013). Engineering based problem solving in the middle school: Design and construction with simple machines. *Journal of Pre-College Engineering Education Research, 3*(2), 1–13.

Freire, P. (1987). *Literacy: Reading the word & the world*. South Hadley, MA: Bergin & Garvey Publishers.

Furuto, L. H. L. (2014). Pacific ethnomathematics: Pedagogy and practices in mathematics education. *Teaching Mathematics and Its Application, 33*(2), 110–121.

Gonzalez, H. B., & Kuenzi, J. J. (2012). *Science, technology, engineering, and mathematics (STEM) education: A primer*. Washington, DC: CRS.

Gromov, M. (1998). Possible trends in mathematics in the coming decades. *Notices of the AMS, 45*(7), 846–847.

Horta, H. (2013). *Education in Brazil: Access, quality and STEM*. Lisbon, Portugal: Instituto Superior Técnico.
Johnson, L., Adams, S., & Cummins, M. (2012). *The NMC horizon report: 2012 higher education edition*. Austin, TX: The New Media Consortium.
Jolly, A. (2014). *5 hot topics in STEM education. STEM by design. STEM challenges*. MiddleWeb. Retrieved from http://www.middleweb.com/17350/five-hot-topics-stem-issues/
Kintgen, E. R., Kroll, B. M., & Rose, M. (1988). *Perspectives on literacy*. Carbondale, IL: University Press.
Lantz, H. B. (2009). *Science, technology, engineering, and mathematics (STEM) education: What form? what function?* Retrieved from https://dornsife.usc.edu/assets/sites/1/docs/jep/STEMEducationArticle.pdf
Lipka, J. (2002). *Schooling for self-determination: Research on the effects of including native language and culture in the schools*. Charleston, WV: Clearinghouse on Rural Education and Small Schools.
Marginson, S., Tytler, R., Freeman, B., & Roberts, K. (2013). *STEM: Country comparisons. Report for the Australian Council of Learned Academies*. Melbourne, Australia: Australian Council of Learned Academies.
Markham, T. (2011). Project-based learning. *Teacher Librarian, 39*(2), 38–42.
Martinez, J. G. R., & Martinez, N. C. (2001). *Reading and writing to learn mathematics: A guide and a resource book*. Boston: Allyn & Bacon.
Mayes, R., & Koballa, T. (2012). Exploring the science framework: Making connection in math with the common core standards. *The Science Teacher, 79*(3), 27–34.
Moon, J., & Singer, S. R. (2012). Bringing STEM into focus. *Education Week, 31*(9), 24–32.
Museus, S., Palmer, R. T., Davis, R. J., & Maramba, D. C. (2011). *Racial and ethnic minority students' success in STEM education*. Hoboken, NJ: Jossey-Bass.
Myers, C. B., & Pavel, D. M. (2011). Underrepresented students in STEM: The transition from undergraduate to graduate programs. *Journal of Diversity in Higher Education, 4*, 90–105.
National Governors Association. (2011). *Building a science, technology, engineering and math agenda: An update of state actions*. New York: NGA.
NCTM. (1996). *Mathematics: An introduction to the NCTM standards*. Reston, VA: NCTM.
NRC. (2011). *Successful K-12 STEM education: Identifying effective approaches in science, technology, engineering, and mathematics*. National Research Council. Washington, DC: The National Academies Press.
NRC. (2014). *STEM integration in K-12 education: Status, prospects, and an agenda for research*. Washington, DC: National Research Council/The National Academy Press.
Roehrig, G. H., Moore, T. J., Wang, H. H., & Park, M. S. (2012). Is adding the E enough? Investigating the impact of K-12 engineering standards on the implementation of STEM integration. *School Science and Mathematics, 112*(1), 31–44.
Rosa, M. (2010). *The perceptions of high school leaders about English language Learners (ELL): The case of mathematics*. Sacramento, CA: College of Education, California State University.
Rosa, M., & Orey, D. C. (2007). Cultural assertions and challenges towards pedagogical action of an ethnomathematics program. *For the Learning of Mathematics, 27*(1), 10–16.
Rosa, M., & Orey, D. C. (2013). Ethnomodelling as a research lens on ethnomathematics and modelling. In G. A. Stillman & J. Brown (Orgs.), *Teaching mathematical modelling: Connecting to research and practice – International perspectives on the teaching and learning of mathematical modelling* (pp. 117–127). Dordrecht, The Netherlands: Springer.
Rosa, M., & Orey, D. C. (2015). A trivium curriculum for mathematics based on literacy, matheracy, and technoracy: An ethnomathematics perspective. *ZDM, 47*(4), 587–598.
Schlechty, P. C. (2011). *Engaging students: The next level of working on the work*. San Francisco: Jossey-Bass.
Seemann, K. (2000). Technacy education: Towards holistic pedagogy and epistemology in general and indigenous/crosscultural technology education. In *Proceedings of improving practice*

through research: Improving research through practice (pp. 60–74). 1st Biennial international conference on technology education research. Brisbane, Qld: Griffith University.
Smith, K. A., Douglas, T. C., & Cox, M. F. (2009). Supportive teaching and learning strategies in STEM education. *New Directions for Teaching and Learning, 117*, 19–32.
Steen, L. A. (2001). *Mathematics and democracy: The case for quantitative literacy*. Princeton, NJ: National Council on Education and the Disciplines.
Stohlmann, M., Moore, T. J., & Roehrig, G. H. (2012). Considerations for teaching integrated STEM education. *Journal of Pre-College Engineering Education Research, 2*(1), 28–34.

Milton Rosa graduated in mathematics, sciences, and pedagogy in Brazil, earned both a Master's and Doctorate in Education at California State University, Sacramento, in the United States, and earned a postdoctorate degree in Education in Brazil. His studies focus on mathematics education, educational leadership, and long-distance learning. Dr. Rosa is a professor in the Centro de Educação Aberta e a Distância of the Universidade Federal de Ouro Preto, Ouro Preto, Minas Gerais, Brazil. In 2010, Dr. Rosa was awarded the Dr. Carlos J. Vallejo Memorial Award, in Multicultural and Multiethnic Education, Special Interest Group in the American Education Research Association (AERA) as an emerging researcher in mathematics education.

Daniel Clark Orey is Professor Emeritus of mathematics and multicultural education at California State University, Sacramento. He has taught and lived in Guatemala, México, and the United States. He is a Fulbright senior specialist with experiences at the Pontifícia Universidade Católica de Campinas in Brasil (1998) and at Kathmandu University in Nepal (2007). He is currently professor of mathematics education in the Department of Mathematics Education and serves as a collaborator in the Master's Program in mathematics education at the Universidade Federal de Ouro Preto, Brasil.

Chapter 12
Enlivening STEM Education Through School-Community Partnerships

Russell Tytler, David Symington, Gaye Williams, and Peta White

Abstract A major response to the growing concern with diminishing engagement and participation of students in STEM pathways, in Australia and internationally, has been the involvement of the STEM community in school outreach activities. In Australia there has been a proliferation of links between scientists and schools, with the aim of engaging students in authentic activities and providing models of what STEM work pathways might entail. This chapter will draw on a series of projects studying partnerships between the professional science/mathematics communities and schools, to explore a range of partnership models, the experience and outcomes for students and teachers, and challenges for crossing the boundary between school and STEM professional communities. Such school/STEM community partnerships are particularly suited to studies related to environmental and sustainability issues, a focus explored in the chapter. Further, we will draw on a recent evaluation of the Australia-wide, CSIRO-led Scientists and Mathematicians in Schools (SMiS) program. That study provided insight into the use and outcomes of the SMiS model. We will explore some of the challenges of working across the school-STEM professional practice boundary, implications for curriculum, and differences in partnerships for mathematics compared to science.

12.1 School Community Partnerships to Enliven STEM Education

12.1.1 Introduction

While there is widespread agreement that a nation's future requires an effective school STEM education program, there is less agreement about how that will best be achieved. There is a growing belief, however, that the involvement of STEM professionals in school programs could enrich them and capture the interest of

---

R. Tytler (✉) • D. Symington • G. Williams • P. White
Deakin University, Burwood, VIC, Australia
e-mail: russell.tytler@deakin.edu.au

R. Jorgensen, K. Larkin (eds.), *STEM Education in the Junior Secondary*,
DOI 10.1007/978-981-10-5448-8_12

students. In New Zealand, for example, it has been suggested at governmental level that:

> If we are to more effectively engage all students in science by using learning opportunities that are of interest and relevance to them, teachers need to have access to information on contemporary contexts that students can relate to. To achieve this we need appropriate planned connections between the science education sector and science communities. (New Zealand Office of the Prime Minister's Science Advisory Committee, 2011, p. B59)

Similarly the Australian National STEM School Education Strategy (Education Council, 2015) includes as one of the five areas for national action: "facilitating effective partnerships with tertiary education providers, business and industry."

In this chapter we will explore examples of how such planned connections between the education and STEM communities have been operationalized and what the research to date has indicated about the value of such connections. While it is recognized that STEM education involves more than science and mathematics, in our research we have mainly focused on connections between scientists and the science curriculum, often with a sustainability focus, as this is common in the Australian context. These partnerships frequently have a strong environmental focus, and so we explore also the strong and particular tradition of community partnerships in sustainability education. Later in the chapter, we give particular attention to connections between schools and mathematicians and other STEM professionals who use mathematics. Readers will be able to identify the relevance of the work done in relation to school science and mathematics to other aspects of STEM education.

Again the focus will be on the Australian experience, since this is where we have been researching. However, our Australian research has been influenced by developments and understandings across the globe.

The picture which emerges from studies in this field is complex since the programs are enormously diverse having a broad range of purposes, organizations, and outcomes. However, it is possible to point to understandings which can be used as guidelines to shape future activity in the field.

12.1.2 The Range of Ways STEM Professionals Connect with Schools

There have been several attempts recently (Office of the Chief Scientist, 2016; Tytler, Symington & Cripps Clark, 2016) to document the range of programs developed to engage the STEM community with school education. The programs range from those in which STEM professionals come to a school with a prepared program to others where the programs are planned between the professionals and schools. The range and reach of these programs show the strength of commitment from the STEM professional communities to the enrichment of school programs and also indicate that many schools are ready to embrace these links with the STEM community.

Particular approaches to partnerships between schools and the science community have been the subject of a number of in-depth studies (Burgin, Sadler, & Koroly, 2012; Forbes, 2014). The focus of most of these is on evaluation of programs in which organizations outside the school are the primary drivers. For example, Husher (2010) has conducted a study with a focus on what she calls "outreach programs," which she defines as "Face to face programs related to Science, Technology, Engineering and/or Maths, which are delivered to K-12 schools actively and regularly, rather than one-off events" (p. 9). One example of this type of program is the Science and Engineering Challenge, a program mounted in Australia primarily by university schools of engineering and designed for students in year 10 with a specific aim of increasing enrolments in STEM studies in the senior secondary level of schooling and subsequently at the tertiary level. Among the findings of the study, Husher found that 2 weeks following participation (in the Challenge), 4.4% of students overall were found to be more likely to consider a career in science and 1.5% were more likely to consider a career in engineering than they were prior to the Challenge. For the group of students who responded 12 months after participation, no overall change was found in the percentage of students likely to consider a career in engineering compared to prior to participation; however, an increase of 3.4% was found in students considering a science career.

Not all of the initiative for programs involving STEM community members has come from outside the schools. There have been many instances where teachers and schools have worked with scientists to develop locally relevant curriculum activities. In this chapter we will draw significantly upon our own research, which has focused on the type of collaborations where the teachers have been involved in a significant way in determining the nature and focus of the collaborative activity. Below are some examples of the sorts of programs we explored, to illustrate the type and variety of curriculum activities that can arise from such partnerships.

In one of our studies (Cripps Clark, Tytler & Symington, 2014), we spoke with a number of professionals whose work involves science relevant to the lives of the students in rural areas and who were approached by local schools to contribute their expertise to the science program. Our interviewees included:

- A forester who met the students in the field for half a day at the plantation where he led them through a number of activities, such as calculating the quantity of timber per hectare. A colleague then led them through biodiversity exercises, and they talked to the landholder about how the plantation was integrated with their enterprise.
- A meteorologist who mentored students in air quality and weather projects.
- An engineer who, in partnership with the local water utility, engaged students in a case study of water movement throughout their school and developed a plan to install wetlands in part of the school grounds. (They have subsequently found money to implement the plan, and it is part of continuing study of the ecology of the area.)

In a later paper Tytler et al. (2016) reported on a scientist who is working on the transfer of genes from algae into oil seed crops. His links with schools began with him taking his expertise to a school in an area where such research is very relevant, a farming district where the oil seed crops are grown. His employer, a scientific research organization, had a major industry partner based in a country town close to the school. In this instance the students were being made aware of the way in which science was working with their local industry. It introduced them to the forefront of scientific knowledge and processes in an area of considerable local relevance.

The focus of an earlier study (Tytler, Symington, Kirkwood & Malcolm, 2008) was on collaborations initiated by teachers in schools in rural settings. This study provided insights into the variety of foci and strategies adopted by the teachers who initiated the collaborations we explored.

> Rose (an alias) is a physics and science teacher who was concerned about the uptake of physics by students. She seized the opportunity provided by a grant to link science with industry and the community. She organised a cluster of schools including her own and neighbouring primary (elementary) schools to run science workshops, supported by students from the local university, and ran science nights for parents and students. She organised a careers event and had 'the optometrist, the chiropractor and even the massage therapist come along'. A local manufacturing company had hitherto been low profile and interacted with schools on an ad hoc basis, but the initiative saw them involved in the careers nights and also running excursions and activities for local students in tours of their plant, emphasising applications of science and opportunities in local industries. These events were very successful and Rose reports an increase in the numbers of students choosing senior science. As well as students having greater understanding of science and of science opportunities in the community the links made between the various players was a very significant outcome. (Tytler et al., 2008, p. 15)

In the case cited above, the teacher who was the driving force for the initiative planned to increase awareness among the students of the range of career options which draw upon the sciences and saw as an indicator of success an increase in the enrolment in physics in senior classes. A later study (Tytler, Symington & Smith, 2011) identified another teacher in a rural setting who was concerned to ensure that the students realized the range of career options available in their own area that drew upon science.

> The GrowSmart project in the Riverland area of South Australia provided an opportunity for pupils to learn about local horticulture and agriculture industries and professional career opportunities which exist where innovative, highly skilled science and technology concepts and processes are crucial to the longevity of the local industry. (Tytler et al., 2011, p. 32)

In a further study Symington and Tytler (2011) explored the operation and impact of a national science investigative competition, organized and run by science teachers through their organization and sponsored by a major international company, in which students singly or in groups undertake their own research projects. In this type of program, the aim is to introduce students to the experience of working scientifically, to enable them to experience the process of developing knowledge using scientific processes and thinking. In quite a few of the schools, the participating teachers linked students with outside STEM professionals who had expertise which could assist with their project. One of the findings that emerged from the study was the value of these linkages with the STEM community. The following extract comes from an interview with a teacher conducted as part of the study.

> It's also good to have people in the community to draw on. We've had great help from CSIRO Marine and Atmospheric Research labs in (state capital), who have given students samples of material to work with and detailed notes to base their methods on. Likewise university departments have been helpful. Also people related to students who have expertise – they may not be people from big industries, but people like bee keepers, or agricultural advisors, or people with farms or people working with the department of primary industries – they might take samples and test them for students. Over the years I've developed a lot of contacts and I use them. These people also act as role models for students in areas in which they might find themselves in a few years time. It really is good for students to have contact with such people, so they can see the value of the sciences. Or it might be that we have people come into class as guest speakers, from organisations such as the Department of Primary Industries or Waterwatch, perhaps taking in crayfish, as was the case back when we were doing the work on them. In this way you can expose the species (then endangered) to as many as 200 students even though the project only involved 6 students. (Alison, Moriah School) (Symington & Tytler, 2011, p. 10)

Research into programs where teachers initiate the partnership is quite limited. However, the work which has been done has identified some key issues which need to be considered when planning such collaborative strategies. These will be introduced in the following section.

12.2 Challenges in Developing Partnerships with STEM Professionals

The programs we have investigated, often where teachers were significantly involved in determining the focus of the partnership and where the science programs reflect the local context, are very diverse. In many cases the collaborations were outstandingly successful. However, our research has also alerted us to significant factors impacting on the outcomes and sustainability of such partnerships. In particular our work has pointed to the relevance of the theoretical constructs of *communities of practice* and *boundary crossing* (Akkerman & Bakker, 2011). Participating STEM professionals bring with them understandings from their own community of practice and generally lack knowledge of the ways in which schools operate and problems can arise when the partners do not appreciate these differences in understandings. A simple illustration of this difficulty was scientists' frustrations in communicating with teachers who were not contactable by phone for most of the working day because of teaching commitments and who have a very different e-mail practice to their own (Tytler et al., 2016).

Curriculum is another aspect of the culture of schools that needs to be understood by STEM professionals willing to work in school programs as it can be a major site of sociocultural discontinuity across the boundary. A number of the scientists interviewed by Tytler et al. (2016) portrayed the school curriculum as a potential barrier to their ability to contribute optimally with the key issue often being lack of flexibility on one or both sides of the boundary. There are those community members whose employment dictates the fields in which the collaboration is to operate. For example, participants employed by water authorities perceived the field of their possible contribution to school science programs constrained by the purposes of their employers: to raise young peoples' awareness of water supply and usage issues. On the other hand one scientist whose work was focused on urban water systems expressed frustration at her partner teacher who insisted she teach a lesson on the "water cycle" rather than negotiate to use her expertise more flexibly. Whether or not curriculum is a barrier depends on the flexibility and capability of both sides of the partnership. Not all volunteering scientists saw curriculum as an obstacle to collaboration. There were cases where collaborating community scientists saw themselves as adding value to the curriculum beyond a particular topic or outcome framework, often focusing on meta-level learning of knowledge and skills. These often focused on the benefit to the students or on the authenticity of what was being offered, addressing curriculum in its broader sense as what students experience, and for what purpose. The key to matching community members' knowledge and skills to the needs of schools depends upon clarity of purpose, understood from both sides of the boundary.

12.3 Sustainability Education Through Partnerships

Many of the partnerships we have explored reflect the particular relevance of school-community links to environmental and sustainability education and its long-term practice in schools. Programs investigating local environmental and sustainability issues lend themselves well to community input and engagement; hence, there is a strong tradition for sustainability partnerships with STEM professionals. This tradition is acknowledged through many opportunities in the curriculum. The strong sustainability education focus through partnerships is illustrated by exemplars from the large Australia-wide Australian School Innovation in Science, Technology and Mathematics (ASISTM) initiative (Tytler et al., 2011) including:

- *Connections*: a one-day a week program focused on connecting students with themselves, the community, and the environment. The projects led to public outcomes such as a reclaimed piece of river or reduced water use at the school (p. 25).
- On Kangaroo Island, South Australia, scientists worked with teachers and students in monitoring fish and other animal populations and environmental conditions, to build up a substantial database of scientific interest (p. 26).
- Clusters of schools worked with university scientists and industries involved in ecotourism on projects focused on Marine Science and Aquaculture, Marine Ecology and Ecotourism, and Bio-remediation (p. 26).

We found a strong theme of promoting environmental literacy in interviews with scientist partners (Cripps Clark, Tytler & Symington, 2014) consistent with this focus on environmental projects.

12.3.1 *The Particular Place of Community Partnerships in Sustainability Education*

In this section we consider some of the important benefits of forming community partnerships to explore sustainability issues aligned with STEM education. These include curriculum links, community network formation, volunteering in environmental work, and students' career and attitude development. Many of these programs operate in quite a different manner to most of those discussed in the earlier sections of this chapter in that the issue of boundary crossing, discussed above, is often handled by education officers employed by companies or instrumentalities, such as water authorities, to run outreach activities. These individuals span the boundary between professional STEM practice and teaching by virtue of their history and can thus act as "brokers" between the two communities.

12.3.1.1 Curriculum Links

The Australian Curriculum learning areas of Science and the Humanities and Social Sciences provide obvious opportunities for engagement in sustainability issues via partnerships.

The learning area of science offers many connections where curricula outcomes can be addressed via STEM professional engagement in local sustainability issues. Learning strategies that have been used successfully include inquiry and investigation into local issues, experimental design, classifying and grouping, problem-solving, communication, history, and change (or change over time). These strategies can be augmented through connection with STEM professionals who use them on a daily basis in their work. While teachers are trained to develop skills that encourage proficiency across a variety of curriculum areas, they cannot be expected to bring a depth of understanding to all areas. Developing community partnerships with STEM professionals, then, becomes an important part of resourcing the curriculum.

Further, the Australian Curriculum features three cross curriculum priorities, one of which is Sustainability, addressing "the ongoing capacity of Earth to maintain all life" (ACARA, v7.5 http://v7-5.australiancurriculum.edu.au/crosscurriculumpriorities/Sustainability). Reading into the statements included in the curriculum documents, it is clear that developing partnerships between schools and community is an important way to establish sustainability-focused STEM educational programs that matter. "It enables individuals and communities to reflect on ways of interpreting and engaging with the world. Actions that support more sustainable patterns of living require consideration of environmental, social, cultural and economic systems and their interdependence" (ACARA, v7.5 http://v7-5.australiancurriculum.edu.au/crosscurriculumpriorities/Sustainability).

Additionally, the Australian Curriculum includes seven general capabilities, one of which is the general capability "Ethical understanding" which provides strong possibilities for partnerships in sustainability issues: "Complex issues require responses that take account of ethical considerations such as human rights and responsibilities, animal rights, environmental issues and global justice" (ACARA, v7.5 http://v7-5.australiancurriculum.edu.au/generalcapabilities/ethical-understanding/introduction/introduction).

Finally, the learning area of Civics and Citizenship also provides opportunities to explore sustainability issues via partnerships. "The Civics and Citizenship" curriculum aims to reinforce students' appreciation and understanding of what it means to be a citizen. It explores ways in which students can actively shape their lives, value their belonging in a diverse and dynamic society, and positively contribute locally, nationally, regionally and globally" (ACARA, v7.5 http://v7-5.australiancurriculum.edu.au/humanities-and-social-sciences/civics-and-citizenship/rationale). Partnerships focused on sustainability provide opportunities for interdisciplinary knowledge, skill, and attitude development.

12.3.1.2 Community Network Formation

Many environmentally related organizations consider the development of education programs and resources as a viable marketing and community awareness strategy. These organizations and government departments (at all levels, local, state, and federal) will also often employ education officers to facilitate their involvement in partnerships. These people provide valuable resources in sustainability education as they are usually capable educators with curriculum understandings as well as knowledgeable in their area of science/environmental expertise, often as practitioners. They can, therefore, be useful conduits between the STEM professional community and teachers/students. These boundary crossers are well placed to navigate the potential gap that may exist between teachers and STEM professionals. Many are trained with qualifications in both teaching and STEM professions, fusing the two with a passion for sustainability education, often focused on local issues.

An interesting example of where the joint credentialing of these boundary crossers is employed is in the "Watch" programs. Australia boasts many "Watch" programs: CoastWatch, RiverWatch, ReefWatch, FrogWatch, and AirWatch, to name a few. The Watch programs host networks that link schools and provide additional support for the implementation of citizen-based environmental monitoring across large areas within states. The accumulated action that these communities undertake as engaged volunteers can be significant. We explore this further in the next section.

There are many resources and teaching tools available related to sustainability issues. However, these tools need to be critically considered as they are often produced to share perspectives that are intended for profit through changing community perception. The potential problem associated with this form of outreach organized by government or industry bodies is the potential for vested interests to capture the curriculum. For example, a parody of The Lorax was developed by the National Oak Flooring Company declaring that some forms of logging are, in fact, appropriate. The title of this targeted resource was "The Truax" (Birkett, n.d.). Critical evaluation of such materials can provide good training for teachers and students. When schools open up their doors to community influence, we must be careful to preserve the integrity of the curriculum and schooling purposes.

12.3.1.3 Volunteering in Environmental Work

While using students to achieve on-ground works (such as revegetation, weed control, native species management, storm drain protection, water monitoring, riparian zone and river management, litter cleanup, or beach cleanup) is common practice, it seems to be best managed when married with a strong educational/awareness program so that the students involved are cognizant of the reasons for taking action.

An example of successful action-orientated (volunteering) programs facilitated by teaching/STEM professional boundary crossers is the Western Australian South West Catchments Council "Dune Dudes" program. The coastal facilitators established a "Dune Dudes" program in 2011, and it has significantly influenced a number of students, schools, teachers, and programs and been significant in sustaining coastal areas. "Captain CoastCare" and "Dune Dudes" run and unpack several adventures with participating students and work to improve the coastal vegetation and promote environmentally sensitive practices. As a result, in a 2-year period, on-ground volunteering works carried out along the southwest coastline were calculated as providing a value of $33,000 (Gibbs, 2013). Such programs can also lead to significant, authentic learning outcomes.

Citizen science programs are another way of facilitating community and school engagement in environmental monitoring. A useful educational program that facilitates science data collection is the Global Learning and Observations to Benefit the Environment (GLOBE) Program (http://www.globe.gov/about/overview). This international science and education program provides students and the public with the opportunity to participate in data collection and scientific processes and to contribute meaningfully to our understanding of the Earth system and global environment (GLOBE, n.d.). In these programs students and teachers often work under the guidance of STEM professionals/education officers using carefully designed scientific processes.

Having students take action in their local community, with support and advice from STEM professionals and facilitators to ensure engagement in the issues behind the action, provides a rich learning environment. The positioning of students as future engaged citizens, who take responsibility for their own and others' actions, affords empowerment to learning. Student interest and awareness of future career opportunities are also often a significant outcome of these partnership and action-focused learning opportunities.

12.3.1.4 Students' Career and Attitude Development

At the heart of connecting students and STEM professionals in meaningful partnerships are actions toward understanding science as a human endeavor. The "understanding science as a human endeavor" strand in the Australian Science Curriculum includes understanding "the nature and development of science" and the "use and influence of science," which span the ways science is practiced and the people and careers involved in science. Through partnerships between school and community, students can interact with STEM professionals and imagine career potential for themselves. Through engagement in local sustainability issues, where community members from a variety of employment pathways interact to develop shared understandings and better ways of practicing, students can develop attitudes that engage with scientific literacy and science epistemologies. This is where a strong social justice agenda becomes embedded in the practices of sustainability education.

With the exploration of sustainability issues, and the exemplification of how community partnerships can be highly effective, comes greater clarity around how these programs are greatly supported by individuals – boundary crossers – who marry teaching and STEM professional skills.

12.4 The Scientists and Mathematicians in Schools Program

Despite the prevalence of STEM partnerships with schools, these are not often subjected to explicit evaluation. Rather people draw on anecdotal evidence and an expectation that having scientists interacting with students must yield positive dividends in terms of providing role models and images of potential careers in the STEM area. However, this leaves unexamined a number of questions concerning: What forms of knowledge can STEM professionals productively bring to the school STEM curriculum? What ways of interacting provide effective use of STEM professionals' expertise and time? and What are the conditions under which such partnerships are sustained?

The Australia-wide Scientists and Mathematicians in Schools (SMiS) program run by CSIRO (Commonwealth Scientific and Industrial Research Organization) is an example of a STEM partnership model with national reach. SMiS was initiated in 2007 and in 2015 involved 1800 active and assigned partnerships. SMiS has taken the issue of evaluation seriously. Performing an evaluation of the model and its impact (Tytler et al., 2015) allowed researchers to build on previous evaluations (Howitt & Rennie, 2008; Rennie, 2012; Rennie & Howitt, 2009) and to examine in more detail the nature of partnerships including the roles of the partners and the outcomes for students, teachers, and the STEM professionals. The evaluation of the program has allowed the extension of insights previously mainly focused on scientists, to explore also the potential of mathematicians working with teachers and what this offers that is similar to or different from the science partnerships. The STEM professional partners contributing to Mathematics in Schools included scientists taking on the STEM professional role to illuminate how mathematics was crucial to their scientific work.

12.4.1 Evaluating the SMiS Model

The SMiS model involves one-on-one partnerships between an individual teacher in a participating school and a partner STEM professional, with details of the activities not prescribed but to be negotiated between these partners. The CSIRO SMiS Team Member matches teachers with STEM professionals; the matching considers expressed school needs and the STEM professional's expertise. The evaluation involved developing, executing, and analyzing a survey of scientists and mathematicians and teachers involved, as well as conducting and analyzing interviews with the SMiS management team, and developing case studies based on interviews with participants in a number of targeted partnerships. In this way some light was shone on the issues raised above concerning the potential of such partnerships to add value to the STEM curriculum, the range of activities that may prove productive, the value of such an arrangement for students, teachers, and STEM professionals, and issues around successfully negotiating and sustaining such partnerships. Importantly this research extended our understandings, mainly gained with programs around science, to include the nature and outcomes of partnerships focused on mathematics.

12.4.1.1 The Nature of the Partnerships

The first thing that became obvious in examining the survey responses was the variety of partnership and activity arrangements. The time spent by the scientists and mathematicians in schools is spread across class presentations, explorations, or discussion, working with individual students or with groups of students within or separate from the class, with presentations across several classes, and planning with teachers, all depending on what the teacher and STEM professional negotiate. Scientists often support individuals or groups of students with investigative projects, help teachers develop activities in content areas outside their immediate expertise, or present about careers. Often activities developed and altered over time as the partners got to know their respective strengths and what was possible.

> My scientist partner and I are always looking at new ways that he can conduct science in my classroom. We try to integrate these into my science unit of work where possible. (Science Teacher)
>
> The activities have become more focused on students' understanding of mathematics and have become more appropriate to their needs. The sessions are more relaxed as we have all become more comfortable with each other. (Mathematics Teacher)

The program allows the STEM professional to adjust their involvement to the needs of the school. The following extract, describing the involvement of one scientist over a number of schools, illustrates the point.

George trained as a neuroscientist and in psychiatry, and has partnered both primary and secondary schools (That STEM professionals can work with a number of schools was noted in our previous research (Cripps Clark, Tytler & Symington, 2014)). One partnership involves having year 10 students work with him on research projects about 2–3 times per year, another involves him, during science week, demonstrating clinical activities, yet another has him giving career guidance to aspiring scientists. George notes the positive effect on student engagement when he brings in equipment from his laboratory for them to use. George believes that he offers some "real" examples providing opportunities for students to develop more in-depth understanding of a curriculum he sees as 'superficial.' (Extract from Tytler et al., 2015, SMiS Evaluation)

12.4.1.2 The Knowledge Brought by STEM Professionals

While knowledge of science or mathematics, brought by the STEM professionals, was an important component, more important, according to both teachers and the STEM professionals themselves, were aspects of thinking and working scientifically and mathematically and science as a human endeavor (these being significant organizers in the Australian science curriculum). When asked in the survey about the significance of outcomes for students from the collaboration, the top category for both science and mathematics teachers was "passion and curiosity," followed by "knowledge of contemporary science/mathematics and the way science builds evidence/mathematicians think and work" (stronger for mathematics teachers), and then in roughly equal measure "knowledge of key concepts," "what it's like to work as a scientist/mathematician as part of a team," and "capacity to tell stories." Knowledge of careers was further down the list. The STEM professionals' nominations echoed this pattern.

We argue from this evidence that the value that STEM professionals bring to the school curriculum is not primarily formal conceptual knowledge but rather that they represent what it is like to be a person involved in science or mathematics through inquiry, problem-solving and reasoning, and appropriate ways of working, with a passion for pursuing this type of knowledge. They are in a position, through living what it is to practice in STEM, to offer a unique contribution to the school curriculum, beyond what is normally possible for teachers.

In the specific area of our Scientist's expertise, the students and teachers involved in the program had access to current thinking and research which enhanced their learning. In addition, the students were very engaged in that area through research and investigation. (Science Teacher)

Interacting with a real life, practicing scientist is an amazing opportunity for the students. My partner is inspiring and down-to-earth which makes her both accessible and relatable. The partnership continues to inspire my teaching practice; it has increased my experience, knowledge and understanding. I LOVE being involved in the SiS program. (Science Teacher)

Comments concerning the mathematicians' contribution often referred similarly to engagement in mathematical activity but rarely to the mathematician as a person:

> The students love the challenge and excitement generated by the interesting problems. (Mathematics Teacher)

The flexibility of the negotiated SMiS model enables partnerships to build on the expertise of the STEM professional while addressing specific needs identified by the teacher.

12.4.1.3 The Outcomes for Students and Teachers

Across the board there was agreement that the program had brought significant benefits to students. The specific benefits for students, identified by teachers (and STEM professionals) as "very significant," were "increased awareness of how scientists/mathematicians think and work," "increased appreciation of scientists/mathematicians as people," "increased interest in and enjoyment of doing science and mathematics," "awareness of relevance of mathematics to society," followed by "knowledge of contemporary science/mathematics," "increased ability to recognize and ask questions about science-related issues," and "increased awareness of the nature of scientific/mathematical investigations and inquiry skills." Again, there is a strong theme of awareness of thinking and working scientifically and mathematically, appreciation of the human aspects of contemporary STEM work, and increased scientific and mathematical literacy. Knowledge of careers was also high on the list for secondary teachers. Many respondents to the survey saw the value of the partnership with STEM professionals as awakening students' curiosity and passion for mathematics.

The overwhelming sense from these survey responses is that for most participants the value of the partnerships is not so much about topping up specialist knowledge or skills, as about introducing students to scientific and mathematical thinking as ways of being in the world. However, we need to note that comments made by some secondary teachers and/or mathematicians revealed a subset of teachers of mathematics who were strongly focused on the STEM professional *delivering* content knowledge.

12.4.1.4 Sustainability of the Partnerships

As described above, there are a number of challenges for scientists and mathematicians working with teachers in schools associated with the very different communities of practice within both STEM and, more broadly, education. Thus, boundary crossing is an issue, and sustainable partnerships require some form of brokerage in the form of a person at the school or in the STEM workplace, or in the SMiS program within the support team, who understands issues from both sides of the boundary. Sometimes this can be the STEM professional themselves if they have a teaching background or the teacher if they have research or professional experience in STEM. In such cases, the "boundary" is not so significant because each has already practiced, to some extent at least, on both sides of the boundary.

This boundary crossing can be aided by teachers and STEM professionals recognizing and respecting each other's expertise. Our evaluation showed that teachers and STEM professionals were remarkably aligned in their views about what helped sustain projects. This included the need for partners to listen to each other and learn to understand their different points of view and develop a partnership that adapts to this.

> I also think another aspect of a successful partnership is just building an appreciation of the requirements that each other has in their careers. (SMiS Team Member)

A member of the SMiS management team reflected on building of successful relationships for Mathematics STEM partnerships as follows:

> Even the teachers that I've spoken to that weren't really sure what they were going to be doing, or what the mathematician was going to be able to do, through the conversation they've obviously both learnt about each other, and they find the spot that suits both, and whether that's taking a session with a few students or just talking to the staff in their staffroom it's really about how they communicate with others and how they can then come up with something together, so that's the really … powerful …. (SMiS Team Member)

Our evaluation showed that this occurred quite often but not always in SMiS (Mathematics) partnerships.

The other factor that supports commitment to partnerships is the existence of a common interest, a joint concern to pursue a goal and an alignment of expertise with genuine purpose (see also Tytler et al., 2011).

> It [the partnership] has been growing steadily to include more students in more classes and year levels, and more interaction with more teachers. It's taken a little time for the notion to percolate through the school. With growing familiarity there has been increased appreciation, demand and support from the whole teaching body. (STEM Professional: Mathematics)

This growth of interest in what the mathematics STEM professional could offer occurred over several years, suggesting that some teachers needed to "see" outcomes before committing.

Less positive, but equally compelling, the evaluation of the SMiS revealed a few instances where the school had no expectation of contributing to the partnership:

> The partner school expected me to come up with the entire format … that we were going to work on together, that there would be no input from them. So, while I didn't mind putting my time in, I wasn't prepared to do so with no input from them. (STEM Professional: Mathematics)

These instances illustrate the importance of teachers' expectations in negotiating productive and sustainable partnerships.

12.4.2 *Mathematicians Contributing to STEM in Schools Through SMiS*

Involvement of STEM professionals in school programs in mathematics does not have the same history as the involvement of scientists in school science programs. Allied with this, the SMiS program began from science and operated for 2 years before partnerships in mathematics were introduced. Even in 2015 there were far fewer partnership programs operating involving mathematics compared to science. Hence, our understanding of partnerships around the teaching and learning of mathematics is much less well developed. Accordingly the evaluation around the SMiS mathematics partnerships is of special significance.

Partnerships in mathematics within the SMiS program took many forms because expectations of what could be gained by participating differed markedly between teachers and between teachers and STEM professionals. However, the most common reason given by teachers of mathematics for participating in SMiS was to increase student engagement in mathematics through providing students with access to a "mathematics" STEM professional. What mathematicians contributed to partnerships in general included showcasing mathematics to add interest and excitement (which sometimes included the mathematicians' own research), raising awareness of possible careers in mathematics, and providing activities to stimulate interest in mathematics. With each of these types of activities, there were sometimes links to the curriculum, and sometimes the activities were more broadly focused. The teacher often played a major part in decisions on whether the mathematical activity had to relate directly to the curriculum. Illustrations are given below of the nature of relationships between partners, activities undertaken, and possible differences in what was capturing student interest.

12.4.2.1 Variations in Who Makes Decisions About the Focus of the Activity

There was significant variation in the parts that the partners played in determining the focus of the classroom activity. In some cases the teacher played the primary role.

> Once my partner teacher asked if I could talk about the search for MH370 – being as it was Bayesian statistics, & recently in the news. That was fun. (STEM Professional: Mathematics)

In other partnerships the STEM professional played a major part in determining the focus of the activity.

> My partner teacher generally leaves it up to me what activity to run, though I do ask what topics they have covered in class recently and try to make the activity relevant. I find the teachers have little idea of what to expect from me or what type of activities we could do in mathematics. (STEM Professional: Mathematics)

In other partnerships the partners reported that the roles changed over time as the STEM professional became more at home in the school context.

> In the first year the teacher suggested after my first couple of lessons that I teach some fundamentals about geometry. /I agreed and did that but ran out of time to present as I wished to present. /Subsequently … In years 2 through 5 of my participation the teachers allowed me to present as I wished. They made suggestions from time to time and I attempted to accommodate those suggestions in how I presented the material. (STEM Professional: Mathematics)

As was the case in science partnerships, there were examples of relationships where there was fruitful discussion between the partners in determining the focus of the activity. In some cases this joint planning was focused on ensuring that the activity meshed with the school mathematics curriculum.

> Our Mathematician participates in sessions once a week, sometimes once a fortnight, depending on the work commitment. We often talk after school, targeting our learning to complement the learning our students are engaged in. (Teacher of Mathematics)
>
> The mathematician [called by first name] has asked for ideas on what we have been covering in class and then he will come up with some ideas, we will discuss those and then he will come with the lesson ready to go. (Mathematics Teacher)

Sometimes partners were willing to explore ideas as they planned and participated together:

> We're a bit more opportunistic. My partner [teacher] came up with a new suggestion, we are trying it out. (STEM Professional: Mathematics)

12.4.3 Factors Determining the Focus of the Classroom Activity

As illustrated in the previous section, the evidence points to a great deal of variation in the roles that the partners play in determining the program in which the students are to be engaged. A variety of factors influenced decisions about the focus of the partnership activities. There was evidence that in some cases the curriculum was the major factor.

> [My] Initial contributions to MiS partnership tended to be not quite in line with the present curriculum, or what is useful to the teacher and their program. Listening to the teacher presentation to the class can assist in design of more pertinent maths content. (STEM Professional: Mathematics)
>
> The aim was not to teach the curriculum but use it in interesting ways. (STEM Professional: Mathematics)

Commitment to reflecting the school curriculum, however, was not universal. Many partnerships appeared to be less constrained by curriculum considerations. There is evidence of the STEM professionals developing activities around what were described as "real math problems."

Have students participate in team efforts to solve real math problems. (STEM Professional: Mathematics)

Again, there were instances of programs being designed to address what the partners saw as appropriate to the students themselves. The reference in the previous quote to "team efforts" suggests a concern that learning to work in a team is seen as important for the students. A focus on the needs of the students was shown by others.

I have a range of activities … [and] tailor them to suit the needs of the particular student cohort and their current learning in mathematics. I am happy to take a class every month, whenever the teaching timetable and my work commitments allow. (STEM Professional: Mathematics)

12.4.3.1 The Mathematical Activities Undertaken

Activities undertaken in partnerships varied considerably. They included work with rules and procedures already known to the students, and applying mathematics from the curriculum in a new context, and were at times about presenting rather than interacting, as seems apparent from the following quote:

(I had to) identify the lesson relevant to the curriculum being taught, and then to flesh out the speaking/writing notes. (STEM Professional: Mathematics)

Mathematicians implemented activities involving unfamiliar challenging problems with varying degrees of guidance and varying degrees of student control over the mathematics used. For example, the following activity which was unfamiliar to the students was heavily guided by the mathematician:

My intention was to have the students engage in making three dimension models and learn how to represent geometry in orthographic projection with projects used as vehicles for learning being the design and making of sundials and sunshades to a performance specification that required insights into conceptualising three dimensional solar geometry. This is a complex topic. All I could expect was that the students follow through the process, make their models and explain to the class the processes of designing and making their models. /I wished the students to see how it was possible to apply solar geometry to the design and making of a product.

On the other hand, the following quotes which focus on *creativity* suggest students had more control over the mathematics they used and the pathways they took.

The creative application of geometry to problem solving. (STEM Professional: Mathematics)

(The partnership) gave the teacher a chance to see students working on different topics and engaging with creative ideas in approaching mathematical tasks. (STEM Professional: Mathematics)

12.4.4 The Expected and Achieved Outcomes

Positive affect, including interest and excitement, was displayed by students, with varying levels of mathematical performance, during and after activities with STEM professionals. It was not always clear whether this increase in interest related to opportunities for creative mathematical thinking, the introduction of new contexts, or the opportunities for students to access mathematics in different ways:

> Students really look forward to the visits of our partnered Mathematician and enjoy the mathematics activities. Students who do not always experience success in formal written mathematical work feel success with hands on activities and can express their understandings verbally. (Teacher of Mathematics)
>
> The students love the challenge and excitement generated by the interesting problems. (Teacher of Mathematics)
>
> Addressing the needs of our high achievers has seen engagement and learning really increase as they have been able to explore their mathematical thinking beyond the realm of the classroom! (Teacher of Mathematics)

There were also benefits for the teachers, who indicated that they did derive "very significant benefit" from the partnership with respect to: opportunity to communicate with a mathematician (52%), enjoyment in working with a mathematician (39%), increased engagement of my students (39%), updated mathematics knowledge (35%), and increased awareness of mathematics-related careers (35%).

12.4.4.1 Reflections on the Activity Within Mathematics Partnerships

As can be seen, there was diversity in the relationships between the SMiS partners: how partners participated in topic selection, the autonomy the mathematician had in working with the students, whether or not partnership activity was directly connected to the curriculum, and who made decisions about whether or not this would occur. There was also diversity in opportunities for students to use mathematics creatively, and there appeared to be differences in whether student interest was elicited by the context within which the mathematics occurred or by the creative exploration of mathematical ideas.

As current mathematics education research literature and policy documents point to the need to nourish creative and innovative thinking in the next generation in Australia and beyond (Cunningham, Theilacker, Gahan, Callan, and Rainnie 2016; Marginson, Tytler, Freeman and Roberts 2013), it could be useful to consider SMiS partnerships from this perspective. To develop creative and innovative citizens, opportunities need to be provided for learners to play an active part in their own learning by working independently or collaboratively, being resourceful, and being effective users of technology who are able to problem solve by drawing upon ideas across learning areas and employ mathematics to make sense of their world (Ministerial Council on Education, Employment, Training and Youth Affairs (MCEETYA), 2008; National Curriculum Board, 2009).

Internationally (e.g., Korea, China, and Singapore), questions have been raised about prevailing teacher-centered and procedural approaches to learning mathematics. These countries recognized that even though their students ranked highly on international benchmark tests at school, many of their graduates in Engineering and Economics could not identify and pose problems but rather only solve those problems identified for them by others and where they were told what mathematical procedures to use (Shimizu & Williams, 2013). Their mathematics education had not developed their skills in "problem-solving, critical thinking, communicating, collaborating, and self-management … [that] have become more important skills and attributes in the modern workplaces" (Masters, 2013, p.24). These countries have subsequently changed their mathematics curricula placing different emphases in doing so, with Korea focusing on creativity, China on group interactions, and Singapore on problem-solving (Williams & Huang, 2015). Galbraith, Stillman, and Brown (2010) provide an illustration of how mathematicians can work with students to develop and use these thinking skills. The research illustrated the high positive affect experienced by students during creative mathematical modeling activity designed to track the spread of cane toads (an introduced pest) in Australia.

The data from SMiS (Mathematics) raises questions about the extent to which the mathematical activity undertaken includes opportunities for creative and innovative mathematical thinking, and whether we should care if it does not, as long as student interest in mathematics is increased. If we do care, we need to address the questions: Are there strategies that could be put in place to increase the prevalence of opportunities for creative, innovative, and critical mathematical thinking? Can we develop models of mathematics partnerships that illustrate for teachers how to make productive use of STEM mathematics professionals' capacity to stimulate such thinking?

There are SMiS partnerships that could inform these questions and the broader question: In what ways can STEM professionals and teachers develop partnerships that increase student interest in mathematics *and* the types of mathematical thinking crucial to future societal participation?

12.4.4.2 How Can STEM Professionals Contribute to Mathematics in Schools?

The data to date suggest that there may be more teachers of mathematics than of science who are unsure, for a variety of reasons, as to what contribution a STEM professional could make to their classroom program. The following extract from one of the case studies tells something of a counter narrative of one teacher who identified an opportunity to involve a scientist who used sophisticated mathematics and how the collaboration developed.

Setting

Patrick is a secondary school mathematics teacher in a school approximately 45 kilometers from a capital city. After reading an article about the CSIRO Mathematicians in Schools Program, Patrick made contact with CSIRO requesting a partner-mathematician. Patrick wanted to be partnered with a mathematician who could speak to his year 12 specialist mathematics students, and work with students in a year 8 and 9 acceleration program. He particularly wanted a mathematician-partner who could let his students know about the "mathematics available in the real world."

Nature of the Partnership

The partner-mathematician, Heather, has visited the school twice since the partnership began less than 12 months ago. She has addressed the year 12 students speaking about, among other things, how mathematics works in her area of employment (astrophysics) and how it is such a big part of getting any job. When Patrick spoke with these students afterwards he noted that this latter point resonated most with them. Patrick hopes that this message might be communicated to all students in the future.

Heather has also spent time with the students in year 8 and 9 speaking about how mathematics relates to astrophysics. Patrick was "blown away" by the impact Heather has had on the students so far and has begun planning with Heather to facilitate a project for the year 8 and 9 accelerated students with a focus on the mathematics involved in astrophysics. Some students have also requested that they have one-on-one time with Heather to discuss her area of expertise. It is unclear whether it was the context, or the mathematics, or both that elicited the high level of interest.

Other teachers in the school have been inspired and requested that Patrick "share" Heather with them. He anticipates that this will also happen as the partnership continues.

Motivation and Influences

Patrick envisaged that having a partner-mathematician was an opportunity to provide his students with a fresh face to answer their questions such as "why do we need to learn this?" or "when will we ever use this?". He hopes to utilise Heather for many students in the school, not just "accelerated" students, but also those disengaged with learning mathematics.

Quality Aspects of This Partnership

Patrick did not know what to expect when first partnered with Heather. As it turned out Heather really wanted to be involved with the students, and offered to help out in many classes during her visits. Heather comes across as someone who "really wants to put in the time and effort" and Patrick feels that the "students pick up on that."

Patrick has observed the difference in how his students ask questions of an outsider, compared to when they ask questions of him, their teacher. Subsequently he has tried to be "a bit more open about how I structure my classes."

Members of the school leadership team are supportive of Patrick's partnership with Heather. Parents of some the students commented on the fact that they like that the school has taken the initiative to pursue such a partnership.

Patrick used the online support materials provided by CSIRO SMiS to prepare himself for the partnership. He acknowledges that quality communication is a key factor in maintaining this partnership. (Extract from SMiS report: Case study 5)

12.5 Implications of These STEM Partnerships

In this chapter we have explored, through findings from a number of research investigations, a variety of models of partnerships between teachers and schools and scientists and mathematicians that provide illustrations of the curriculum enrichment possibilities of this increasingly prevalent practice. Activities in this area vary from single visits of STEM professionals representing their work, to programs brought by community STEM organizations to deliver in schools, to locally designed programs where teachers and STEM professionals form a partnership to pursue science and mathematics activity that is often embedded in the school community context and which extends and enriches the curriculum in ways not easily achieved by teachers and schools on their own. The STEM professionals' expertise offers a genuine addition to curriculum resources.

Although these partnership activities are rarely locally evaluated, where evaluations have taken place, the evidence points to significant gains for students in terms of interest and enjoyment and knowledge of contemporary STEM practices and of STEM professionals as people. In the SMiS evaluation, understanding the way mathematicians think and work was rated by teachers of mathematics as more important than specific knowledge and expertise the mathematician brought to the partnership. This aligns with big picture views of what learning mathematics should achieve.

The SMiS evaluation also provided evidence that such partnerships generate significant outcomes for teachers and for scientists and mathematicians (Tytler et al., 2015). Tytler et al. (2011) found that the pedagogies associated with partnership activities tend to be more inquiry focused and student centered than prevailing classroom practice, thus indicating significant teacher learning flowing from the partnership. These partnerships offer a chance to break the cycle of orthodoxy in STEM disciplinary teaching and encourage enlivened pedagogies and curriculum vision.

We argue that the real potential strength of having STEM professionals in schools does not primarily lie in their substantive knowledge but rather in their modeling of scientific and mathematical ways of working including their orientation to knowledge generation – their passion and curiosity. Thus, there is significant potential for partnerships to offer models for bringing school STEM practices closer to disciplinary practices in STEM and that this should be a core aspect of a contemporary STEM curriculum that genuinely represents a version of mature practice in the discipline area. There is a strong indication that such a principle pays dividends in terms of student interest and engagement with significant learning.

Many of these STEM professionals in their working lives represent cross-disciplinary practices, such as scientists with significant mathematics knowledge and commitments, or engineers, or mathematicians working in cross-disciplinary areas. These partnerships (e.g., Patrick and Heather) can in these cases represent contemporary practice in interdisciplinary STEM activity and provide authentic models of people who customarily move across disciplinary boundaries. Along with employing known mathematics in unfamiliar contexts, raising awareness of careers that rely heavily on mathematics (in applying mathematics, and in creating new ideas), and using mathematics as a tool to further other disciplines, emphasis needs to be placed on illuminating mathematics as a domain in which creative activity can occur with mathematical insights developed as a result.

Of course, there are challenges associated with setting up and sustaining such partnerships, associated with the difficulties of crossing the boundary between STEM and school communities of practice. In these cases there is a need to make use of boundary objects that exist at the interface between the two communities

(Akkerman & Bakker, 2011). The curriculum can operate as such a boundary object if it encourages activities and principles aligned with authentic disciplinary practices, and examples of this are the "inquiry skills" and "science as a human endeavor" and "mathematical problem-solving and reasoning" dimensions of the Australian curriculum. A key requirement for pursuing such partnerships, however, is the existence of space and freedom in the curriculum to allow teachers and schools to enter into these partnerships with confidence. A restrictive, locked down curriculum inhibits genuine local project work.

Many of these partnerships are supported by brokers, for instance, the SMiS management team, who operate at the boundary to encourage and support partners to understand each other's practices. There are also many instances of the community STEM person themselves sitting across the boundary, such as the education officers from government instrumentalities who run outreach activities, who themselves have a teaching background. The advantage of an education officer running outreach activities is that they can understand the school curriculum and culture. However, they are generally less able to represent the variety of STEM professional practice, and provide the insight into working and thinking scientifically and mathematically, that usually occurs with locally inspired partnerships or those of the SMiS program.

12.5.1 Suggestions for the Way Ahead

This chapter has provided some snapshots of some of the activity we have explored in Australia involving partnerships between schools and the STEM community. Space prevents us from representing here the full range of models which have proved to be effective or even to fully describe the ones we have discussed. However, the literature enables readers to pursue such details. Furthermore, our explorations have suggested that there is no single organizational model which outperforms all others. What we are able to do, and have attempted to do above, based on our understanding of the field, is to make some strong recommendations about the characteristics of a system which will allow productive partnerships to develop.

A key feature is the curriculum framework within which the schools work. Our research suggests that it should not be totally focused on content to be learned but needs to encourage exploration of the disciplinary practices in STEM: a curriculum which provides space for STEM professionals to model, for the benefit of students, ways of working within and across disciplinary boundaries.

Again, research in this field suggests that the system-wide curriculum should not be unnecessarily prescriptive with respect to content so that teachers have the space to introduce topics which have relevance to the daily lives of the students and the local community. In our research (e.g., Tytler et al., 2011; Tytler & Symington, 2015), we have found a disproportionately high representation of schools in rural areas developing exemplary programs involving collaboration between school and STEM professionals. An earlier section of this chapter pointed to the opportunities to involve students with STEM professionals in sustainability projects. While these abound in the countryside, there is a need to support schools in urban areas to engage in such local activity.

Finally, our research suggests that the success of collaborative activities does not depend upon who initiated the ventures. Our investigations have identified very successful school/STEM professional collaborations which have been initiated by a single school, by clusters of schools, by education systems, by individual STEM professionals, and by STEM organizations of various types. The critical characteristic of successful programs is that there is mutual respect for the expertise that the partners, both teachers and STEM professionals, bring and opportunities for them to solve any boundary-crossing issues.

References

ACARA (v 7.5). *Links to the Australian curriculum*. Retrieved from links at each reference.

Akkerman, S. F., & Bakker, A. (2011). Boundary crossing and boundary objects. *Review of Educational Research, 81*(2), 132–169.

Birkett, T. (n.d.). *The Truax*. Retrieved from woodfloors.org/truax.pdf

Burgin, S. R., Sadler, T. D., & Koroly, M. J. (2012). High school student participation in scientific research apprenticeships: Variation in and relationships among student experiences and outcomes. *Research in Science Education, 42*, 439–467.

Cripps Clark, J., Tytler, R., & Symington, D. (2014). School-community collaborations: Bringing authentic science into schools. *Teaching Science, 60*(3), 28–34.

Cunningham, S., Theilacker, M., Gahan, P., Callan, V., & Rainnie, A. (2016). *Skills and capabilities for Australian enterprise innovation*. Melbourne, Vic: Australian Council of Learned Academies. www.acola.org.au

Education Council of Australia. (2015). *National STEM school education strategy*. http://www.educationcouncil.edu.au/site/DefaultSite/filesystem/documents/National%20STEM%20School%20Education%20Strategy.pdf

Forbes, G. A. (2014). *MyScience: Communities of practice around the teaching and learning of primary science*. Unpublished doctoral thesis. Southern Cross University, Lismore, NSW. http://epubs.scu.edu.au/theses/366/

Galbraith, P., Stillman, G., & Brown, J. (2010). Turning ideas into modeling problems. In R. Lesh, P. Galbraith, C. Haines, & A. Hurford (Eds.), *Modeling students' mathematical modeling competencies: ICTMA 13* (pp. 243–283). New York: Springer.

Gibbs, C. (2013). *Dune Dudes – A school program. Southwest Catchments Council*. Retrieved from http://swccnrm.org.au/wp-content/themes/blanktheme/inc/casestudy/CST-%20Dune%20Dudes%20-%20C.%20Gibbs.pdf

GLOBE Education. (n.d.). *The Global Learning and Observations to Benefit the Environment (GLOBE) Program*. Retrieved from http://www.globe.gov/about/overview

Howitt, C., & Rennie, L. J. (2008). *Evaluation of the scientists in schools Pilot project*. ACT: CSIRO. Available online at http://www.scientistsinschools.edu.au/evaluation.htm

Husher, K. (2010). *Evaluation framework for Australian science and maths outreach programs in schools*. PhD, University of Newcastle, Newcastle, Australia.

Marginson, S., Tytler, R., Freeman, B., & Roberts, K. (2013). *STEM: Country comparisons*. Melbourne: The Australian Council of Learned Academies. www.acola.org.au

Masters, G. (2013). *Australian education review. Reforming educational assessment: Imperatives, principles, and challenges*. Melbourne, Vic: Australian Council for Educational Research.

Ministerial Council on Education, Employment, Training and Youth Affairs. (2008). *Melbourne declaration on educational goals for young Australians*. Retrieved June 26, 2016, at: http://www.curriculum.edu.au/verve/_resources/National_Declaration_on_the_Educational_Goals_for_Young_Australians.pdf
National Curriculum Board. (2009). *Shape of the Australian Curriculum: Mathematics*. Barton, ACT: Commonwealth of Australia.
New Zealand Office of the Prime Minister's Science Advisory Committee. (2011). *Looking ahead: Science education for the twenty first century*. Wellington, NZ: New Zealand Office of the Prime Minister's Science Advisory Committee.
Office of the Chief Scientist. (2016). *STEM Programme Index 2016*. Retrieved from http://www.chiefscientist.gov.au/2016/01/spi-2016-stem-programme-index-2016-2/
Rennie, L. J. (2012). *"A very valuable partnership". Evaluation of the Scientists in Schools Project, 2011–2012*. Dickson, ACT: CSIRO Education. Retrieved from http://www.scientsitsinschools.edu.au/evaluation.htm
Rennie, L. J., & Howitt, C. (2009). *"Science has changed my life!" Evaluation of the scientists in schools project*. Dickson, ACT: CSIRO.
Shimizu, Y., & Williams, G. (2013). Studying learners in intercultural contexts. In K. Clements, A. Bishop, J. Kilpatrick, F. Leung, & C. Keitel (Eds.), *Third international handbook in mathematics education* (pp. 145–168). New York: Springer.
Symington, D., & Tytler, R. (2011). Schools and teachers supporting student open investigations. *Teaching Science, 57*(1), 8–12.
Tytler, R., & Symington, D. (2015). Science learning in rural Australia: Not necessarily the poor cousin. *Teaching Science, 61*(3), 19–25.
Tytler, R., Symington, D., & Cripps Clark, J. (2016). Community-school collaborations in science: Towards improved outcomes through better understanding of boundary issues. *International Journal of Science and Mathematics Education, 15*, 643–661. doi:10.1007/s10763-015-9711-9.
Tytler, R., Symington, D., Kirkwood, V., & Malcolm, C. (2008). Engaging students in authentic science through school – Community links: Learning from the rural experience. *Teaching Science, 54*(3), 13–18.
Tytler, R., Symington, D., & Smith, C. (2011). A curriculum innovation framework for science, technology and mathematics education. *Research in Science Education, 41*, 19–38.
Tytler, R., Symington, D., Williams, G., White, P., Campbell, C., Chittleborough, G., Upstill, G., Roper, E., & Dziadkiewicz, N. (2015). *Building productive partnerships for STEM Education: Evaluating the model and outcomes of the Scientists and Mathematicians in Schools program*. Melbourne, Vic: Deakin University. Available at: http://www.scientistsinschools.edu.au/downloads/SMiSEvaluationReport2015.pdf
Williams, G., & Huang, H. M. (2015). Learning and cognition in mathematics. In S. J. Cho (Ed.), *The proceedings of the 12th international congress on mathematics education* (pp. 497–502).

Chapter 13
Inserting Critical Mathematics into STEM Education

Mark Wolfmeyer, John Lupinacci, and Nataly Chesky

Abstract The chapter begins by asking whether STEM education is a friend or foe to the field of critical mathematics education (CME) by reviewing how mainstream STEM conflicts with CME but also provides spaces for critical work. Tensions between CME and STEM include mainstream STEM's emphasis on human capital, inattention to environmental degradation, and soft-critical orientation to social justice issues. However, STEM's emphases on interdisciplinarity can provide opportunities for critical mathematics education to take place. We argue that STEM education as policy can be an opportunistic space to simultaneously resist and reconstitute in line with the values and goals of CME. We extend CME's goals with deeper theoretical consideration to the nature of the ecological and social crises, in so doing we draw on ecofeminism and EcoJustice Education. The chapter concludes with a model "critical STEM" unit plan sketch that is appropriate for the Junior Secondary level. CME, ecofeminist theory, and internationally benchmarked content standards provide the foundation for our STEM unit plan titled "A Story of Incarceration." By this example, we intend to show that critical STEM projects can be transformative for learners as well meet the content goals of standard STEM education.

Science, technology, engineering, and mathematics (STEM) education is primarily linked to "vital preparation for today's high-tech information economy" (Drew, 2011, 1) and as such has become an educational priority of many nations and organizations. Much of today's writing on STEM education is framed by workforce readiness, including research efforts addressing diversity (e.g., Hrabowski, 2016)

M. Wolfmeyer (✉)
Kutztown University of Pennsylvania, Kutztown, PA, USA
e-mail: wolfmeyer@kutztown.edu

J. Lupinacci
Washington State University, Pullman, WA, USA

N. Chesky
State University of New York, New Paltz, NY, USA

R. Jorgensen, K. Larkin (eds.), *STEM Education in the Junior Secondary*,
DOI 10.1007/978-981-10-5448-8_13

that attempt to open up more pathways for increasing access to STEM jobs. However, in this chapter, we position ourselves within a critical frame that simultaneously opposes STEM education for human capital and proposes working within it for educational motives inspired by critical examination. Our unique line of inquiry is to draw heavily from the wide research program of critical mathematics education (CME) as a means to both critique STEM education and propose critical work within it. Doing so will give us opportunities to explore the question: Should CME consider STEM education a friend or foe?

To begin, we provide a comprehensive review of our framework, CME. As it turns out, this expansive field includes a variety of orientations and programs, and these are brought together for our critique and for our proposed reconceptualization of STEM education. We next complement CME with a tradition of ecocritical curricular scholarship, a theoretical contribution not yet fully applied to mathematics education but one which can bring to light some of the specific concerns that CME would have with STEM education. With a thorough synthesis of these fields at hand, we are equipped to critique STEM education's dominant objectives and do so through an analysis of some of the more prominent research agendas in STEM education. Such work provides the pitfalls and cautions that CME must avoid if it were to engage with STEM. The last third of the chapter provides sketches of three STEM lessons at the Junior Secondary level informed by CME and ecocritical considerations.

13.1 Orientations Toward Social Justice: Critical Mathematics Education

Alive and well since the early 1980s, the field of critical mathematics education (CME) is our starting point in critiquing STEM education. In this section we review CME and especially note particulars that help us to reconsider STEM education in the next section. Initial work in CME rests on two domains: critical pedagogy and critical theory. As to the former, Frankenstein's (1983) application of Freirian praxis provides the starting point to thinking deeply about teaching mathematics for social justice. Frankenstein makes clear that mathematical literacy should be taught in the Freirian efforts to raise consciousness: her work motivates the teaching of mathematics as an essential knowledge for those to understand class hierarchy and one's place within it. A discussion of her "own experience teaching urban working-class adults basic mathematics and statistics for the social sciences demonstrates ways in which Freire' s theory can illuminate specific problems and solutions in critical teaching, and ways in which mathematics education can contribute to liberatory social change" (p. 316).

Set up by Frankenstein, the trajectory of applying Freire to mathematics education continues with many examples, including Gutstein's (2006) call to "read the world with mathematics" by using "mathematics to understand relations of power, resource inequities, and disparate opportunities between different social groups and to understand explicit discrimination based on race, class, gender, language and other differences" (p. 26) and to "write the world with mathematics" by developing a sense of social agency to make change in the world, doing so in part with mathematical argumentation. This relates to what the mathematical community refers to as quantitative literacy. The Mathematical Association of America, comprised of mathematicians, published a volume describing the relationship between quantitative literacy and democracy (Steen, 1997).

> Quantitatively literate citizens need to know more than formulas and equations. They need a predisposition to look at the world through mathematical eyes, to see the benefits (and risks) of thinking quantitatively about commonplace issues, and to approach complex problems with confidence in the value of careful reasoning. Quantitative literacy empowers people by giving them tools to think for themselves, to ask intelligent questions of experts, and to confront authority confidently. These are skills required to thrive in the modern world. (p. 2)

Coming from a different initial starting point, Skovsmose (1985) intersects the New Frankfurt School of critical theory with mathematics education, finding that the literature to date has hardly made such connections and calls on the mathematics education community to do so. Skovsmose's own work has certainly continued this trajectory, including a mathematics education that engenders equitable distribution of resources (Skovsmose, 1994) to his more recent landscape of the field presented in Skovsmose (2011). Here he provides several CME questions bearing relevance to our present inquiry into STEM education. For example, CME critiques the "school mathematics tradition" for its relationship with the global economic and political elite. Calling the project "prescription readiness," school math assimilates children into the behavior of taking long sequences of commands. Accordingly, adults subjected to this education are more ready to uncritically accept their roles of obedience to corporate profit.

> Could it be that such a prescription-readiness is serviceable for very many job functions in our society and that the school mathematics tradition serves society perfectly well in exercising this readiness? Could it be that a prescription-readiness, including submission to a regime of truths, cultivates a socio-political naivety and blindness that is appreciated at today's labour market? Could it be that a prescription-readiness fits perfectly well the priorities of a neo-liberal market, where hectic and unquestioned production serves the economic demands? (pp. 9–10)

Research efforts that critique mathematics education for its service in developing an obedient workforce abound (e.g., Apple, 1992; Gutstein, 2006; Wolfmeyer, 2014).

Skovsmose offers a counterproposal to "prescription readiness," what he refers to as "landscapes of investigation." In this case, "a landscape can be explored in different manners and through different routes. Sometimes one must proceed slowly and carefully and sometimes one can jump around and make bold guesses" (p. 31). Skovsmose's examples span mathematical behaviors, from classic math procedures (what he calls "references to pure mathematics") to applied mathematics (what he calls "references to reality"), such as his example of 7-year-olds designing a playground (Skovsmose, 2011, pp. 46–47). As well, Skovsmose's "mathematics in action" suggests mathematics to model and solve problems.

We might say that these applications of critical theory and critical pedagogy lay a foundation for CME. In our use of CME, we do not suggest that critical mathematics education is a defined branch of study following entirely from these applications or with definitive participants and outsiders. Instead, we are presenting a variety of research topics that align with these orientations and that help to frame critical inquiries into STEM education. The remainder of this review of CME calls forward critical conceptions of race and mathematics education and mathematics education and the environment, chosen as relevant for our reframing STEM education.

An abundance of research exists on race and mathematics education and runs the gamut from radical reconstructions of mathematical curriculum to changes within the teaching and learning of mathematics aimed at increasing the performance of students of color. Providing a thorough review of the landscape of this research, Larnell, Bullock, and Jett (2016) orient us toward the most critical of these within a tradition of "Teaching and learning mathematics for social justice." They identify two camps: (1) critical mathematical literacy and (2) social justice as access; however, these are not mutually exclusive in their participants, orientations and products. The first comprises research on theory and practice along the Freirian framing described earlier and includes reference to Frankenstein and Gutstein. In addressing critical conceptions of race, a critical mathematical literacy teaches mathematics to raise consciousness about racial injustice. An example might be lessons on proportion in which students learn about disproportionate incarceration rates. The second camp of teaching and learning mathematics with a critical ear bears an interesting influence on STEM education discourse. Moses and Cobb (2001) call mathematics a "civil right" because success in school mathematics has a positive correlation to job opportunities in later life. This means that culturally relevant mathematics teaching *of traditional content* is essential for steps toward racial justice. In this sense, mathematics education that prepares students to think critically about their world might at the same time limit their access to job opportunities.

Reviewing these two camps in race and mathematics education helps to think through the opportunities for a critical conception of STEM education. On the one hand, critical STEM teaching and learning can open up questions and further inquiries that use STEM disciplines to read and write the world in a Freirian sense. On the other hand, the teaching and learning of STEM must prioritize an understanding of STEM as access to economic opportunity. Our view echoes that of Larnell et al. (2016) in which an either/or approach would stifle critical education goals achieved through STEM education. Recognizing the material consequences of STEM education and teaching to raise consciousness about social injustice are necessary for teaching critically.

Other work in CME that addresses race theorizes more deeply the racialized nature of mathematics education and the school mathematics experience. Martin (2013) claims mathematics education as a "white institutionalized space" in which the interests expressed, and for whom the project serves, reinforce racial hierarchies. For example, in reviewing the history of mathematics education in the United States, he looks at the Cold War era's ushering in of a new math:

> The New Math reform project was not an antiracist vessel in the sea of racial discord characterizing that time. With its emphasis on the 'best and the brightest,' it was just another mechanism for maintain White (male) privilege…The 'prominent persons' involved in the political project of the New Math movement identified mostly White males, from various backgrounds, as the key leaders and decision makers of the movement, a finding that is common for White institutional spaces. If the nation had minimal will to integrate Black children into their schools and other public institutions or the voices of Blacks into its policymaking circles, it was certainly no more willing to integrate their needs into the mathematics education reforms of the day. As a result, it could be argued that the New Math movement to educate a generation of students who would protect the U.S. from the Soviet intellectual threat did not include Blacks (or Native Americans, Chinese, Japanese, or Latina/os). Rather, mathematics education in the United States served to help maintain the prevailing racial project. (p. 326)

To what extent can the same be said for STEM education? Martin's accurate description of new math reform efforts describes a "color-blind" world in which "the best and the brightest" are referenced and assumed to be white. Today, with STEM education there is an agenda to increase student success and thus diversify the STEM workforce. The article referenced in this chapter's introduction (Hrabowski, 2016) takes up the issue front and center. He points to the reality that in the United States, nonwhite populations will increase their share of the population, and without a proper education in STEM, the US population will fall behind in competition for global STEM jobs. Although these efforts are no longer "color-blind," how might we think more critically about these racialized goals of STEM education? Did the creation of these policy goals involved nonwhite communities and leaders? In whose interest are these goals formed?

Abundant work on race and mathematics education aligns with CME, and, although more limited, some CME-inspired work is attending to issues of environmental sustainability. Coles, Barwell, Cotton, Winter, and Brown (2013) make these connections explicit by pointing to CME's emphasis on social issues but also in its relevance to environmental issues as well:

> Skovsmose, however, emphasizes the role of mathematics itself in *creating* our world. As a result, he argues that mathematics teaching can include a critical analysis of this role. Mathematics is not simply a powerful way of interrogating the world around us; it is part of the structure of our society. A critical mathematics education offers students some insight into how mathematics is part of their lives and the consequences it can have. (pp. 12–13)

The authors provide several examples of applying such a CME to sustainability issues and suggest that these link mathematical content that is taught in school mathematics to the environmental topics of agriculture, climate change, the economy, and biodiversity. In some cases, such as the chapter by Jan Winter's (Coles et al., 2013) discussion of food, the interconnection of environmental issues with social concerns comes front and center. It is these conversations, linking environmental sustainability to social justice, that we find productive in thinking through a critical STEM education.

As we are discussing the application of CME to STEM, we realize that the disciplinary content of CME, as mathematics, is narrower in scope than STEM, as science, technology, engineering, and mathematics. Given its interdisciplinary nature, we expect STEM to provide a more productive space for CME's projects. CME routinely applies mathematics and, in this sense, is often interdisciplinary to start. As it has done already, integrating social and environmental issues into mathematics teaching and learning requires interdisciplinary approaches. Our drafted unit and lessons in this chapter aim to demonstrate how to operationalize this. We expect that the STEM space, as *adding disciplinary content* to mathematics units, will further legitimize CME work.

We find that the ground laid by Coles et al. (2013) sparks an important conversation about the relationship between environmental and social crises, and this is something that has not been taken up too much in the mathematics education literature. We do briefly note Khan's (2011) productive discussion of this linkage. He explains of the potential for ethnomathematics to address "decolonization, liberation, justice, and sustainability" (p. 17). Khan's contribution is highlighted here as a question to consider: can ethnomathematics and the ethical-aesthetic-mathematical experience (Khan, 2010) find potential in STEM education as well, in particular for achieving goals of social justice and sustainability? Although this is an excellent consideration that will play a complementary role in a critical STEM, for the time being, we are drawing primarily on the work of CME and its application to social justice (with special attention to race) and sustainability. In our efforts to expand CME's role in STEM education, we turn for a moment away from the work of mathematics education and toward a field of ecocritical scholarship in education, where the connections between social justice and sustainability are theorized more deeply and have, as of yet, to be applied to mathematics and STEM education.

13.2 Ecocritical Conceptions of Education with Direct Implications for CME and STEM

Picking up from the previous section, we draw from work in education that has not yet received sufficient attention from the mathematics and STEM education communities. A trend in curriculum studies now attends to the consistent interrelationship between social justice and environmental catastrophe, a pattern that one of us (Lupinacci & Hapel-Parkins, 2015) has referred to as "ecocritical." This is a theoretical and practitioner-focused approach that draws from a variety of disciplines and most notably ecofeminist and poststructural feminist theory. A variety of key terms are associated with the groups working on this, including EcoJustice, ecopedagogy, and posthumanist education. These all have direct implications on how CME and STEM education are conceived of and how they are taught in schools.

One of these, EcoJustice, is defined by Martusewicz, Edmundson, and Lupinacci (2015) as: "The understanding that the local and global ecosystems are essential to all life; challenging the deep cultural assumptions underlying modern thinking that undermine those systems; and the recognition of the need to restore the cultural and environmental commons" (p. 20). Central to an EcoJustice framework is the importance of recognizing the differences between ecological cultures and Western dominant individual-centered cultures. Central to this work is acknowledging the role that language and culture plays in shaping our Western habits of mind, what has been referred to as "discourses of modernity." Martusewicz et al. (2015) draw from postmodernism and ecofeminism to define these as "the specific set of discourses that together create our modern, taken-for-granted value hierarchized worldview" (p. 86).

The critical examination of these discourses, or shared cultural meanings, is complex and allows for the multidimensional analysis of language and culture in connection with taken-for-granted assumptions regarding what is valuable, what is worthless, and how these concepts are applied. The analysis of superior/inferior dualisms allows EcoJustice theorists to identify a powerful group of discourses that form metaphors that dominate how we, as subjects in a modern era, interpret difference and construct meaning. These discourses of modernity consist of individualism, mechanism, progress, rationalism/scientism, commodification, consumerism, anthropocentrism, androcentrism, and ethnocentrism (Martusewicz et al., 2015). For those of us disciplined by modernist assumptions of human superiority and individualism, the analysis of the aforementioned discourses allows for the examination of the relationships between our language, how we think, and our behaviors that undermine living systems. These powerful discourses contribute to the ever-growing ecological crisis—a crisis that EcoJustice educators identify and understand as a cultural crisis. This cultural crisis is responsible for both the environmental catastrophes and social injustice that, thus far, CME (critical mathematics education) has taken up as separate entities.

EcoJustice educators recognize how language shapes culture and that culture is understood by how we interpret the "differences that make a difference" (Bateson, 1972, p. 315). In other words, we are bound by the metaphors of our language. This distinguishes EcoJustice Education from other pedagogical approaches that engage in a deep analysis of culture without consideration of language and the historical roots of the patterns shaping how we think and act. Language is a process that carries forward ways of thinking from the past. This is significant in that all languaging processes, which include past ways of thinking, are framed by and reproduce the assumptions of the culture. For example, Bateson (1972) writes about the way Cartesian thinking and Occidental—or Western—assumptions create the illusion of a separation existing between mind and environment. Bowers writes about root metaphors and the master metaphorical templates in reference to how metaphors in an industrial culture differ from metaphors for a sustainable culture; and Martusewicz et al. (2015) explain how the ways that we identify and behave are created though discursive patterns rooted in language that "are complex exchanges of meaning that use metaphor" (p. 66).

Western culture is defined by the languaging processes being passed on and includes deeply embedded assumptions like anthropocentrism, ethnocentrism, androcentrism, and other life-threatening centric discourses that come from mythopoetic narratives and prominent "attitude" changing experiences—to draw from Bateson's criteria for naming major historical cultural events (Bateson, 1972). The codes of these mythopoetic narratives and prominent experiences are embedded into metaphors—and more specifically, root metaphors. These root metaphors work together to shape discourses that provide the framework of a culture. They are passed on generation to generation, having great influence on values, problem-solving, habits, and traditions.

It is important to address the ways in which we are shaped by language because of its role in discourse as influencing what is marginalized or silenced by dominant root metaphors. Educators using an EcoJustice Education framework emphasize how industrialized Western thinking, and the habits it shapes, contributes to a culture of social violence and ecological destruction. By examining the ways in which language works, EcoJustice educators suggest that we ought to work toward alternative root metaphors that replace modern discourses with life sustaining discourses that are rooted in ecology rather than Cartesian individualism.

The linkages between social injustice and environmental catastrophe are further developed by ecofeminist theory, to which we now turn. Ecofeminism provides important insight into helping us look at how we might engage STEM teaching that interrupts the discourses of modernity. Ecofeminist scholars connect the unjust suffering inflicted upon women with the subjugation and destruction of nature in patriarchal cultures. Karen Warren (2000) offers a specific starting point for the ecofeminist philosophy influencing EcoJustice Education: "The dominations of women, other human Others, and nonhuman nature are interconnected, are wrong, and ought to be eliminated" (Warren, 2000, p. 155). The importance of learning about other cultures resides in the need to not only understand ourselves as subjects but also to gain consciousness of how we exercise or submit to power relations. EcoJustice educators engage in this ethical process in order to understand how from positions of power, others—and even sometimes ourselves—get excluded, homogenized, backgrounded, incorporated, and instrumentalized (Plumwood, 2002).

In many ways, for many of us as subjects, this is the historical understanding of how we think and act. In order for us to heal from both atrocities we have experienced and inflicted upon each other and on the "more-than-human world" (Abram, 1996), we first accept some often silenced historical truths followed closely with humility and an authentic reconciliation. We must know our history—understand how and why we think and act the way we do—in order to cease doing evil. Then we must learn to do good. Ecofeminist scholars, like historian Carolyn Merchant and philosophers Val Plumwood and Karen Warren, bring a well-rounded feminist perspective to the EcoJustice Education framework. In other words, if we are to critically and ethically understand our history, we must consider how patriarchy has provided a painfully obscure bias in favor of androcentric versions of human history.

Carolyn Merchant's historical work to trace mechanism and rationalism to specific events and thinkers coming from the Enlightenment, the Industrial and Scientific revolution, and the rise of capitalism in Western Europe brings a perspective and insight to EcoJustice Education that traces modern dominant Western culture. Merchant's scholarship debunks the myth that domination is the natural evolution of humanity. Merchant (1983) in *The Death of Nature: Women, Ecology, and the Scientific Revolution* writes:

> An ecosystem model presents an earth's-eye view of history. By looking at history "from the ground up," factors having an impact on the earth's resources can be analyzed and a new and different interpretation of historical change developed, based on the assumption that the natural and human environments together form an interrelated system. (p. 42)

Merchant (1983) further explains: "An ecosystem model of historical changes looks at the relationships between the resources associated with a given natural ecosystem (a forest, marsh, ocean, stream, etc.) and the human factors affecting its stability or disruption over historical time periods (p. 43)." Merchant details a historical transformation in language and thought from organic metaphors for living systems to mechanized metaphors of domination that reduce living systems to lifeless machines and calls this transformation the "death of nature" (Merchant, 1983).

Val Plumwood's approach offers insight into how important a feminist perspective is to the male-dominated field of environmental philosophy. Plumwood (1993) writes:

> People suffer because the environment is damaged, and also from the process, which damages it, because the process has disregard for needs other than those of an elite built into it… As the free water we drink from the common streams, and the free air we breathe in common, become increasingly unfit to sustain life, the biospheric means for a healthy life will increasingly be privatised [*sic*] and become the privilege of those who can afford to pay for them. The losers will be (and in many places already are) those, human and non-human, without market power, and environmental issues and issues of justice must increasingly converge. (pp. 13–14)

This statement from Val Plumwood may be one of the strongest descriptions of the context within which EcoJustice Education is situated. Plumwood's philosophical tools provide the necessary logics to deepen the linkages between the environmental and social justice, thereby reinforcing works in education like Gruenewald (2003), and Bowers (2001). While several environmental and social justice-oriented educators offer arguments for the inseparability of social and environmental justice issues, we find these philosophical tools and linkages as particularly helpful to this project. Her articulation of the role of ecological feminism as it contributes to male-dominated environmental philosophy goes deeper than simply casting a positive version of woman as nature. She links the insight of feminism's ability to cast the likening of woman to nature in connection with a culturally constructed negative value for woman that hinges on a negative value for nature and seeing them as together less than human, or as less than the fully human male, as the basis for women's inferiorization and oppression. Most importantly, she does this with a historical understanding of how forms of domination emerge and shape our modern perceptions of relationships. She introduces an ecologically oriented feminism that acts as a promising lens through which we might illuminate not only the domination of women but also the domination of the nonhuman world. Since the oppressed in modern society are in almost all cases feminized and naturalized, Plumwood (1993) suggests that through ecological feminism, we can perceive how value-hierarchized dualisms—superior/inferior dualisms like culture/nature, reason/emotion, mind/body, and man/woman—work discursively to marginalize women, other human groups, and nature. Plumwood maintains that these dualisms are inseparable from each other and from the root discourses that create and recreate oppression and unsustainable relationships. She examines how forms of centric thinking work to exclude, homogenize, background, incorporate, and instrumentalize life to create what Warren (1990) calls "a logic of domination." Plumwood calls for ecofeminist philosophy to help guide us toward an ecological ethic and, drawing from the words of Rosemary Radford Ruether, shares: "An ecological ethic must always be an ethic of ecojustice that recognizes the interconnection of social domination and domination of nature" (Ruether in Plumwood, 1993, p. 18).

EcoJustice Education requires a commitment to ecological ethics in order to engage in recognizing the interconnectedness of both social and environmental suffering. Val Plumwood's work brings to EcoJustice Education a framework for understanding how important an ecological ethic comprised of mutuality and relationality is to a cultural ecological analysis. Plumwood and ecofeminists, such as Ruether, Warren, and Merchant, offer perspectives that serve as guidance in how to navigate dominant discourses undermining life. Their work seamlessly weaves through multiple historical and androcentric philosophical attempts to address human and more-than-human suffering on the planet, highlighting strengths and weaknesses or flaws in those attempts as they illustrate how dominant discourses, often in contradictory and hidden ways, work to shape approaches responding to environmental and social degradation.

Reviewing the deep theorization of the social and environmental crises, as *one cultural crisis*, we feel it provides more promising directions in a critical STEM education that interrupts the perpetuation of discourses of modernity. These theorizations extend the work of the two critical mathematics education (CME) camps, that of social justice and environmental sustainability, and address more deeply the issues CME tackles. As an interdisciplinary space, we feel that STEM education is an ideal means by which such CME work can take place. In the second half of this chapter, we provide a sample STEM education unit emerging from the CME tradition and with the conceptual underpinnings that CME and EcoJustice Education provide.

13.3 Initial Considerations Leading to "A Story of Incarceration"

This section introduces the remainder of the contributions in this chapter, in which we lay out our process in designing a critical STEM education unit for the Junior Secondary level. At the heart of our work throughout is the notion of appropriating a mainstream policy space for critical work. We aim to insert the spirit and motivations of CME into STEM education, and we will begin by carefully considering the STEM space as it has been laid out. We thus turn our attention to the STEM content standards defined for us, and given our current context, we look to the US content standards in mathematics, science, and technology as well as secondary source materials based on these standards. We are headed toward the goal of repackaging these standards as a critical STEM unit. However, our unit can still claim to teach the content goals of STEM education as set forth by policy. We find this is an important consideration for educators doing critical work today. This is not to say that we find mathematics, science, and technology exempt from controversy. We do not have the space to discuss these issues at length, but we briefly mention that the process and implementation of these standards are part and parcel of the STEM education policy phenomenon that, as we suggested at the opening of the chapter, is essentially motivated by concerns with human capital.

Even though some states have not adopted these officially, mathematics instruction in the United States is steered mostly by the *Common Core State Standards for Mathematics (CCSSM)* (2010). Although a bit younger and not fully in implementation, the national science standards are *Next Generation Science Standards (NGSS)* (2013). Finally, we considered the National Science Foundation's *Standards for Technological Literacy: Content for the Study of Technology*, which, like the NGSS, are not yet officially adopted standards. All three sets of standards claim to be "internationally benchmarked" meaning that they were developed with consistent reference to the content standards of countries whose students perform well on standardized tests. *CCSSM* provides some suggestions of interdisciplinary focus, whereas *NGSS* consistently provides these opportunities by referencing mathematics standards throughout that link up to the science standards at hand. *NGSS* also links in engineering and technology concepts when appropriate. In this sense, *NGSS* is a wonderful resource in developing STEM projects. We provide a word of caution, however, that the linkages they provide are not the only ones that can be made. These are examples of linkages between content areas that can inspire educators to search for others. In our view, *NGSS* reads as if these are the interdisciplinary links that need to be made, where we feel that educators can generate new ones, and in particular ones that link in social and environmental issues as well.

Out of *CCSSM* have grown several curricular projects, some of which are effective in maintaining the mathematical content of *CCSSM* and especially by promoting a few of teaching and learning in which meaning making through connections to big ideas is encouraged. One of these curricular projects is Engage NY, a *CCSSM* content-based curricular project to be used by teachers and students in the state of New York. One of the exciting features of Engage NY is the "curriculum overview" documents that help educators to navigate standards by looking at the connections between content and process standards for mathematics. Their curriculum overview for grades 6–8 provides a helpful survey of the mathematics to be learned by students approximately aged 12–15. "A story of ratios: A curriculum overview for grades 6-8" (Common Core, Inc., 2013) provides a comprehensive curriculum map that links together the content to be learned. The title reflects this resource document's approach in using the important concept of "ratio" as a unifying theme to the work at hand. This is an exceptional strategy for teaching to develop meaning for students and provided our inspiration for inserting an alternative theme in its place that can together link the mathematical contents and science standards chosen.

Ratio is an important topic and does nicely link the mathematical standards in grades 6–8. We focused on the mathematical topics in grade 7, which include ratios and proportional relationships, percent and proportional relationships, statistics and probability, and geometry. We will outline the details for these topics at the grade 7 level again when later discussing the particulars of the unit. For now, the basic nature of the topics at this level includes a rich exploration of ratio and proportionality. Learners are expected to conceptually understand ratios and proportions as well as apply these concepts, for example, in determining when objects are, or are not, in proportion and by connecting proportionality to percentage. As for geometry, the content addresses 2D and 3D properties of figures and measurement skills including computation of area, surface area, and volume, with applications to real-world situations. As well, this geometry content is linked to ratio and proportion.

Given our orientation to CME and EcoJustice Education, the “story of ratio” immediately made us think of incarceration. The mathematics taught in the unit affords students the opportunity to digest information about disproportionate incarceration rates, by race and class, and we feel that 13-year-olds are developmentally ready to tackle this material. In fact, we suggest that the mathematical contribution to the conversation about prisons greatly enhances our knowledge about incarceration. The next step was to think through the multiple facets of incarceration and link them to content both within the mathematical topics of 7th grade as well as make connections to STEM-related content, including incorporating science content standards from *NGSS*. These included the engineering and design of prisons which links to engineering, technology, and the geometry standards.

These were all very exciting linkages, and the details of these connections will be explained shortly. However, with these topics and content areas, we felt the critical STEM unit as such was mathematics-heavy and social-justice-heavy. The mathematics content for 7th grade was addressed in its entirety and supported by a few engineering and technology standards. The topic of incarceration rates lays more in social concerns without clear links to the logic of domination existing elsewhere, especially anthropocentrism and human supremacy. As well, science standards seemed lacking in our STEM unit. And then (this was spring 2016), the news of Harambe the gorilla and the Cincinnati Zoo just broke, and it all came together for us. As people operating under Western culture’s habits of mind, the very word of incarceration had only applied to people. Yet, the news story jolted us into our theoretical predispositions, reminding us that, of course, a great many nonhuman animals are incarcerated throughout the lands. Looking back at the *NGSS* standards, we found appropriate links to science content, specifically biodiversity, evolution, and human interactions with ecosystems. Our ecocritical STEM unit titled “A Story of Incarceration” had thus emerged. We felt it contained the linkages necessary to problematize Western culture’s habits of mind while fulfilling several STEM content standards.

We decided to share this quasi-narrative of the process by which we arrived at coming up with the idea for a critical STEM unit so that educators can apply and adapt the approach. In so doing, we suggest the following: start with content standards that you are required to teach. Think through what might be a good ecological or social issue that relates to the topic. Once this has been identified, list out a variety of subtopics and link these to STEM content. All the while, make sure you are thinking about two goals: (1) bring in as many of the STEM official standards in order to legitimize your unit according to the powers that be and (2) keep in mind both social and environmental issues and the linkages between these; attempt to problematize as many of the discourses of modernity as possible in the given unit.

With an initial plan at hand, the next step is to dig into the social and environmental issues, assuming you have a solid STEM disciplinary knowledge base (otherwise you would need to do that as well). In the next section, we present our efforts in the research we undertook to develop a more realizable unit plan. Keep in mind this is slightly different than a research literature review, where we are seeking publications that can readily apply as teacher content resources as well as resources to use with the learners themselves.

13.4 A Survey of Incarceration: From Rates to Open Prisons, Zoos to Farms

In this section we review the variety of information coming to bear on our unit plan "A Story of Incarceration." In this plan we will be connecting the mathematical content of ratios, percentages, and proportions to the science content of biodiversity and evolution. The linkages for these content areas are through the story of incarceration in all its forms, including imprisonment of people and its related inequities, the engineering and design of prisons and related ethical issues, and finally extending the notion beyond nonhuman animals by considering the enslavement that occurs in agribusiness, experimental testing, and animal entertainment. The discussions contained here refer to resources that can readily apply as teacher friendly materials for use in a similar unit as well as some that can be used as student resources.

To begin, rates of incarceration are at an all-time high. The World Prison Population List (Walmsley, 2015) indicates that 10.35 million people are held in penal institutions worldwide. Given the world population of 7.4 billion, that makes a percentage of .144 of the world's population that is imprisoned. Given the large numbers, it's helpful to discuss these as rates per 100,000, and the worldwide imprisonment rate would thus be 144 imprisoned people per every 100,000. Incarceration rates are typically prepared as such which allows for comparisons by country, a necessity given that differing laws and regulations dictate imprisonment. The United States has the second highest rate of incarceration at 698 per 100,000 people; Australia is at 151. The data readily applies the concept of proportion and would generate more meaning for the students as they examine the data and compare incarceration rates across the globe. Based on their previous knowledge of prisons and the judicial system, learners will start to ask and answer questions about the discrepancies within the data. Initial conversations might lead to students thinking that "well, this country has a lot of crime and criminals." It will be important for educators designing this unit to frame the unit with broader questions about imprisonment that help students to dig deeper, beyond more superficial explanations of the data.

After initial rates of incarceration across the globe are compared, it is equally important to examine carefully the incarceration rates disaggregated by social identity. Mauer and King (2007) provided their policy brief for a group called Sentencing Reform and highlight these discrepancies. Keeping in mind the world incarceration rate of 144 per 100,000, we consider their discussion of incarceration rates for Hispanic people living in the United States at 712 and black people living in the United States at a 2290 per 100,000. Viewing the data on the overall incarceration rate of the US population, 0.4% of all whites are incarcerated, whereas 11.7% of the African American males between the age of 25 and 29 are incarcerated in a prison or jail. The report provides incarceration rates broken down by states in the United States as well with very interesting trends to be discussed. The ratios of black-to-white and Hispanic-to-white prisoners are also revealed and would be fruitful real-world data for learning the concept of ratio more fully. The authors readily apply these concepts and will provoke interesting inquiry into the story of incarceration:

> States with the highest black-to-white ratio are disproportionality located in the Northeast and Midwest, including the leading states of Iowa, Vermont, New Jersey, Connecticut, and Wisconsin… States exhibiting high Black or Hispanic ratios of incarceration compared to whites fall into two categories: 1) those such as Wisconsin and Vermont which have *high rates of black incarceration* and *average rates of white incarceration*; and, 2) states such as New Jersey and Connecticut which have *average rates of black incarceration* and *below-average rates of white incarceration*. In both cases, the *ratio* of incarceration by race is higher than average. (Mauer & King, 2007, p. 5)

The data on incarceration rates, disaggregated by country as well as demographics, provide a wide range of discussions and conversations where learners can apply the mathematical concepts of ratio, proportion, and percentage. The ways to analyze the topic by comparing proportions and ratios are widespread and would spark several questions worthy of further investigation.

Conversations will likely center on crime, criminality, and punishment, and learners might begin to wonder what are the factors involved that relate to the discrepancies. Depending on learner readiness, it might be appropriate to tie in a statistics regression topic on countries average income and their rate of incarceration. The data within countries can be further disaggregated by race as well. This is not to suggest that economics are the entire cause of incarceration rates, but this would be an illuminating graph that students can use to begin to see the relationship between a country's economic structure and its incarceration rates.

Another conversation worthy of exploration that will tie in the geometry standards for the unit is the differences in the culture of crime and punishment across the globe. Research and media outlets alike have lately been dissecting cultures and approaches in Northern Europe, which have consistently low rates of incarceration. Larson (2013) highlights the lack of media that sensationalizes crime in Scandinavian countries and in particular the efforts to design new consequence structures termed "open prisons." Here he describes one example:

> Suomenlinna Island has hosted an 'open' prison since 1971. The 95 male prisoners leave the prison grounds each day to do the township's general maintenance or commute to the mainland for work or study. Serving time for theft, drug trafficking, assault, or murder, all the men here are on the verge of release. Cellblocks look like dorms at a state university. Though worse for the wear, rooms feature flat-screen TVs, sound systems, and mini-refrigerators for the prisoners who can afford to rent them for prison-labor wages of 4.10 to 7.3 Euros per hour ($5.30 to $9.50). With electronic monitoring, prisoners are allowed to spend time with their families in Helsinki. Men here enjoy a screened barbecue pit, a gym, and a dining hall where prisoners and staff eat together. Prisoners throughout Scandinavia wear their own clothes. Officers wear navy slacks, powder-blue shirts, nametages and shoulder bags; but they carry no batons, handcuffs, Tasers or pepper-spray. (Larson, 2013, n.p.)

As another striking example, the Norwegian citizen Anders Breivik, who killed 77 people, received a 21-year sentence by the justice system in Norway. This news story provides a stark contrast for people living and operating within a system, such as in the United States, with a very different outlook on crime and punishment. The approach and philosophy of reentry into public life is at the heart of these approaches and recidivism rates can be compared as well.

By incorporating geometry and engineering standards in the story of incarceration unit, learners will come to learn about both the traditional and open concept to prison design. They will learn the term "panopticon" as the ideal in traditional prison design, in which the prisoner feels under the constant threat of surveillance. This concept came to be of great interest to famed social theorist Michel Foucault, applying it to prisoners and non-prisoners alike and, given time, discussion can move in that direction. Sticking primarily to the standards, however, learners will have the opportunity to analyze the geometry of the panopticon floor plans and understand how prison engineers optimize particulars in their design. They will learn firsthand of the conditions under which prisoners experience life and compare and contrast aspects, such as area and volume measurements, to their experiences in schools and elsewhere. A good resource to begin to understand the engineering of traditional prisons is provided by the US Department of Justice (Kimme, 1998) titled "Jail design guide: A resource for small and medium-sized jails."

We do not suggest that learners be assigned to design their own prison but instead suggest that they analyze floor plans and data they can find on prisons that already exist and, in doing so, come to understand the engineering that goes into prisons as discussed in design guides. These analyses should connect with differing approaches to crime and punishment as discussed earlier (open vs. traditional prisons). We caution educators not to require that students design their own prisons, in line with the stance of the Architects/Designers/Planners for Social Responsibility, an advocacy group that speaks directly to human rights and sustainability issues in design. On their website, adpsr.org, they present the following stance against the design of new prisons:

> It is time to stop building prisons: Our prison system is both a devastating moral blight on our society and an overwhelming economic burden on our tax dollars, taking away much needed resources from schools, health care and affordable housing. The prison system is corrupting our society and making us more threatened, rather than protecting us as its pro-

> ponents claim. It is a system built on fear, racism, and the exploitation of poverty. Our current prison system has no place in a society that aspires to liberty, justice, and equality for all.
>
> As architects, we are responsible for one of the most expensive parts of the prison system, the construction of new prison buildings. Almost all of us would rather be using our professional skills to design positive social institutions such as universities or playgrounds, but these institutions lack funding because of spending on prisons. If we would rather design schools and community centers, we must stop building prisons. (n.p.)

The ADPSR group also takes strong opposition to the design and build of solitary confinement within jails. Several discussions among architects have called for a ban on their creation; however, and in our view unfortunately, the American Institute of Architects did not approve a ban on their member architects from designing them, instead letting individual architects make decisions on what they will design.

The concept of open prisons begins to unpack some of the deeper assumptions related to this story of incarceration. It makes us question the very nature of incarcerating individuals for their crimes, where an approach to restoring justice (see restorative justice movement) through reconciliation and reentry into society promotes the rights of individuals and calls to question the social identities (e.g., race and class) that "make other" the criminals that we incarcerate for long sentences. Such thinking will open the door for learners to think about the incarceration of human animals, and with some careful placement of questions, learners will naturally turn toward the imprisonment of nonhuman animals as well.

Martusewicz et al. (2015) classify the incarceration of nonhuman animals according to three primary occurrences: agribusiness, animal experimentation, and animal entertainment. These three aspects of animal incarceration will incorporate the science standards of biodiversity and evolution by asking the questions about which animals are enslaved and why. To begin, agribusiness refers to the industrialization of food production, and here we focus on the enslavement of nonhuman animals to produce animal-based food. "While television and other advertising still paints images of the family farm in bucolic rural settings, this is far from the reality of agricultural business today" (Martusewicz et al., 2015, p. 114) with a shift to the Concentrated Animal Feeding Operations (CAFO) that have found their way as the main producer of animal flesh for consumption.

> In the dairy industry, for example, cows no longer graze on fields but rather are held and fed in large barns, usually on concrete floors, waiting to be milked by large machines that attach to each cow's udder. Their manure, accumulated throughout the day of standing in one place, is sprayed off the floors into 'lagoons' outside the barn. The result is highly toxic liquidized manure that is then pumped into trucks that spray this concoction onto the surrounding fields. (p. 115)

Animal incarceration that produces milk will always be incarceration, but these authors call us to consider how we can make more ethical decisions as, "If you have a commitment to the species that gives its life in order to provide sustenance, warmth, or shelter, then in return you honor and respect that animal or plant's right to live with dignity and reproduce itself" (p. 116). These considerations of a life with dignity put forth a more ethical story of incarceration not unlike the open prisons of Scandinavia, in which the goal is to make prisoners feel free and to reconnect them with their communities.

The second mass incarceration of animals comes in the form of animal testing. "An entire animal industrial complex has been built around supplying animals for research in the medical, space, military, and food and drug fields," with victimization rates at over 115 million per year (Martusewicz et al., 2015, pp. 117–118). Examples of common procedures include forced exposure to chemicals and infectious diseases, genetic manipulation, mutilation for identification, prolonged physical restraint, and infliction of wounds and pain.

Finally, the third mass incarceration of animals is animals enslaved for the entertainment of people. These include circuses, zoos, and marine life amusement parks like SeaWorld:

> As members of a culture framed by anthropocentric beliefs and practices, we are encouraged to see animals in captivity as somehow happier or protected from their difficult lives in the wild, and thus their captors as benevolent. We are taught that animal imprisonment is 'for their own good.' And, we learn that humans are superior to and have the right to determine the conditions of life for other creatures according to our own interests, whether those be observation, amusement, or profit. (Martusewicz et al., 2015, p. 122)

Taking the three aspects of animal slavery together points to the diversity of species that we enslave and for these differing reasons. We suggest that the science standards included in a unit "A Story of Incarceration" relate to biodiversity and species taxonomy. Looking specifically at the diversity of life on the planet allows for inquiry into the species we select for enslavement within our human supremacist frameworks.

This section has reviewed the themes of incarceration that come to bear on our sketch of "A Story of Incarceration," our critical STEM unit inspired by critical mathematics education (CME) and ecocritical curriculum scholarship. These resource materials reviewed in this section are user-friendly for educators, and we suggest educators use these to unleash a network of materials for use in such a unit plan. Several of these will likely be useful as student resources as well, but we hesitate to suggest a definite list given that unit materials selected will offer a local context for learners as well as appropriate to their readiness. Along these lines, in the final section to follow, we offer an outline of what lessons, questions, and topics this unit might cover, but do suggest that educators approach this as a sketch to be adapted for their particular learners.

13.5 "A Story of Incarceration" Unit Plan Sketch

In this section, we provide a sketch of a critical STEM education unit plan appropriate to the Junior Secondary level. Primarily we aim to teach STEM content as outlined by internationally benchmarked standards under a theme of incarceration that opens learner inquiry into social justice and sustainability themes. These efforts are inspired by critical mathematics education (CME) and ecocritical education scholarship that we reviewed earlier in the chapter, and the organization for our work comes from the "Understanding by Design" (UBD) tool (Wiggins & McTighe, 2005) that makes unit planning straightforward and clear to present.

A brief introduction to the elements of this design helps to navigate our sketch of "A Story of Incarceration." UBD is often referred to as backward design, because the tool requires educators to begin with the established goals of a unit, next develop the assessments, and finally look to the specific lessons of the unit. Thus there are three stages to the UBD tool, and in our sketch of the unit here, we include the first stage. Planning assessments and lesson planning, we feel, should be left to the educators who are specifically tailoring instruction to their learners. This sketch might even require some adaptation for a particular group of learners, but we offer it as a place to start to get thinking about a critical STEM unit that can work for this level. Specifically, in UBD stage 1, the "established goals" are the standards set forward by policy. By linking these together, we connected the standards to formulate "big ideas" that interrelate these concepts. These big ideas also generate the "essential questions," expressed in learner-friendly language that will motivate student inquiry throughout the unit. From there, the details of the unit plan include the specific concepts and skills that will be covered in the unit and throughout lessons. We have expressed these efforts in the standard UBD display format, as follows (Table 13.1):

With these initial sketches at hand, educators can design a unit that is suited to their context and student readiness. Perhaps learners require more explicit development of the concepts, in which case they will need a week or two on the mathematics, science, and engineering standards before linking them together as a unit on prisons. For learners who have some prior knowledge of some or all of these disciplinary ideas, the unit can begin with questioning incarceration and discussion of what this includes and motivations for how to apply mathematical and scientific knowledge to this context. Whatever these particular needs, we encourage the use of questions to spark inquiry, debate, and student invention of knowledge, rather than teachers modeling the knowledge for students to copy.

13.6 Conclusion

This chapter began by asking the question: Can the space afforded by STEM education take up critical orientations in the ways that critical mathematics education (CME) has in the past? We feel that STEM education, as an interdisciplinary space, legitimates CME work because STEM units must link content across the disciplines with strong themes. Our example, "A Story of Incarceration," intends to show these possibilities. We have sketched the beginnings of a unit plan that targets internationally benchmarked STEM standards but links them with a critical approach that questions the nature and practice of incarceration. Our work continues the trajectory of CME and is enhanced by ecocritical curriculum scholarship as reviewed earlier. Thus, we answer in the affirmative: STEM education can and should be used as a

Table 13.1 Standards, goals, and content for "A Story of Incarceration" unit plan

A Story of Incarceration, unit plan sketch	
Established goals (standards):	
1. Compute unit rates associated with ratios of fractions, including ratios of lengths, areas, and other quantities measured in like or different units (CCSSM, 7.RP.1)	
2. Recognize and represent proportional relationships between quantities (CCSSM, 7.RP.2)	
3. Solve problems involving scale drawings of geometric figures, including computing actual lengths and areas from a scale drawing and reproducing a scale drawing at a different scale (CCSSM, 7.G.1)	
4. Biodiversity describes the variety of species found in Earth's terrestrial and oceanic ecosystems (NGSS, LS2-C)	
5. The fossil record documents the existence, diversity, extinction, and change of many life forms throughout the history of life on Earth (NGSS, MS-LS4–1)	
6. Define the criteria and constraints of a design problem with sufficient precision to ensure a successful solution, taking into account relevant scientific principles and potential impacts on people and the natural environment that may limit possible solutions (NGSS, MS-ETS1)	
7. The development of technology is a human activity and is the result of individual or collective needs and the ability to be creative (STL Standard 1-G, grades 6–8)	
8. Knowledge gained from other fields of study has a direct effect on the development of technological products and systems (STL, Standard 3-F, grades 6–8)	
What essential questions will be considered?	**What understandings are desired?**
1. How do we compare rates and ratios?	1. Mathematical concepts of rates, ratio, proportion, and disproportion illuminate our understanding of the real world and augment our ability to solve problems
2. What do the terms proportional and disproportional mean and how do we recognize these relationships?	2. Scientific concepts of biodiversity, taxonomy of species, and evolutionary relationships describe the variety of life on the planet and promote respect by humans for living things
3. How do incarceration rates and proportions differ across populations?	3. The engineering process applies scientific and mathematical concepts and is deeply intertwined with ethical considerations
4. What do we know about human incarceration by studying the geometry and engineering of prisons?	4. Ethical applications of science, engineering, and mathematics reveal that the mass incarceration of human and nonhuman animals may rest on unethical assumptions of domination and subordination. As well, such applications present steps toward eradicating these practices of incarceration
5. How do we classify life on earth and what does this tell us about evolutionary relationships?	
6. Which species among the classifications are incarcerated and for what reason?	
7. Is it possible to design an ethical incarceration?	

(continued)

Table 13.1 (continued)

What specific knowledge will learners acquire?	What specific skills will learners acquire?
1. Mathematical concepts of rate, ratio, and proportion	1a. How to compute rates and compare proportions
	1b. How to use a scale drawing to compute actual lengths and area
	1c. Apply the above to incarceration as (i) computing rates of incarceration across sectors of the population, (ii) identifying disproportionate groups that are incarcerated, and (iii) studying prison floor plans and scaling to real-life dimensions
2. Scientific concepts of species taxonomy, biodiversity, and evolutionary relationships	2a. Identify the levels in the taxonomy of species
	2b. Distinguish species within the animal kingdom and the mammal class
	2c. Apply knowledge of evolutionary relationships and taxonomy to incarceration of animals existing in agribusiness, entertainment and experimental testing
3. Ethics as a constraint on engineering design	3a. Ethically analyze the existing design of incarceration
	3b. Debate the possibility of designing ethical incarceration systems, such as open prisons and humane zoos

space for critical work. It provides a bounty of opportunities to address critical issues today and can harness the collective power of STEM disciplinary content.

We do not, however, commit to STEM education as the triumphant answer to these problems. As we (Chesky & Wolfmeyer, 2015; Wolfmeyer, 2013) and others (Bowers, 2016a, 2016b) have claimed, STEM, or more properly, the conflation of reason, science, and technology as an objective, always-good project, has in fact created much of the world's problems. However, we continue the CME tradition that does not, on the whole, reject STEM content. Without these knowledges, we would not know the extent of the problem of incarceration, for example. And without them, we will not find viable solutions either.

References

Abram, D. (1996). *The spell of the sensuous*. New York: Pantheon Books.

Apple, M. (1992). Do the standards go far enough? Power, policy and practice. *Journal for Research in Mathematics Education, 23*(5), 412–431.

Bateson, G. (1972). *Steps to an ecology of mind*. Chicago: University of Chicago Press.

Bowers, C. A. (2001). *Educating for eco-justice and community*. Eugene, OR: Eco-Justice Press.

Bowers, C. A. (2016a). *In the grip of the past*. Eugene, OR: Eco-Justice Press.

Bowers, C. A. (2016b). *A critical examination of STEM: Issues and challenges*. New York: Routledge.

Chesky, N., & Wolfmeyer, M. (2015). *Philosophy of STEM education: A critical investigation*. New York: Palgrave Macmillan.

Coles, A., Barwell, R., Cotton, T., Winter, J., & Brown, L. (2013). *Teaching secondary mathematics as if the planet mattered*. New York: Routledge.

Common Core, Inc. (2013). *A story of ratios: A curriculum overview for grades 6–8*. Available at: https://www.engageny.org/file/8761/download/a-story-of-ratios-a-curriculum-overview-for-grades-6-8.pdf

Common Core State Standards Initiative. (2010). *Common core state standards for mathematics*. Available at: corestandards.org

Drew, D. (2011). *STEM the tide: Reforming science, technology, engineering and math education in America*. Baltimore, MD: Johns Hopkins University Press.

Frankenstein, M. (1983). Critical mathematics education: An application of Paulo Freire's epistemology. *Journal of Education, 165*(4), 315–338.

Gruenewald, D. (2003). The best of both worlds: A critical pedagogy of place. *Educational Research, 32*(4), 3–12.

Gutstein, E. (2006). *Reading and writing the world with mathematics: Towards a pedagogy for social justice*. New York: Routledge.

Hrabowski, F. (2016). Broadening perception in the American STEM workforce. *BioSience, 62*(4), 325–326.

Khan, R. (2010). Performing oneself differently: A mathemaesthetician's responsibility. *Educational Insights, 13*(1): http://einsights.ogpr.educ.ubc.ca/v13n01/articles/khan/index.html

Khan, R. (2011). Ethnomathematics as mythopoetic curriculum. *For the Learning of Mathematics, 31*(3), 14–18.

Kimme, D. (1998). *Jail design guide: A resource for small and medium-sized jails*. Champaign, IL: Kimme and Associates.

Larnell, G., Bullock, E., & Jett, C. (2016). Rethinking teaching and learning mathematics for social justice from a critical race perspective. *Journal of Education, 196*(1), 19–30.

Larson, D. (2013). Why Scandinavian prisons are superior: Open prisons, in which detainees are allowed to live like regular citizens, should be a model for the U.S. *The Atlantic*.

Lupinacci, J., & Happel, A. (2015). Recognize, resist, & reconstitute: An EcoJustice response to neoliberalism. In M. Abendroth & B. Porfilio (Eds.), *Schools against the neoliberal rule* (pp. 269–287). Charlotte, NC: Information Age Publishing.

Martin, D. (2013). Race, racial projects and mathematics education. *Journal for Research in Mathematics Education, 44*(1), 316–333.

Martusewicz, R., Edmundson, J., & Lupinacci, J. (2015). *EcoJustice education: Toward diverse, democratic and sustainable communities* (2nd ed.). New York: Routledge.

Mauer, M., & King, R. (2007). *Uneven justice: State rates of incarceration by race and ethnicity*. Washington, DC: The Sentencing Project.

Merchant, C. (1983). *The death of nature: Women, ecology and the scientific revolution*. New York: Harper and Row.

Moses, R., & Cobb, C. (2001). *Radical equations: Civil rights from Mississippi to the algebra project*. Boston: Beacon.

NGSS Lead States. (2013). *Next generation science standards: For states, by states*. Washington, DC: The National Academies Press.

Plumwood, V. (1993). *Feminism and the mastery of nature*. New York: Routledge.

Plumwood, V. (2002). *Environmental culture: The ecological crisis of reason*. New York: Routledge.

Skovsmose, O. (1985). Mathematical education versus critical education. *Educational Studies in Mathematics, 16*(4), 337–354.

Skovsmose, O. (1994). *Towards a philosophy of critical mathematics education*. Boston: Kluwer Academic Publishers.

Skovsmose, O. (2011). *An invitation to critical mathematics education*. Boston: Sense.

Steen, L. A. (Ed.). (1997). *Why numbers count: Quantitative literacy for Tomorrow's America*. New York: College Entrance Examination Board.

Walmsley, R. (2015). *World prison population list*. Available at: http://www.prisonstudies.org/sites/default/files/resources/downloads/world_prison_population_list_11th_edition.pdf

Warren, K. (1990). The power and the promise of ecological feminism. *Environmental Ethics, 12*(2), 125–146.
Warren, K. (2000). *Ecofeminist philosophy: A western perspective on what it is and why it matters*. Lanham, MD: Rowman & Littlefield Publishers, Inc..
Wiggins, G., & McTighe, J. (2005). *Understanding by design*. Alexandria, VA: ASCD.
Wolfmeyer, M. (2014). *Mathematics education for America? Big business, policy networks, and pedagogy wars*. New York: Routledge.
Wolfmeyer, M. (2013). STEM, meet ecojustice. *Green Theory and Praxis, 7*(1), 47–52.

MIX
Papier aus verantwortungsvollen Quellen
Paper from responsible sources
FSC® C105338

If you have any concerns about our products,
you can contact us on
ProductSafety@springernature.com

In case Publisher is established outside the EU,
the EU authorized representative is:
Springer Nature Customer Service Center GmbH
Europaplatz 3, 69115 Heidelberg, Germany

Printed by Libri Plureos GmbH
in Hamburg, Germany